1996
INTERNATIONAL SEMICONDUCTOR CONFERENCE

19th Edition
October 9-12, 1996, Sinaia, Romania

CAS'96 PROCEEDINGS

VOLUME 2

IEEE Catalog Number:	**96TH8170**	
ISBN:	**0-7803-3223-7**	**Softbound Edition**
	0-7803-3224-5	**Microfiche Edition**
Library of Congress:	**96-75375**	

CAS'96

organized by
RESEARCH INSTITUTE FOR ELECTRONIC COMPONENTS (ICCE)

Str. Erou Iancu Nicolae 32B, 72996 Bucharest, Romania
Phone: +401-2121448 / +401-2121553 / +401-6333040
Fax: +401-3127519 E-mail: cas@icce.ro / rusu@gef.pub.ro

IEEE Catalog Number:	96TH8170	
ISBN:	0-7803-3223-7	Softbound Edition
	0-7803-3224-5	Microfiche Edition
Library of Congress:	96-75375	

Session 12

STUDENT PAPERS

THEORY AND PRACTICE OF A FULLY CONTROLLABLE WHITE NOISE GENERATOR

Florin C. Ionescu
Department of Electronics and Telecommunication
University "Politehnica" Bucarest, Romania

This paper describes a method and an apparatus generating white noise using feed-back shift registers. We center on the problem of the minimum structure necessary for producing a white noise by multiplexing several registers with different initial states. We present a first model of this generator realized with FPGAs.

INTRODUCTION

The present work is a part of a more important -by the size- project consisting in the implementation of microscopic charge transport models using the Monte Carlo method on ASICs. The reason for this approach, somewhat different from what is usually done with softwares, is to reach very high speeds of computation with respect to what is commonly achievable with general purpose computers, avoiding continuous transfer of data from/to static memory blocks to/from dynamic computation units. Moreover, parallelism is rendered more easy using datapaths on data transfer BUS and a unique controller [NEG]. In such Monte Carlo simulators, we need white noise generators delevering binary pseudo random numbers (PRN). Moreover, the series of numbers must be perfectly reproducible as many times as we wish, otherwise it would be impossible to debbug the simulation code. This is easily done using feed-back shift registers (FSR). The only problem here is to achieve a perfect independence of two successive numbers, so as to get a time covariance as a Kronecker function giving a white noise spectrum (Wiener-Khintchine theorem). To do so, we use multiplexing of a sufficient number of FSR. In the next paragraph, we remind the reader on the way pseudo random number generators are constructed using FSR. The operation of a proposed structure is then simulated and how independence is obtained is discussed. In the last paragraph, we describe the implementation of the proposed structure as a stand-alone white noise generator.

PRINCIPLES OF PSEUDO RANDOM NUMBER GENERATORS

A pseudo random number generator is most easily realized with a shift register fed-back through exclusive OR's (EXOR). Let a register be made of N+1 bits numbered from 0 to N. We choose a limited number of these and send the result J of their mutual EXOR to the least significant bit (LSB) after the register has been shifted once towards the most significant bit (MSB). For instance with N=31 we make $J=I_0+I_1+I_{21}+I_{31}$, where the index is the bit number. At each clock pulse, the two operations are performed in sequence, and the content of the register is just the binary PRN to deliver. In the example, the characteristic polynomial used was $P_{32}(x)=x^{32}+x^{22}+x^{2}+x+1$, where 32 is the MSB. In his book, Max discussed the conditions by which the longest period of PRN can be obtained [MAX]. The condition for a period of $T=2^{32}-1$, the maximum possible, is that P(x) be irreducible. Max gives a list of irreducible polynomials and the optimum feed-backs to be used for maximum periods. The P_{32} above is just one of them. The process was simulated with a software in order to study the statistical independence of the generated numbers. One of the most efficient ways is via the computation of the covar-

0-7803-3223-7/96/$5,00 © 1996 IEEE

iance. The binary PRN is then converted to a decimal, divided by T and treated as a real PRN X(t) in the interval]0,1]. Here t is not time, properly speaking, but rather the occurence of the real PRN. For practical reasons, as we will see later, we only analyze the PRN appearing in the 16 MSB (then X(t)=0 is accepted). This PRN is used for the covariance $C_x(\tau)$ where τ is the occurence. We have simulated up to 16 FSR placed in parallel. All of the 16 registers were initialized with widely different numbers spanning in the range $[1,2^{32}-1]$. The capability of the registers in terms of numbers of different PRN is thus reduced by a factor of 16 to $2^{28}-1$, but the period of the block of registers, is at least bigger than $2^{32}-1$. At each clock pulse the content of a different register is transfered to the calculation of the covariance, in sequence, as in a multiplexer. We can use all or only some of the registers. By way of illustration, in figure 1a we show the covariance when we use only one register and in figure 1b, when we use 16 registers. The difference in nature between the two is obvious. Using all the occurences of one FSR, the covariance exhibits strong memory effects, decreasing slowly with the occurence, whereas using 16 FSR the covariance is maximum at the first occurence and then flat at all next occurences. If the PRN occur with a uniform probability whatever the value (as it is expected) we must find $C_x(0)=1/3$ and $C_x(\infty)=1/4$. Although the

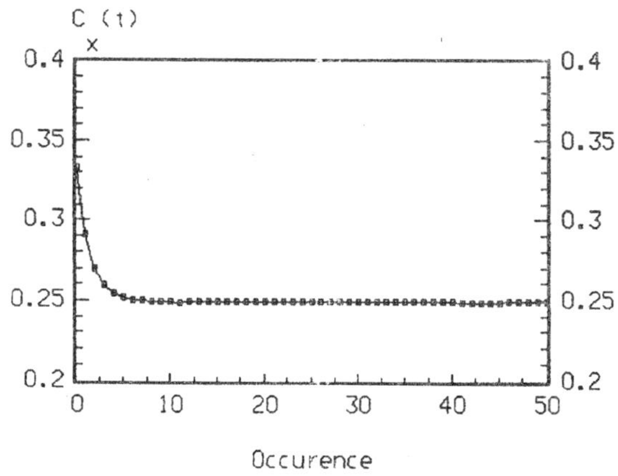

Figure 1a. *Covariance obtained with one FSR only.*

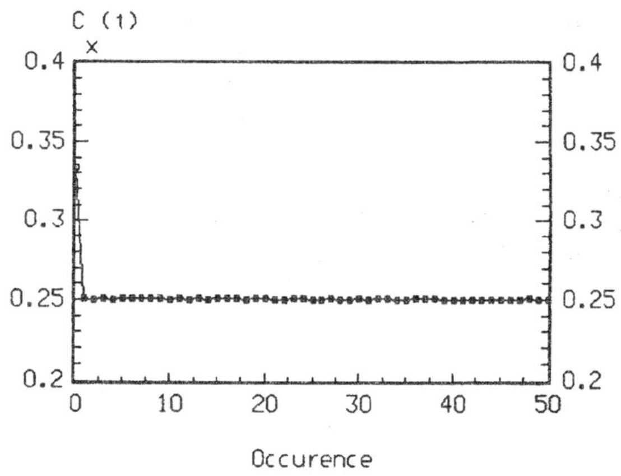

Figure 1b. *Covariance obtained with 16 FSR.*

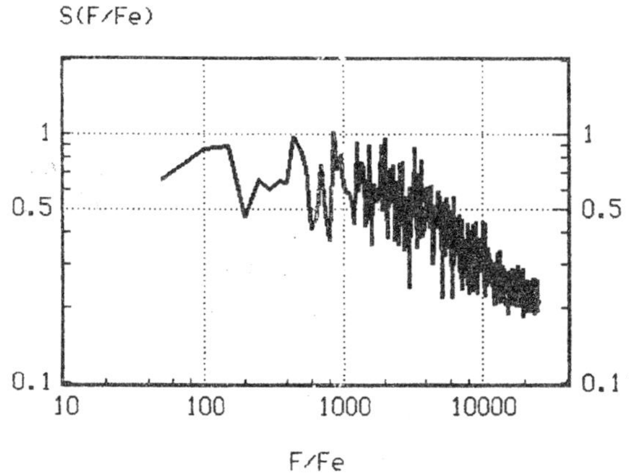

Figure 2a. *Spectrum corresponding to figure 1a.*

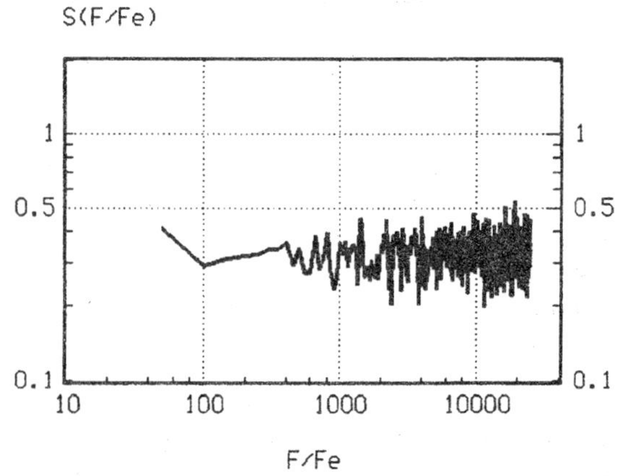

Figure 2b. *Spectrum corresponding to figure 1b.*

covariance is shown here up to the 50th occurence, it remains flat up to occurence 50000 at least. Moreover, the computations here used 1000000 PRN at least. When using only one FSR, it was seen that at least 8 occurences were necessary to obtain statistical independence. It is obvious that, for a 32 bit register, capacity overflow begins when the content has been shifted at least 32 times and the original content has disappeared completely. But as shown in figure 1b, mutual independence does not require such a stringent condition and with 16 FSK only, a high degree of independence has been achieved yet. The noise spectra (obtained using FFT) corresponding to the two cases of figure 1 are shown in figure 2. It is clear that the desired results have been obtained. In figure 2a we notice a spectrum decreasing as $f^{1/2}$, while the spectrum of figure 2b remains flat. Notwithstanding this result, we decided to build a generator with 32 FSRs of 32 bits fed-back according to the P32 above and transfer the result of the 16 MSB to the analogue output. We present it now.

PRACTICAL REALIZATION AND MODE OF OPERATION

We now glance at the practical implementation of this algorithm into a stand-alone apparatus and how it can be used in practice. The generator stands into three main blocks.

1. The heart of the PRN generator is a FPGA (see figure 3). It incorporates 32 FSR of 32 bits connected to a 32 branch DEMUX at the input. This is used for the initialization of the 32 FSR made by transfer of the content of external memories, the DEMUX being activated with an address selector SEL. All the FSR are connected to another multiplexer MUX and the number of active FSR driven by another selector. The two selectors have 5 bits (0-31) at the input and the output, but 16 bits of the MUX only are connected to the DAC. This block was implemented using a description of data flow in VHDL and simulated with SYNOPSIS for XILINX. Using a FPGA was preferred to commercial circuits because of its compactness and lightness. A hand-held apparatus was then possible.

The schematic diagram of the apparatus is described in figure 4.

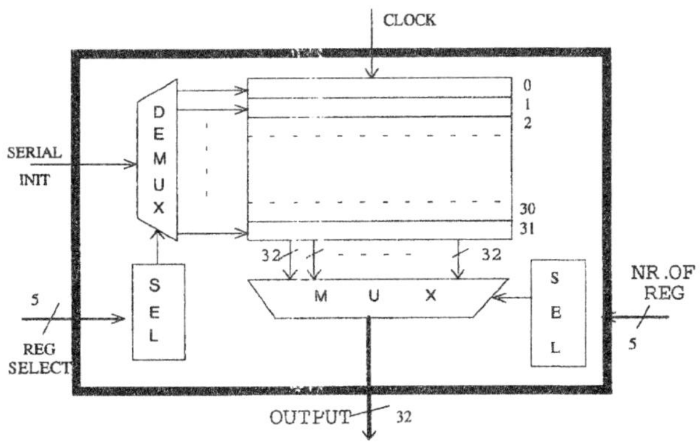

Figure 3. *Block diagram of the FPGA.*

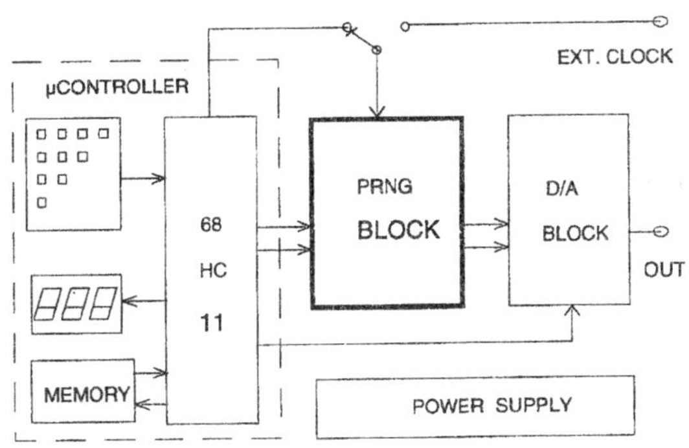

Figure 4. *Block diagram of the generator.*

2. A second block incorporates the digital-to-analogue converter (DAC). It includes a 16 bit AD7846 DAC with all the auxiliary circuits needed for normal functioning. It might be placed in parallel with a binary output for a signal analyzer providing direct binary acquisition inputs.

3. A third block is the control block. We used a Motorola 68HC11 micro-controller (μ-C) for the coordination. It is connected to a keyboard pad and a LC display as the user interface on one side and to a block of programmed EPROM on the other side. This block has a multiple role. It interacts with the user by a menu. It loads the FPGA with the initial values and synchroni zes the DAC. It also provides the internal clock. However, an external clock can be connected to the FPGA. Using this, we could study how the spectrum at the output is affected by the clock frequency. By itself, the μ-C will supply the maximum clock frequency supported by the DAC (related to the minimum time required for full conversion).

Although all of the 32 bits are available at the output of the FPGA, 16 only are transfered to the DAC, but this does not affect the period of the generated sequences. In the final stage the generator will be characterized with a signal analyzer connected at the DAC output. Characterization will show the differences between practical and theoretical spectra. At the time being, the precise way the generator will be programmed is still premature, but the main steps are defined yet. After power-on reset, the μ-C performs self-test and begins the dialogue with the user. It asks whether the generator would work as a stand-alone device or whether programming the apparatus would be necessary. In the first case, the generator is loaded with memorized values assuring a white noise at the output automatically. In the second, the user is asked for the number of FSR to be active, and for the initial values for each of them. It will be an option in the menu that available initial values can be read in the PROM block of the μ-C. This option and the possibility of entering initial values in hexadecimal from keypad assures easy handling and minimum programming time. In all cases it would be necessary to specify the length of the output sequence in seconds. This limitation, precisely controlled by the μ-C clock along with the possibility of having the same initial states in the generator registers justify the fact that the generator is fully controllable.

CONCLUSION

We have described a digital procedure permitting practical implementation of a fully controllable white noise generator. The generator will be realized in practice using feed-back shift registers and multiplexers incorporated into a FPGA. The rest of the apparatus is made of commercial circuits. The conditions by which white noise can be obtained have been discussed in terms of number of registers and digital-to-analogue conversion.

REFERENCES

[NEG] Andy Negoi and Jacques Zimmermann, "Monte Carlo Hardware Simulator for Electron Dynamics in Semiconductors" (These proceedings).

[MAX] J. Max, "Méthodes et Techniques du Traitement du Signal et Applications aux Mesures Physiques", Masson Ed., Paris, 1972.

This work has been realized at the LPCS-ENSERG of the Institut National Polytechnique de Grenoble (INPG) under the auspices of ELESA, a federative laboratory of INPG. The valuable help of Andy Negoi and Jacques Zimmermann is sincerely acknowledged.

Tree Grouping for HDLA-models

Niculae Sebe
"Politechnica" University of Bucharest

Abstract

This article presents a data-structure for the representation of behavioral and structural time-continuous-systems, with an emphasis on VLSI-integrated circuits, which allows the synthesis of VHDL-A-like code.

Introduction

Simulation of heterogeneous systems on the system level implies the simulation of the whole system and of some particular devices with a special emphasis on their functional behavior.

Functional simulation can only be done efficiently by using a single system description language suitable for a single simulator. Such languages exist for the domain of digital microelectronics - VHDL or VerilogHDL. The advent of Analog HDLs like VHDL-A by Anacad offers the possibility to create behavioral models of electrical and non-electrical devices without having the limitations of Spice.

Generation of a "flat" VHDL-A-like model

In this section we will describe the possibility to generate a single VHDL-A-like model starting from two different models, as well as their interconnections. The models have to be available in the source format in order to parse their code. Simple operations as connection of two models in series and parallel (with different signal directions) are sufficient to be used recursively in order to obtain a single homogeneous VHDL-A-like model. We will call this operation *grouping,* whereby modularity is the primary advantage offered by this approach.

The accuracy of the global model reflects those of the initial ones. Despite the fact that mapping of VHDL-A-like code into the simulator-kernel or equation solver is proprietary to the CAD-vendors it can be assumed that the "flat" code will improve the speed, accuracy and stability of the performed simulation.

The grouping algorithm consists of three stages (Fig.1):

- The models are analysed lexically and syntactically, using Lex and Yacc, resulting a parse-tree.
- The models are grouped: Kirchoff's Current Law for the common pin results in a supplementary equation; Kirchoff's second law is reflected by a new state, representing the potential difference (i.e. voltage) at the common pin.
- The supplementary state and equation are eliminated.

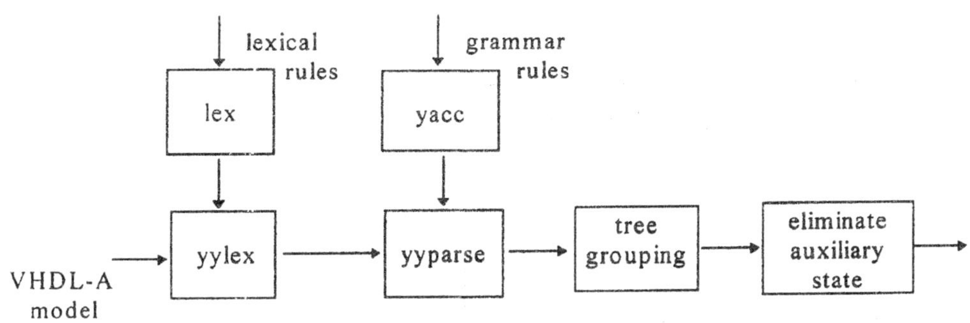

Fig.1: Diagram block

0-7803-3223-7/96/$5,00

Lexical and syntactical analysis of VHDL-A-like models

Lexical and syntactical analysis of VHDL-A-like models enables an efficient and modular construction of the associated tree structure; the check for syntactical correctness of the VHDL-A-like description is another effect of this approach.

The software utilities used are Lex and Yacc. Lex is a program generator designed for lexical processing of character input streams. Lex accepts a high-level, problem-oriented specification for character string matching and produces a C routine that recognizes regular expressions. Lex divides the input into meaningful units (which are usually called *tokens*) realizing a lexical analyzer. In this case the tokens are specific words from VHDL-A-syntax as: ARCHITECTURE, ENTITY, PROCEDURAL, EQUATION, etc.

As the input is divided into tokens, one needs to establish the relationship among the tokens. This task is known as *parsing* and the list of rules that define the relationship that the program understands is a *grammar*. Yacc takes a concise description of a grammar and produces a C routine that can parse that grammar, a parser.

In our case the constructing grammar must follow the syntactic rules of VHDL-A-syntax.

Example: We consider the VHDL-A syntax having following top-view:

```
entity       ::=  ENTITY entity-name IS
                        GENERIC         (declarations)
                        COUPLING        (declarations)
                        PIN             (declarations)
                        PORT            (declarations)
                        END [entity-name];
architecture ::=  ARCHITECTURE name OF entity-name IS
                                 {local-declaration}
                        BEGIN    {concurrent-instruction}
                        END [name];
local-declaration ::=
             external-function | constants | variables | signals | states
concurrent-instruction ::=
      RELATION {procedure | implicit-algorithm}END RELATION;
```

The following rules describe a simplified form of the grammar:

```
entity_body :   ENTITY entity-name IS
                          entity_declarations
                END  entity-name ';'
                          ;
entity_declarations:   generic_declarations
                       pin_declarations
                          ;
generic_declarations: GENERIC '(' name_list ':' analog_indications ')'
                          ;
analog_indications:    ANALOG
                          ;
 name_list :             NAME
```

```
                    | name_list ',' NAME
                        ;
pin_declarations:       . . .
```

The Yacc parser automatically detects whenever a sequence of input tokens matches one of these rules and performs a specific action; it also detects a syntax error whenever its input do not match any of the rules.

The main action that is performed is to build a tree structure associated with VHDL-A-model. This *parse tree* can be built by supplying actions such as:

```
expr :          expr 1 '+' expr2
                      {   $$ = node ( '+' , $1 , $2 ) }
```

in the specifications. That means we can call the function *node* ('+', expr1,expr2).

Composing the tree structure associated with each model

The second stage consists of composing the tree structure associated with each model, taking into account the first of Kirchoff's laws for the common pin and adding a new state that corresponds to the voltage of the eliminated pin. When eliminating this pin from the pin-list we add a supplementary equation for the new internal state. Building tree structures has the advantage that the operation, otherwise applied to VHDL-A-text, is done using a tree (it is easier to add and delete a node, to modify node parameters, etc.).

Example: Suppose we have to link two RC-circuits (Fig.2):

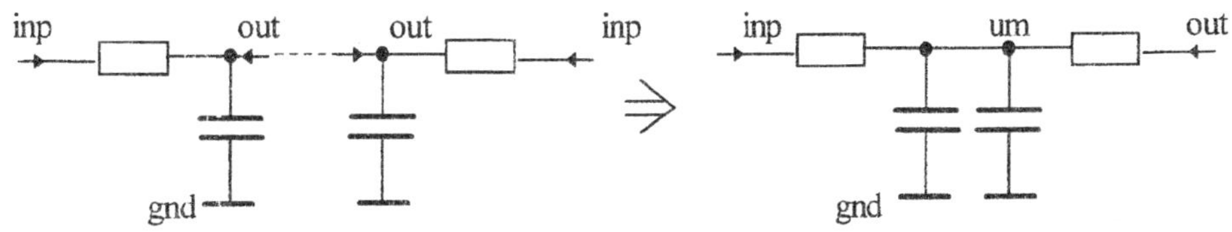

Fig.2: Grouping RCs

The initial model is described by the following VHDL-A-like description:

```
ENTITY RC IS
        GENERIC (R,C : analog);
        PIN (inp, out, gnd : electrical );
END ;

ARCHITECTURE a OF RC IS
 STATE ic: analog ;
 BEGIN
   RELATION
      PROCEDURAL FOR init =>
           ic:= 0.0 ;
```

```
PROCEDURAL FOR dc, transient =>
        [out,gnd].i %= ic-[inp, out].v/ R;
    EQUATION ( ic ) FOR transient =>
        C * [out , gnd].v == integ( ic );
  END RELATION ;
END;
```

After this stage we obtain the following VHDL-A-text:

```
ENTITY RC2 IS
        GENERIC (R,C : analog);
        PIN (inp1, out2, gnd : electrical );
END ;

ARCHITECTURE a OF RC2 IS
STATE ic1,ic2,um: analog ;
BEGIN
  RELATION
    PROCEDURAL FOR init =>
          ic1:= 0.0 ;  ic2 :=0.0 ;
    EQUATION ( ic1,ic2,um ) FOR transient =>
        ic1- (inp.v -um )/R + ic2 -(out.v - um )/R == 0.0;
        C * um == integ( ic1 );
        C * um == integ( ic2 );
  END RELATION ;
END;
```

A supplementary state *um* has been introduced and a new corresponding equation.

Elimination of the auxiliary internal state and the corresponding equation

This stage is imposed by obvious complexity reasons.

We use Mathematica's facilities for symbolic calculus. We have to solve symbolically a system of equations to eliminate the state and the corresponding equation. Therefore we use MathLink to make some literal operations in Mathematica and to receive the result.

Conclusion

We have developed a simple method to obtain a tree structure for a global model consisting of the two models that are grouped. The obtained tree structure can be used to generate the text of the resulting VHDL-A model and also to build a structure to be used as a pre-processing part in VHDL-A model compilation. The gain of this approach is that the model-generation is decoupled from writing and learning VHDL-A-syntax, therefore a graphical model entry would be a straight-forward continuation of this work.

Bibliography

J.Levine, T.Mason, D.Brown: "lex & yacc", O'Reilly&Associates, 1992

HDL-A Language Reference Manual , ANACAD ,March 1994

S.Wolfram: Mathematica , Addison Wesley 1992

MODELING OF SiC VERTICAL POWER DMOSFET STRUCTURE USING THE FUNDAMENTAL MOSFET'S SET OF EQUATIONS

Paul Daniel Lungu

"Politehnica" University of Bucharest, Electronic and Telecommunications Department, Romania

SUMMARY

The most important parameters for 6H-SiC and 3C-SiC are extracted from the literature and implemented in an original Vertical DMOSFET modeling program using the fundamental set of equations that account for the MOSFET characteristics. A comparison of current-voltage characteristics and specific on resistance for different values of breakdown voltage at two significant temperatures for 6H-SiC, 3C-SiC and Si is provided in this paper.

1. INTRODUCTION

Silicon carbide (SiC) has been recently given renewed attention as a potential material for high-power and high-frequency application requiring high-temperature operation. Some of the possible applications of SiC as a material for power electronics is for advanced turbine engines, propulsion systems, automotive and aerospace electronics, and applications requiring large radiation-damage resistance. Properties such as large breakdown electric field strength, large saturated electron drift velocity, large bandgap, small dielectric constant, and high thermal conductivity, as is observed from the table 1, make SiC an attractive candidate for fabricating power devices with reduced power loss and die size /1,2/.

Table 1. Comparison of parameters for 6H-SiC, 3C-SiC and Si

Parameter	6H-SiC	3C-SiC	Si
Dielectric constant: ε	9.66	9.7	11.8
Bandgap: E_g(eV)	$3-3.3*10^{-4}$(T-300K)	$2.3-5.8*10^{-4}$(T-300K)	$1.17-(7.02*10^{-4}T^2)/(T(K)+1108)$
Saturation drift velocity: V_{sat}(cm/s)	$2*10^7$	$2.5*10^7$	$1*10^7$
Electric field E_c(V/cm) at breakdown voltage V_B:	$10^6(1.2+18*V_B^{-0.43})$	$0.8*E_c$(6H-SiC)	$0.125*E_c$(6H-SiC)
Thermal conductivity at T=300K: λ(W/cm°C)	4.9	4.9	1.5

2. VERTICAL POWER DMOSFET STRUCTURE; BRIEF DESCRIPTION AND MODELING PARAMETERS.

MOSFET's are the mostly fabricated unipolar semiconductor devices. In order to reduce the area of the device, MOSFET's for high power - high voltage applications are normally designed with source and drain at opposite sides of the wafer. Such a typical vertical double-diffused MOSFET (DMOSFET), schematically shown in fig.1 , is fabricated by using planar diffusion technology with refractory gate such as polysilicon. In these devices, the p-base and n^+ -source regions are diffused through a common window defined by the edge of

0-7803-3223-7/96/$5,00 © 1996 IEEE

polysilicon gate /1/. The surface channel region is defined by the difference in the lateral diffusion between p-base and n^+ -source region. The forward blocking capability is achieved by the p-n junction between p-base region and the drift region. During device operation, a fixed potential to the p-base region is established by connecting it to the source metal by a break in the n^+ -source region. By short circuiting the gate to the source and applying a positive bias to the drain, the p-base/n^- -drift region becomes reverse-biased and this junction supports the drain voltage by the extension of a depletion layer on both sides. However, due to the higher doping level of the p-base layer (in order of 10^{17}-$10^{18}cm^{-3}$), the depletion layer extends primarily into the n^- -drift region (with a doping level, depend of breakdown voltage, in order of 10^{15}-10^{17} cm^{-3}). On applying a positive bias to the gate electrode, a conductive path extending between the n^+ -source region and the n^- -drift region is formed. The application of a positive drain voltage results in a current flow between drain and source through the n^- -drift region and conductive channel. The conductivity of the channel is modulated by the gate bias voltage and the current flow is determined by the specific on resistance ($R_{on,sp}$) of various resistive components. The total $R_{on,sp}$ is determined as

$$R_{on,sp} = R_{n+} + R_{Channel} + R_{Acc} + R_{JFET} + R_{Drift} + R_{Substrate} + R_{cont} \quad (1)$$

where R_{n+} is the resistance of n^+ -source, $R_{Channel}$ is channel resistance, R_{Acc} is the resistance of the accumulation layer in the drift region below oxide, R_{Drift} is the resistance of the drift region given by (4), R_{JFET} is the resistance of the of the JFET-like region indicated in fig.1 below the accumulation layer. Due to current spreading from the channel into the JFET region, R_{JFET} is assumed to be $1.5R_{Drift}$.

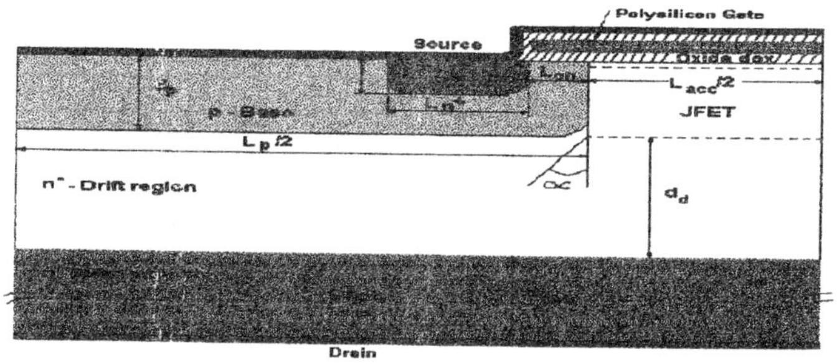

Fig. 1. Schematic cross section of a vertical double-diffused MOSFET. All design dimensions are given in the table 2. For modeling of on-resistance, the current spreading angle α=45° was assumed.

Table 2. Design parameters for 6H-SiC, 3C-SiC and Si Vertical power DMOSFET used for the calculation of the specific on resistance and the current-voltage characteristics. All lengths are given in μm. All concentrations N in cm^{-3}.

d_{Ox}	L_p	L_{n+}	L_{Ch}	L_{Acc}	d_p	d_{n+}	d_{drift}	d_{sub}	N_{n-}	N_{n+}	N_{gate}	$N_{p\text{-}base}$
$50*10^{-3}$	20	3	2	10	1.5	0.3	11	287.5	$9.39*10^{15}$	10^{20}	10^{18}	$5*10^{17}$

The drift region analysis for an ideal DMOSFET structure can be performed by approximating the depletion layer in the drift region as an abrupt one-dimensional junction fabricated in a uniformly doped semiconductor. The doping level $N_{n-}(cm^{-3})$ required to support a given breakdown voltage V_B and depletion width W_d(cm) at the breakdown can be calculated as follows /1/:

$$N_{n-} = \varepsilon E_c^2/(2qV_B) \quad (2)$$

$$W_d = 2V_B/E_c \quad (3)$$

The specific drift region resistance (Ωcm^2) associated with the drift layer to support V_B is

$$R_{Drift} = W_d/(qN_{n-}\mu_n) \quad (4)$$

Usually, both the bulk mobility μ_n and the breakdown voltage are dependent on N_{n-}. The dependence between breakdown field and doping level of the drift region is easy to understand from the table 1 and (2). The mobility dependence on doping level in a n or p type semiconductor is given in /2/ as follows

$$\mu_n=\mu_n^{min}+(\mu_n^{delta}/(1+((N_D+N_A)/N_n^{\mu}))))(T/300K)^{\alpha n} \quad (5)$$

with the constants listed in table 3 and N_D+N_A the total doping of n type semiconductor. For 3C-SiC, mobility is given from /2/ as follows:

$$\mu_n(3C\text{-}SiC)=2*\mu_n(6H\text{-}SiC)$$

Table 3. Set of constants for mobility modeling 6H-SiC and Si /2/

Semiconductor	μ_n^{delta}(cm²/Vs)	μ_n^{min}(cm²/Vs)	α_n	γ_n
n-type 6H-SiC / Si	380 / 1268	20 / 92	-3 / -2.42	0.45 / 0.91

3. ELECTRICAL CHARACTERISTICS FOR 6H-SiC AND 3C-SiC POWER DMOSFET.

Because of hard dependency of the drain current density on specific on resistance ($R_{on,sp}$) the most important parameter in designing of power DMOSFET structures is the parameter $R_{on,sp}$. Fig. 2 shows R_{on} versus DMOSFET breakdown voltage for 6H-SiC, 3C-SiC and Si at two temperatures. To calculate $R_{on,sp}$ was used (1)-(5) in which an incomplete ionization was used, $R_{cont}=1*10^{-4}$ Ωcm^2 /3/, and surface mobility field dependency was neglected.

The most important observation from fig. 2 is the increasing with about an order of magnitude of specific on resistance for 6H-SiC and 3C-SiC at T=800K, but even in these conditions $R_{on,sp}$ is less than specific on resistance for Si at T=300K.

From fig. 2 it is observed that for $V_B \in$(1000-10 000V) the increasing of $R_{on,sp}$ is less than an order of magnitude. A spectacularly decreasing of $R_{on,sp}$ (with about 30% of $R_{on,sp}$ from fig. 2) can be obtained with a good geometrically design of the DMOSFET structure.

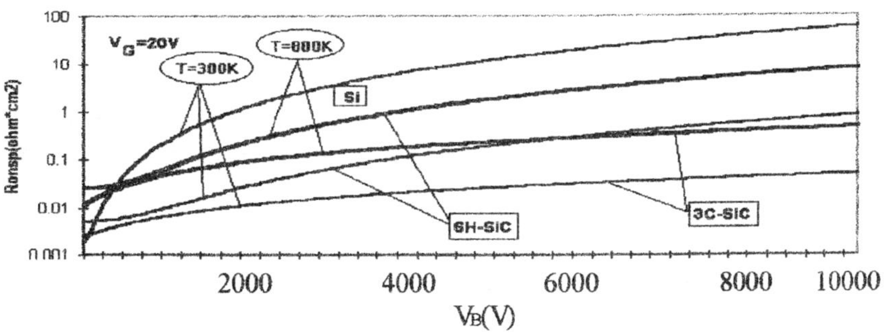

*Fig. 2. The dependence of the specific on resistance on breakdown voltage for 6H-SiC, 3C-SiC and Si. At V_B=1200V, $R_{on,sp}$=10.7*10^{-3} Ωcm^2 at T=300K*

Figs. 3-4 present J_D-V_{DS} calculated characteristics for 6H-SiC and 3C-SiC vertical power DMOSFET at three different values of the gate voltage and two significant temperatures with equations /4/:

$$J_D=K_n((V_{GS}-V_T)(V_{DS}-R_{on,sp}J_D)-(V_{DS}-R_{on,sp}J_D)^2/2) \quad , \; V_{DS}-R_{on,sp}J_D<V_{DS,sat} \text{ linear region} \quad (6)$$

$$J_D=(K_n/2)(V_{GS}-V_T)^2(1+(V_{DS}-R_{on,sp}J_D-V_{DSsat})/V_E) \quad , \; V_{DS}-R_{on,sp}J_D\geq V_{DSsat} \text{ saturated region} \quad (7)$$

where V_T is the threshold voltage, V_E is the Early voltage, $V_{DSsat}=V_{GS}-V_T$ is the saturation voltage, $L_{drain}=L_p+L_{Acc}$ is the drain lenght and $K_n=(1/L_{ch}L_{drain})\mu_n C_{ox}$.

From fig. 4 are observed great values of J_D for 6H-SiC and 3C-SiC. At the same voltage, V_{DS} and V_{GS}, for Si was obtained J_D less than an order of magnitude than for SiC. With increasing of the temperature J_D decrease. Usually, for both 6H-SiC and 3C-SiC channel mobility increase with increase of the temperature and for a common MOSFET structure the drain current increase with increase of the temperature.

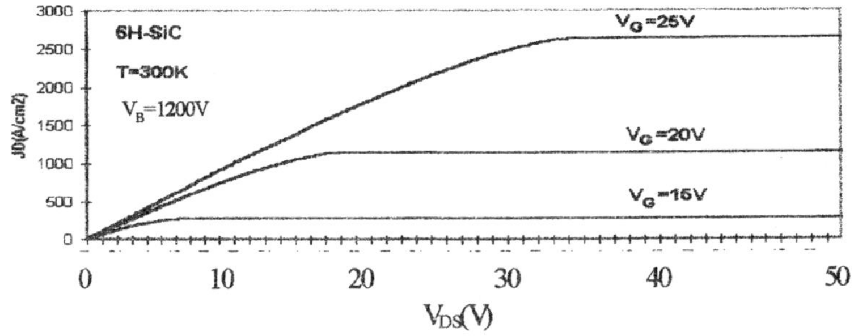

Fig. 3. $J_{DS}(V_{DS})$ characteristics for a n-channel VDMOSFET structure at diferent values of the gate voltage.

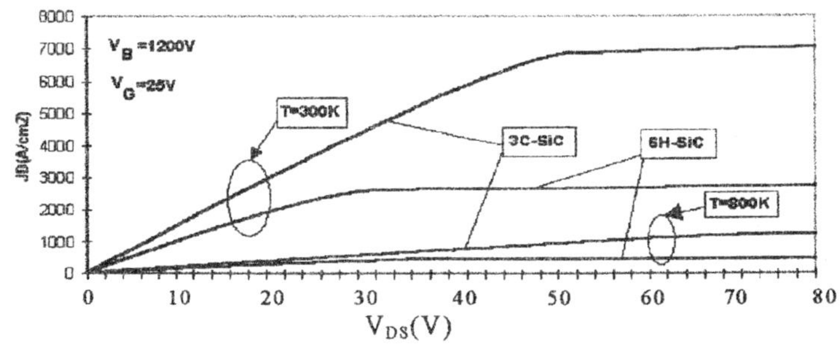

Fig. 4. Drain current density versus drain source voltage for 6H- SiC and 3C-SiC at T=300K and T=800K.

It is observed that the 6H-SiC specific on resistance from fig. 2 has the same value ($R_{on,sp}=10.7*10^{-3}\Omega cm^2$) with the resistance obtained from the slope of the characteristic $J_D(V_{DS})$ at V_{GS}=20V and T=300K (fig. 3).

The calculations performed using the simple MOSFET model but introducing the specific on resistance effect, are in good agreement with PISCES simulation results reported in /2/. In PISCES was obtained a $R_{on,sp}=9.9*10^{-3}\Omega cm^2$ and from the measured data was obtained a $R_{on,sp}=9*10^{-3}\Omega cm^2$. In the model considered in this paper was not take into account the field mobility dependence and the differences between the electrical parameters can be explained.

4. CONCLUSIONS

An original and simple modeling program was performed in order to analyze electrical behavior of the vertical power DMOSFET structures. The usual set of equations for a common MOSFET can be used for modeling power structures if it is take into account the important contribution of the specific on resistance and the capture current surface is assumed to be the drain area. The results obtained from this technique are in good agreement with PISCES simulation results and measured data reported in /2/. The calculation time using the original modeling program in considerably shorter than the PISCES simulation time and the results are comparable.

REFERENCES

/1/ M. Bhatnagar and B.J. Baliga, IEEE TRANS. EL. DEV. (40) p.647 No.3 (March 1993)
/2/ M. Ruft, H. Mitlehner, and R. Helbig, IEEE TRANS. EL. DEV. (41) p.1040 No.6 (June 1994)
/3/ J. Crofton, P.A. Barnes, J.R. Williams and J.A. Edmond, Appl. Phis. Lett. 62 (4), 25 January 1993
/4/ G. Brezeanu, Electronic Circuits and Devices, Course, 1995

OPTICAL PROPERTIES OF $GaAs/GaAs_{1-x}P_x$ SUPERLATTICES

Alexandru Korotcov
State University of Moldova,A.Mateevich str.60, Kishinev,2009, Moldova

Summary

Time resolved photoluminescence of the $GaAs/GaAs_{0.76}P_{0.24}$ superlattice (SL) have been investigated at $90K$ temperature and different excitation levels. The transitions from electrons minibands to holes ones are explained and it is shown that experimental results are described by SL model of II type with valence-band offset larger than conduction band one.

1.Introduction

The $GaAs/GaAs_{1-x}P_x$ strained layers superlattice is one of the most intensively studied strained system. However, the band offsets at the heterojounction is still disputed [1, 2].

In this work a time resolved photoluminescence (PL) of the $GaAs/GaAs_{0.76}P_{0.24}$ SLs (layerwidths $\sim 60\AA$) grown by CHVPE method [3] have been investigated at $90K$ temperature and different excitation levels (*Fig.1a and 2*). It is shown, that location and halfwidth of PL maxima yields information about the energy levels and layers qualities, while a time dependence of PL peaks intensities gives information about a charge density space distribution. The model, presented in [1] (*Fig1b*) permits us to calculate the transition energies, which are in an agreement with experimentally obtained ones.

2.Results and discussion

For PL excitation the $He - Ne$ pulse laser ($\lambda = 632,8nm$) was used , what assured the absorbtion of exciting photons in the SLs region of the sample. The low excitation levels was carried out by beam disfocusing.

At low excitation level and the delay $0ns$ two radiation bands are revealed (*Fig.1a*). The maximum ($1.54eV$) corresponds to the transitions from E_1 to LH_1 and maximum ($1.582eV$) from E_2 to HH_2. The delay up to $40ns$ transforms the spektrum by sharply changing the intensity of low energetic band in comparison with the intensity of high energetic band , which witnesses about a higher life-time values of charge carriers in E_2 and E_1 state. Additional bands , corresponding to ($1.526eV$) from E_1 to HH_2 and ($1.594eV$) from E_2 to LH_1 appear.

At high excitation level (*Fig.2*) a large quantity of holes appears and at the delay $0ns$ transitions from miniband E_1 into all holes minibands in the vicinity of the valence band top are participating. At these the curve is displayed in the region of low energies relative to *Fig.1a* , and also a large half-width of radiation band is observed , which is stipulated as by the band edge spread at the boundary as well as by the state density increas in the direction of low energies. At the delay of $50ns$ maximum of ($1.501eV$) band remain at the same place , which is stipulated by high concentration of charge carriers. The delay increase up to $100ns$ leads to the appearance of the next radiation bands with the maxima : ($1.494eV$) transition from E_1 miniband to HH_1 ; ($1.527eV$) from E_1 to HH_2 ; ($1.545eV$) from E_1 to LH_1 and from E_2 to HH_1.

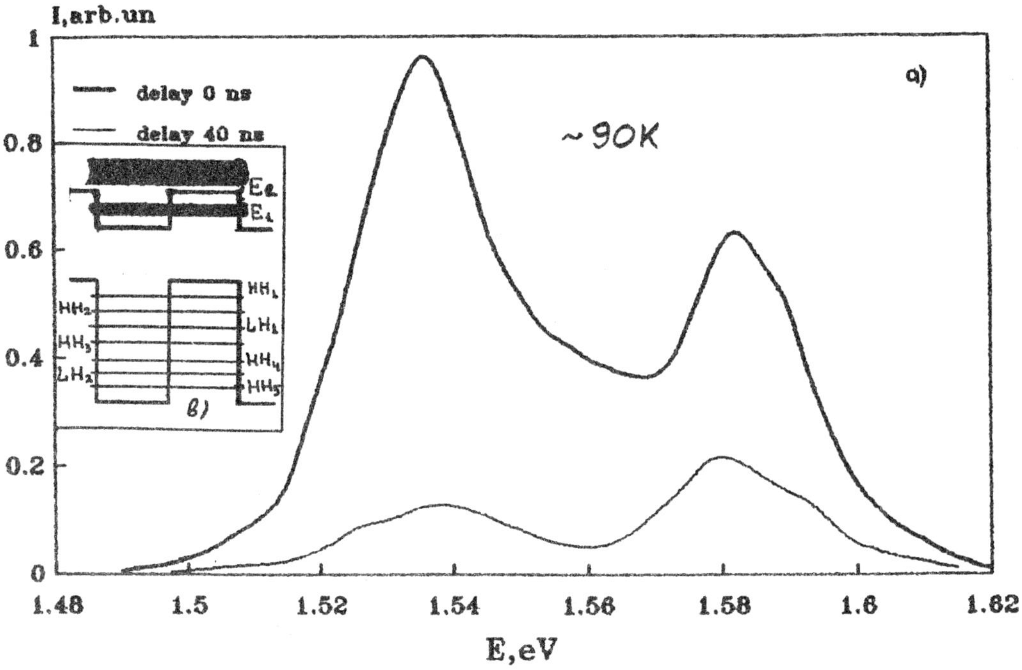

Fig.1 a)Kinetics of the PL at low exitation levtl.b)Energy band diagram.

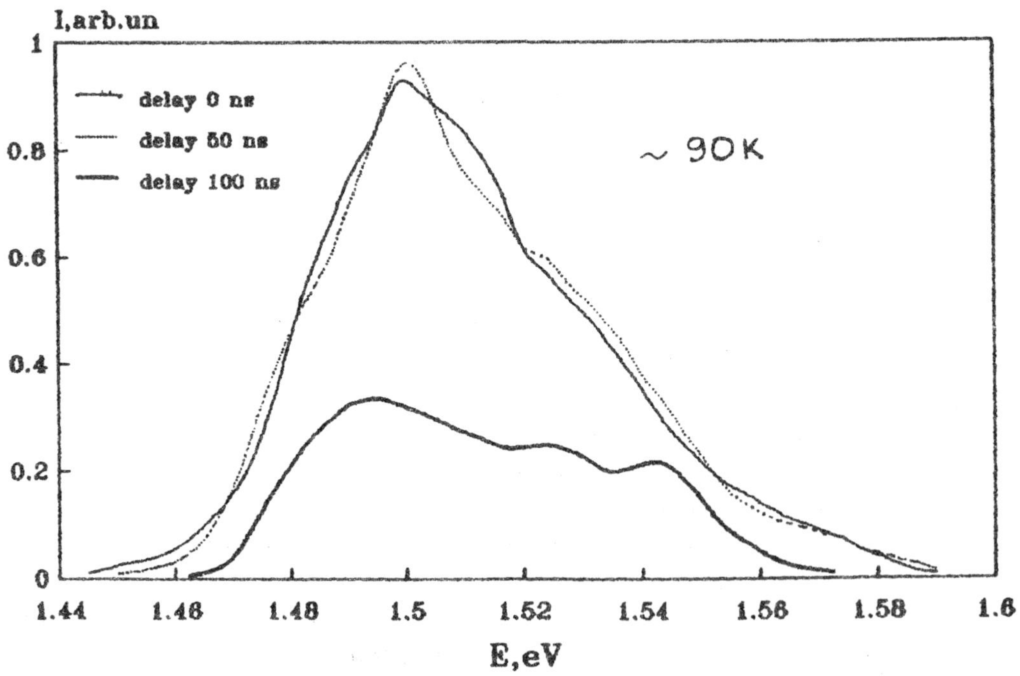

Fig.2 Kinetics of the PL at high exitation level.

In both cases ($Fig.1a$ and 2) one can see that with the delay time increase the miniband character of PL began to be well displayed , moreover these transitions are not limited by the parity selection laws. The conformity of the calculated and experimental transition values is the confirmation of the applicability of the model [1] to the $GaAs/GaAsP$ SLs.

To compare band structure models from [1] and [2] the transition energies for $GaAs/GaAs_{0.77}P_{0.23}$ SLs experimentally investigated in [2] were calculated with the band offsets from [1]. It is shown that all peculiarities of the photoreflectance spectra [2] are well represented in the model [1].

3.Conclusions

All the results obtained are in a good agreement with a theoretical prediction that conduction band in A_3B_5 semiconductors is formed mainly from an III-group atomic orbitals and the fact that a valence-band offset in $GaAs/GaAsP$ heterostructures is significantly larger than conduction band one.

References

1. Gorea O.S. et al.,J.Appl.Spectr.,1995,**V62**,N3,p.160

2. Hiroyuki Yaguchi et al., Jpn.J.Appl.Phys.,1993,**V32**,p.544

3. Gorea O.S. et al., Cryst.Prop.and Prep.,1991,**V32-34**,p.503

ANALOG ASIC FOR EDUCATIONAL PURPOSES*

Marius PĂDURE, «POLITEHNICA» University of Bucharest

ABSTRACT

This work presents a dedicated CMOS chip intended to be used as an efficient and practical teaching support tool. The chip contains typical building blocks (individual transistors, current mirrors, differential pairs) which can be externally configured to realize different typical circuits topologies including amplifiers. Some layout techniques are used such as centroid designs, waffle-shaped (closed) transistors, regular layout structures.

INTRODUCTION

The chip contains the following basic analog building blocks: 6 current mirrors, 2 differential pairs, 2 pairs of common source MOS transistors, 2 pairs for MOS devices characterisation. Each type of cell is implemented with NMOS and PMOS transistors. Because of the complementary symmetry, are implemented the same number of cells (for a given version) geometrical parameters for a specified type (e.g. differential stage) being the same. The configured circuits are biased using an external programmable instrument (e.g. HP4145). This chip is designed only for DC measurements. It is aimed for understanding some basic circuit topologies and allows to implement relatively complicated circuits with minimal number of blocks. Different circuits can be realized with external connections. After the practical measurements the students can compare the practical results with the simulation results. A lot of MOS circuits can be built using the aforementioned basic blocks: cascode current mirrors, regulated cascode current mirrors, OTA and folded cascode amplifiers.

The final chip is tailored for a 44 pins package and has a die area of about 3,1 × 3,1 mm^2. All pads are protected because the intended measurements are only in DC (the protection diode's parasitic capacitances being negligible in DC). The final layout is presented in figure 1.

BASIC BUILDING BLOCKS DESCRIPTION

1. Current mirrors

There are 3 NMOS (1:1 current ratio) current mirrors and the same number of PMOS mirrors. The mirroring ratio can be modified using another block that is integrated in the same

* This work was carried out during a stage in the frame of TEMPUS SJEP 08180-94 at Centro Nacional de Microelèctronica (CNM), Barcelona, SPAIN.

0-7803-3223-7/96/$5,00 © 1996 IEEE

chip such as one of the device with common source. For instance, 1:2, 2:1, 1:3 and 3:1 ratios can be achieved using this approach. This is only one of the multiple possibilities that can be used to accomplish different ratios using only the blocks implemented on the chip.

The NMOS and PMOS current mirror's layout is implemented using 2 basic cells all of these having the same dimensions (170µ × 170µ) for all the blocks (differential stages, MOS devices or common source devices). That means the dimensions are the same and the layout has the same structure, minimal modifications being needed to achieve one particular cell (for instance cell for the differential block). This versatility is the result of the fact that all the blocks have only two transistors (see figure 2).

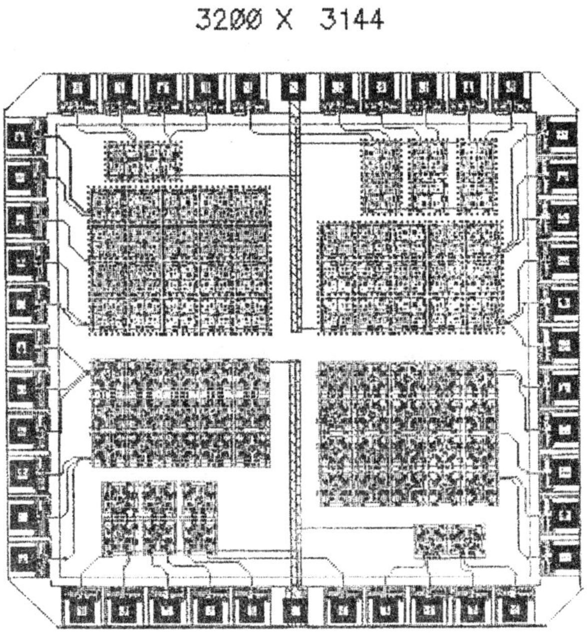

Fig. 1 Layout of the chip.

Geometrical parameters:

Each cell has 24 transistors each of these having W/L=6µ/6µ.

Because each block has two cells the resulting transistors of the current mirrors has 48 transistors in parallel. So W/L=48 × (6µ/6µ).

2. Differential stages

There is only one differential pair of every type (NMOS and PMOS). The transistors of these stages are larger than the transistors from the current mirrors or common source pair. The reason of that choice is the possibility to keep (with proper external bias) the transistors of the differential pair in weak inversion while the transistors of the current mirrors are in strong inversion. This bias will achieve maximum gain with a given active load transistors (fixed geometrical parameters and variable bias current from the external tester).

Because the width of these transistors is an order of magnitude greater than the current mirror's transistors, the measurements will indicate the amplifier's gain reduction as the differential's transistors entry in the strong inversion region (bias current will be increased).

Geometrical parameters:

Each block has 20 cells. This provides W/L=480 × (6µ/6µ).

The cell used in the differential stage layout has the same structure and dimensions as the cells used in current mirrors or common source blocks (the modifications at layout level are very easy to be made because of the common centroid structure which

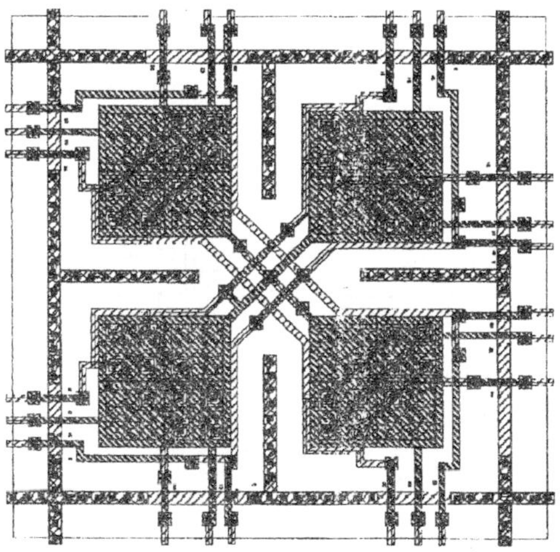

Fig. 2 A cell that implements the current mirror.

permits to make simple connections in the middle of the cell final block being just an a array of these cells). With that approach, no extra-connections are made, every cell being a differential stage.

3. Common source devices

There are two blocks (one PMOS and one NMOS) each of them having two transistors with their sources tied to VDD (PMOS) or VSS (NMOS). This topology can be used as active load as well as inverting amplifiers. Like the current mirrors these transistors are smaller than the differential stage transistors.

Geometrical parameters:

Because there are implemented 2 cells at each block, the geometrical parameters are the same with current mirror transistors:

$W/L=48 \times (6\mu/6\mu)$.

4. Block for MOS devices characterisation

These blocks are used in DC characteristics measurements (output and transfer characteristics). It can be used also as cascode topology amplifier (because of the floating source). The bulk is tied to VSS (NMOS) respectively VDD (PMOS) so one can measure the bulk effect impact on DC characteristics.

Geometrical parameters:

The layout is an array implemented using 15 cells and that will provide a $360 \times (6\mu/6\mu)$ MOS transistor aspect ratio.

THE LAYOUT

There are two basic cells (one for PMOS blocks and one for NMOS blocks). These basic cells have a rectangular layout that permits very easy to make regular blocks using that pattern alternatively mirrored in different directions. No intra-cells connections are needed because each of these cells has external connections. Final layout will be an array of these cells.

To obtain other cells (for differential blocks, common source devices or current mirrors) the cells mentioned above can be easily modified. The final layout of these blocks is achieved with the same approach (mirroring and repeatability). Finally, each of these cells will inherit only one structure that is the basic cell structure.

Two layout techniques are used:

1. Common centroid geometry is chosen because of the cancelation of random gradients of concentration;

2. Each part of splited transistors has a "waffle-shaped" geometry. That means there are 12 "closed" transistors (sources and drains are surrounded of polysilicon). That will provide a maximum W/L on unity area.

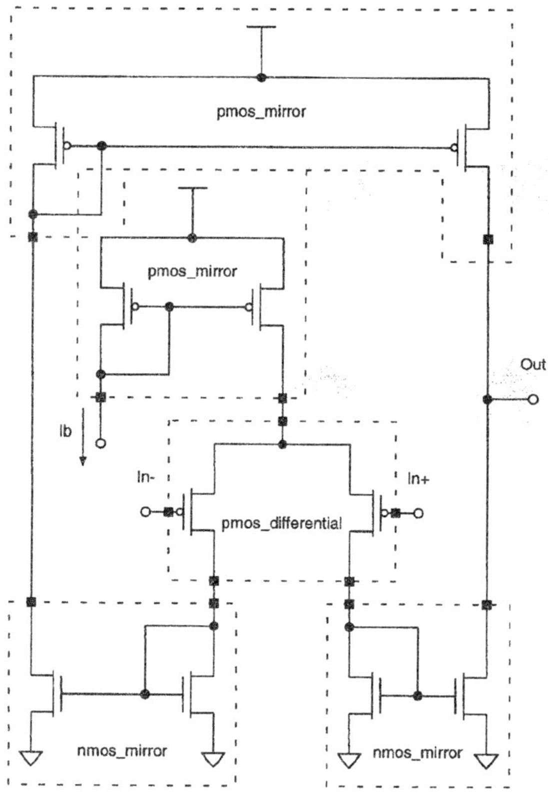

Fig. 3 Basic OTA schematics.

CONCLUSIONS

One CMOS chip for educational purposes was presented, containing basic analog building blocks. Complex circuit's layout can be generated very easy in a full custom design using standard and parametrized cells.

Some basic analog circuits can be implemented. For instance a simple OTA can be implemented and the schematic is presented in figure 3.

ACKNOWLEDGEMENTS

I would like to acknowledge the support provided by CNM Barcelona team and especially by dr. Lluis Téres i Téres and dr. Francesc Serra i Graells.

Bibliography

1. Vittoz, E., "Analog design for low power", *EUROCHIP Training Course*, 1992.
2. Gray, P.R. and R.G. Meyer, *Analog Integrated Circuits Analysis and Design*, 3rd edition, McGraw Hill, New-York, 1991.

THE EQUIPMENT FOR MEASURING OF CONCENTRATION PROFILES USING THE SPREADING RESISTANCE METHOD

J.Hybler and V.Švagr

Czech Technical University in Prague

Czech Republic

ABSTRACT

The equipment for measuring of concentration profiles in semiconductor structures using the spreading resistance method has been developed Device is designed for measuring local contact resistance in range from 1Ω to 10MΩ with high level of accuracy. Resistance is measured under conditions of dc constant current with measured contact voltage below 10 mV. The stepping of the sample holder and the measuring pin movement (down and up) and data collection and processing is fully controlled by the PC.

INTRODUCTION

In semiconductor component technology ,a good knowledge of dopant profiles is very important for effective controlling of technological processes. The spreading resistance method is one of methods commonly used for determination of dopant profiles in the whole extent of thicknesses coming into account including very thin layers. Using this method , we actually determine the local value of specific resistivity which is derived from the measured value of spreading resistance.

PRINCIPLES OF MEASUREMENTS

The spreading resistance can be in case of round shaped plane contact area expressed [1] as

$$R_s = \frac{\rho}{4a} K \qquad (1)$$

where ρ is the local resistivity

a is the radius of contact area

K is a correction factor depending on the layer thickness and resistivity gradient.

0-7803-3223-7/96/$5,00

As the sample is used a small piece of investigated structure bevelled under a small angel, as shown in fig.1. We can see, that depth of the point under the sample surface can be derived from a number of steps after crossing the boundary between the bevelled and unbevelled parts of the sample, providing that we know the bevelled angel and the length of one step. The configuration shown in the fig.1 consists of one measuring point contact and two Ga-Sn and Ga-Al contacts can be called the one point probe. Often is used two point probe measuring two contact resistencies(spreading resistencies)in series. The contact area is usually determined experimentally on homogeneous samples of well known resistivity.

To avoid high electric field effects and temperature increase in the contact area, that the d.c. measured voltage on the contact should be kept less than about 15mV [1]. Measuring the spreading resistance , local resistivity and consequently, the local dopant concentration can be determined from the Eq. (1). One of the very important problem is to determine the correction factor K for each measured point , particularly in case of very thin layer. Supposing that a layer with a given resistivity distribution can be replaced by the superposition of a finite number of thin layers with uniformly distributed resistivities of given constant values ρ_i and equal thicknesses w , the correcting factor K can be calculated from measured data by numerical methods [2], [3]. For that reasons, it is useful all the measured data be collected in a digital form.

For reliable evaluation of dopant profiles, there is necessary to carry out a larger amount of particular measurements under well defined conditions on the sample under test followed by subsequent numeral processing. Therefore the full control of measurements and data processing by a PC is desirable.

INSTRUMENTATION

For measuring concentration profiles by the spreading resistance method , highly specialized instruments by a few producers are available. The well known system made by the Solid State Measurements Inc. [4] were based on a principle of an analogue logarithming measuring circuit. The main disadvantage of this principle is given by the fact that the analogue logarithming circuit can cover the necessary extent of seven or eight decades, over which the measured values of spreading resistance can be spread, with a limited accuracy only. Therefore a measuring equipment was developed using a quite different principle of operation explained by the block diagram shown in fig.2.

The measurements of spreading resistance are performed under conditions of d.c. constant current. The current source is so designed that it can deliver into the measuring circuit of the probe , one of seven, fixed, exactly pre-set values of measuring current. For a particular measurement , the measuring current is automatically so chosen, that the probe voltage taken from the measuring circuit is in interval 1mV to 10mV. The local resistivity is derived from the measured voltage and the measuring current. This configuration enables a good accuracy of measuring in the interval of resistances under $10^7\Omega$. Higher resistances can be also measured, but without required accuracy. Very important is the mechanical part which ensures reproducible contact area (constant pressure on a pin and a well determined velocity of the pin moving down to sample surface). The movement of the sample by defined steps is realized by a step motor and micrometric screw pushing the X-Y table.

For communication and for the measuring is used ACL-812PG card. It is a multifunction, high performance high speed data acquisition card, containing A/D, D/A, DIO. Communication is realized by two parallel ports of the card and the voltage is measured by 12 Bit A/D converter. There are measured high resistances using low currents so the high input impedance of measuring probe is required. Accordingly we designed the high input impedance preamplifier using OA. All functions of the device are controlled by the PC : manual adjusting of the measuring probe position using the control panel situated on the body of the sample holder(mechanical part of system) and moving of the measuring pin (down and up).

For controlling of the measuring equipment was developed a program "MOS". A simple preview of the operation is shown in fig. 3. At first -input data about the sample: the angle of bevelled part , number of steps. There must be manually adjusted the position of the measuring pin on the edge of bevelled part of the structure surface. The measuring pin goes down, the lowest current is set in the source and voltage is measured. The measured voltage is tested if it is lower than 10 mV. The result can change the settings of the current source. The current is chosen so that the voltage is in the range from 1mV to 10mV. After writing data into memory the measuring pin goes up and motor moving X-Y table with the sample holder makes a step. The next measuring begins. After all data were collected, the sample holder goes back on the first position. After collection of all the set of data of the measured profile, the data processing (computation of correction factors K and active impurities concentration) is perform. For computation of the correction factor K, the program „KOPRO“ [5] is used.

CONCLUSIONS

The equipment for measuring concentration profiles using the spreading resistance method including software for device control and data processing has been developed. Its function was verified.

ACKNOWLEDGEMENT

Authors gratefully acknowledge Doc.Ing. Vitězslav Benda Csc. for his support, help and useful discussions.

REFERENCES

[1] Mazur R.G. ,Dickey D.H. : A spreading resistance technique for resistivity measurements on silicon.1966, Journal of the Electrochemical Society,Vol.113,No 3, p.255-259

[2] Piessenc, R. :Journal of the Electrochemical Society,130,1983,p. 468

[3] Choo, S.C. and others :Solid State Electronics, 19,1976, p. 561

[4] Solid State Measurements, Inc. :The ASR 100c/2 Technical Data and other manufacturer's publications

[5] Ringelhan, R. : Diploma theses, CTU Prague, 1994

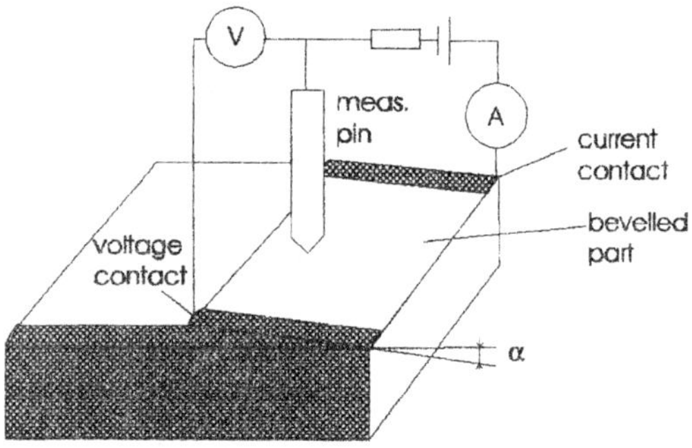

fig. 1.

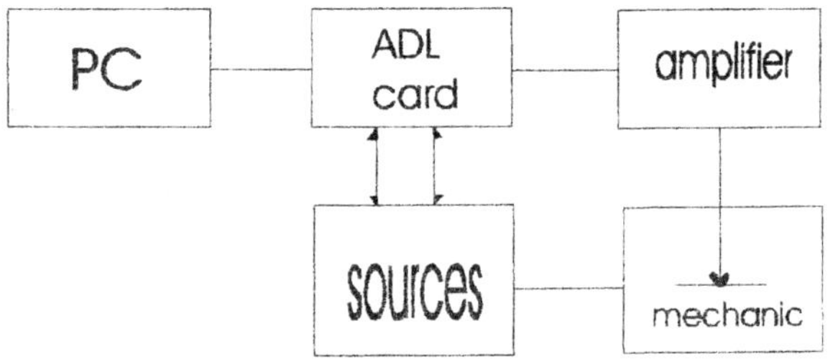

fig. 2.

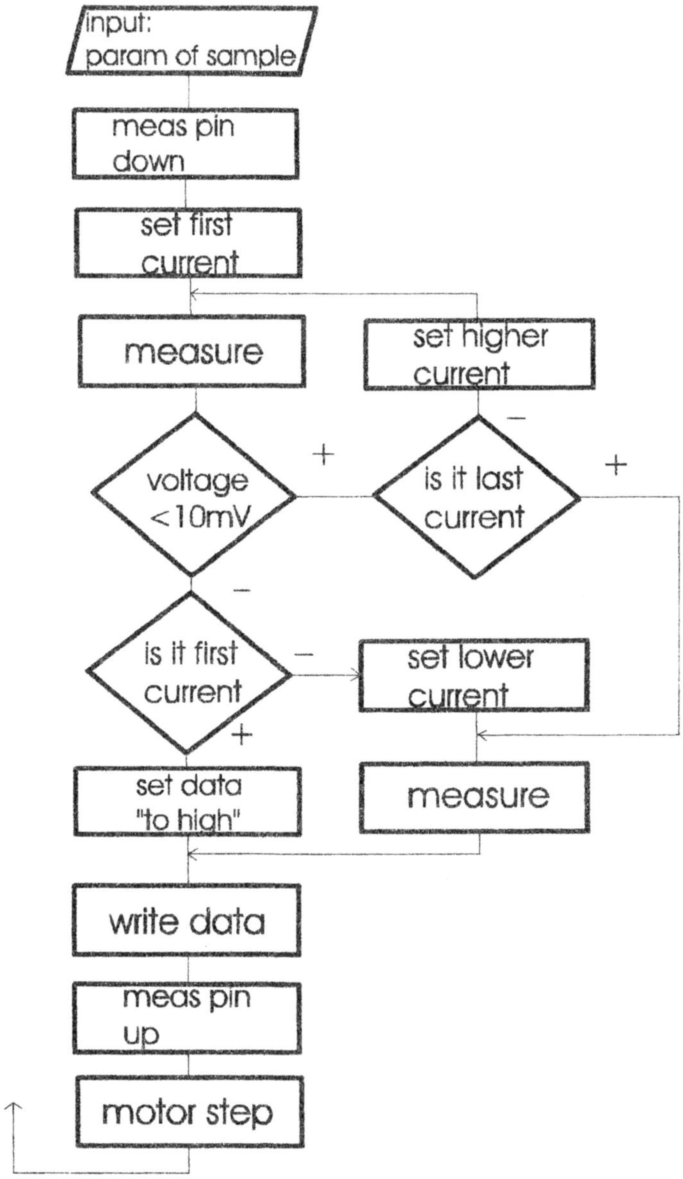

fig. 3.

Session 13

SEMICONDUCTOR PHYSICS 2 (Poster)

OPTIMIZATION OF THE SILICON-BASED TUNNEL MIS STRUCTURES AS HOT ELECTRON INJECTORS.

I.V.Grekhov, A.F.Shulekin and M.I.Vexler.

A.F.Ioffe Phys.-Tech. Institute, 194021 St.Petersburg, Russia

SUMMARY- Hot-electron Auger ionization in Al/ tunnel-thin oxide/silicon structures was shown to be used most efficiently in the devices with high n-substrate doping (Nd) and/or relatively large insulator thickness. The structures with high Nd were also found to be quite appropriate for observation of electroluminescence.

1. INTRODUCTION.

Si-based tunnel MIS structures are very interesting as hot electron injectors since they may operate at high current density (up to 1000 A/cm2 [1,2]) and the electron energy control is very simple. Such structures may be exploited as components of several devices, in particular as heteroemitters of Auger transistors [2] or as sections in thyristors [3].

After being injected into Silicon, hot electrons loss their energy not only via phonon scattering but also through impact (Auger) ionization and (as we will show) different radiative transitions (Fig.1). It is essential to decide - how to fabricate a tunnel MIS structure so that two latter mechanisms of energy losses would be used most efficiently?

In this paper we summarize our new results on the optimization of parameters of Al/SiO2/n-Si tunnel structures for different device applications.

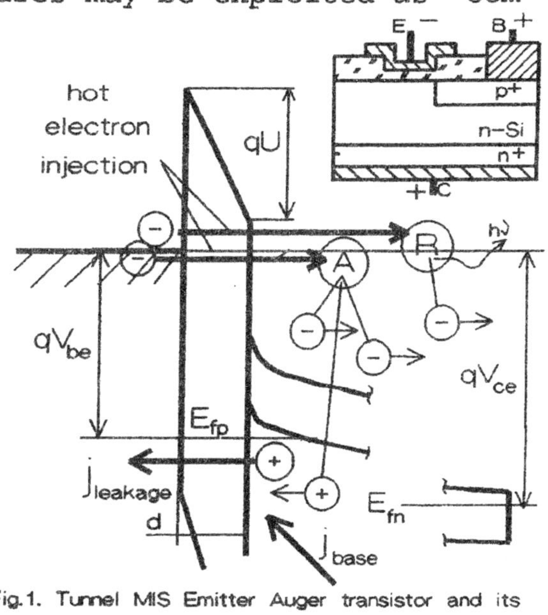

Fig.1. Tunnel MIS Emitter Auger transistor and its energy band diagram. "A" inside the circle denotes "Auger process", "R"– interband radiative transition.

2. HOT-ELECTRON TUNNEL INJECTORS IN AUGER TRANSISTOR.

Auger transistor [2] (which is probably the most promising application of studied structures) is a bipolar transistor

0-7803-3223-7/96/$5,00 © 1996 IEEE

with a tunnel MIS emitter (hot electron injector) and inversion base (Fig.1, insert). This device exhibits high current gain and demonstrates the bistability in its output characteristics (Fig.2). The nature of these features is positive feedback due to Auger ionization.

An important advantage of such a device is that the scaling down the emitter size is not restricted by the base thickness as in conventional bipolar transistors.

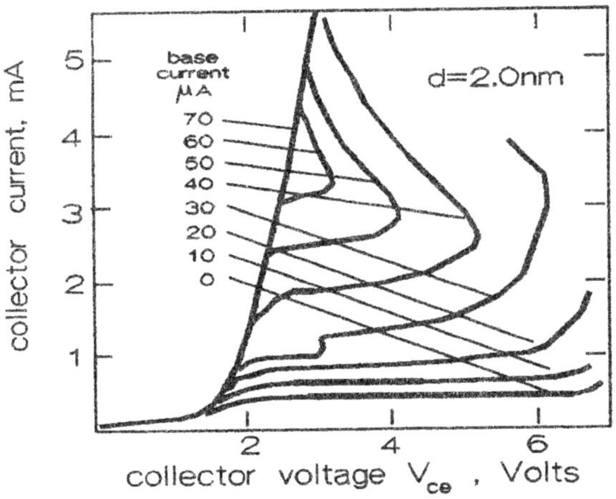

Fig.2. Common emitter DC characteristics of Auger transistor. Emitter area is 20×20 μm^2. Collector doping is 10^{16} cm^2.

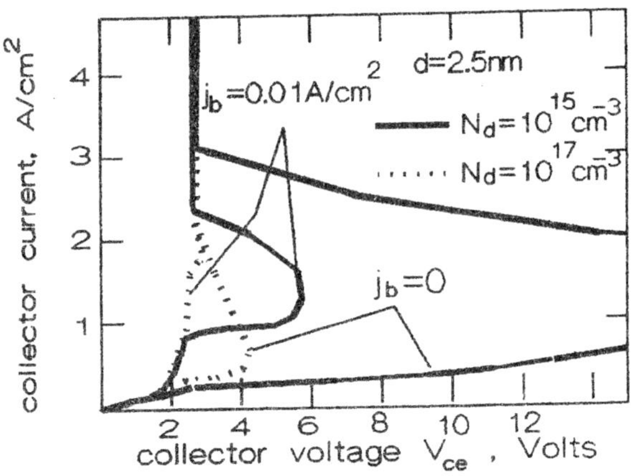

Fig.3. Effect of substrate doping (results of calculation. Experimental data are very similar.

2.1. Substrate doping effects.

As we have shown, the "width" of S-shape segment in DC characteristics of Auger transistor decreases with the increase of substrate doping Nd (Fig.3). There is no S-shape segment (and no active mode) for Nd>5*10^{18} cm^{-3}. Two effects are responsible for such a transformation: first, at high Nd the insulator voltage enough for activation of hot electron injection may be provided by depletion layer charge. Secondly, Auger process probability (and therefore current gain) increase due to additional electron heating in electric field in Si. Since the S-shape segment restricts the active region of Auger transistor operation, Nd cannot be higher than ~10^{17} cm^{-3} if this region should be several volts (in collector voltage).

2.2. Configuration of the base contact area.

It is of critical importance that there were no overlap of the p+ base contact area with the thin oxide area. As we have found out, even 5%-overlap lead to dramatic decrease of current gain, because the emitter current density is much higher namely in this 5%- area and injected electrons recombine with holes in this heavy doped p+ layer. To avoid this problem, we used in our devices the configuration of base contact area shown in Fig.4 (right), instead of a ring contact.

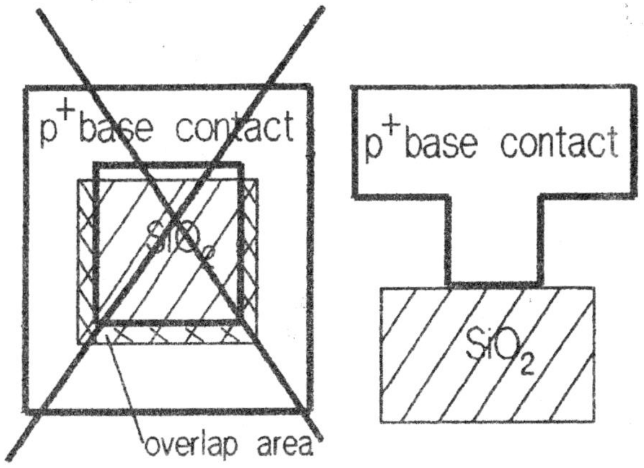

Fig.4. Configuration of the base area. Left: the overlap of emitter metallization (thin oxide area) with p^+-base contact area (such devices have poor current gain). Right – more appropriate configuration used in our devices

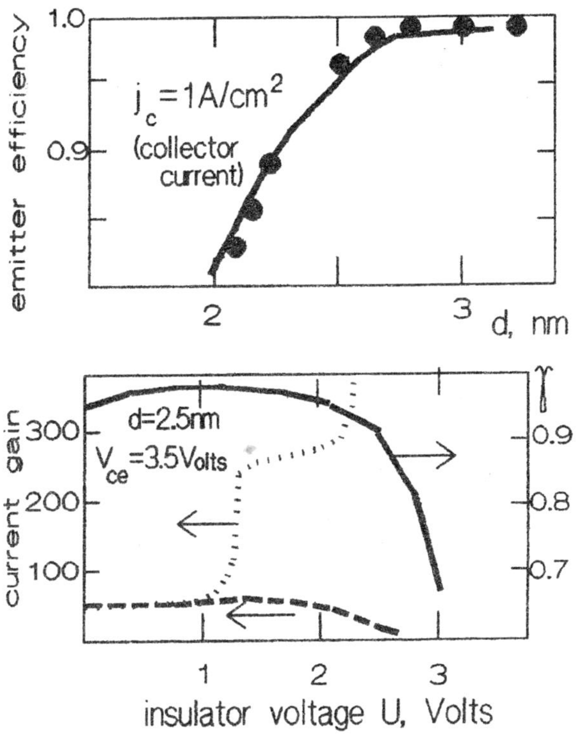

Fig.5. Upper – emitter efficiency vs oxide thickness. Line-calculaiton, dots-experiment. Lower: solid line emitter efficiency, chain curve – calculated current gain (disregarding Auger effect), dotted line – experiment: enhancement of gain due to Auger effect.

2.3. Emitter efficiency.

The efficiency of Al/SiO2/n-Si tunnel injector was found to decrease with the reduction of the oxide thickness d (Fig.5). The higher is emitter efficiency, the more narrow is the range of currents where the transistor may operate in active mode. Therefore, if we are interested in obtaining the highest possible current gain of Auger transistor, we should use the devices with high emitter efficiency (relatively high d (~3.0 nm)). Conversely, if it is desirable that the range of base voltage for active mode was very wide, the structures with small d (~1.5-2.0 nm) are preferred.

3. LIGHT-EMITTING STRUCTURES.

Another kind of interest to Al/SiO2/n-Si tunnel structure arises from its capability of visible hot-electron light emission which we have first observed (Fig.6). We expect that tunnel Si-based MIS structures with hot carriers injection may be successfully used for excitation of luminescence centers implanted into Si.

The most appropriate structure for observation of light-emission effect is the device with high Nd (its advantage is that it can be two-electrode since the applied voltage drops

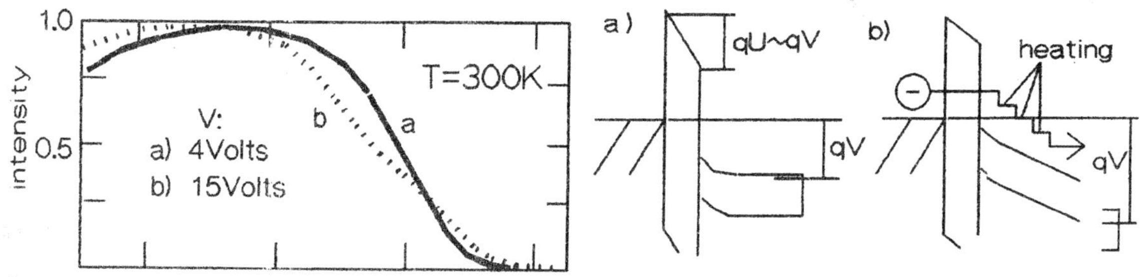

Fig.6. Measured electroluminescence spectra of Al/SiO_2/n-Si structures.
a) $N_d=10^{19}$ cm^{-3}, b) $N_d=10^{15}$ cm^{-3}.

primarily across the insulator). Another interesting operation mode is that where not only insulator bias but also electric field in depletion layer are high. In this case the efficiency of light emission increases, because the electron heating in Si is competitive to the phonon losses and the electrons remain hot in a large distance into the bulk of Si.

CONCLUSIONS.

The optimization of the tunnel MIS structure to be used in Auger transistor should be based on some compromise between the enhancement of small-signal gain and simultaneous "squeezing" of the active region with the increase of Nd and d. For light-emitting devices, the structures with high Nd are favorable.

REFERENCES.

[1] K.M.Chu, D.L.Pulfrey // IEEE Trans. Electr. Dev., v.ED-35, p.188 (1988).
[2] I.V.Grekhov et al. // Solid-State Electr., v.38, p.1533 (1995)
[3] W.K.Choi, A.E.Owen // J.Appl.Phys., v.68, p.6447 (1990)

NON-LINEAR TRANSPORT PROPERTIES OF PbTe<Ga> SEMICONDUCTORS

A. Nikorich, V. Kantser, V. Nikorich*, D. Constantinescu**
International Laboratory HTSC SSE, Kishinev, Moldova
*Moldova State University, Kishinev, Moldova
**ICP "Electromecanica", Ploiesti, Romania

SUMMARY

The results of the I-V characteristics (IVC) investigation on PbTe<Ga> semiconductor samples in the temperature range 67-100 K are given. IVC was found to be N-shaped, and at temperatures below 77 K the electric current oscillations through the sample were observed.

1. INTRODUCTION

One of the ways to change the free charge carriers concentration in semiconductors is their doping with various impurities. However, the influence of doping with elements of the III group on compounds A^4B^6 leads not only to lowering of the background concentration, but also to appearance of qualitatively new properties. Thus, in particular, PbTe doped with Ga in a certain range of concentration shows high photosensitivity at temperatures $T<T_c \approx 80$ K, and photoconductivity has a long-time relaxation and very unusual kinetics. Besides, in the given samples non-linear transport properties are often observed. In the present work the non-ohmic behaviour of IVC of these bulk materials and the current instabilities revealing whereupon are studied.

2. EXPERIMENTAL RESULTS

The measurements of the IVC were performed on the samples n-PbTe<Ga> with the charge carrier concentration $n \sim (1 \div 3) \cdot 10^{13} cm^{-3}$ and resistivity $\rho > 10^4 Ohm \cdot cm$ at 77 K.

The typical curves of IVC at the temperature 77 K are shown in Fig. 1. The characteristic is built in the coordinates voltage on the sample (U) - current through the sample (I). The significant feature of the ascending curve is that in a

0-7803-3223-7/96/$5,00

certain range of controlled current values the applied field is a multi-valued function of the current density, i. e. IVC is N-shaped. At the voltage descending with the same velocity the current value decreases, but the descending part of IVC does not repeat the form of the ascending one, indicating that we deal with the hysteresis effect. Thus, IVC is N-shaped at the voltage ascending and overlinear at its descending. At the increase of velocities of the voltage variation the IVC behaviour remains unchanged. It should be noted that the ohmic

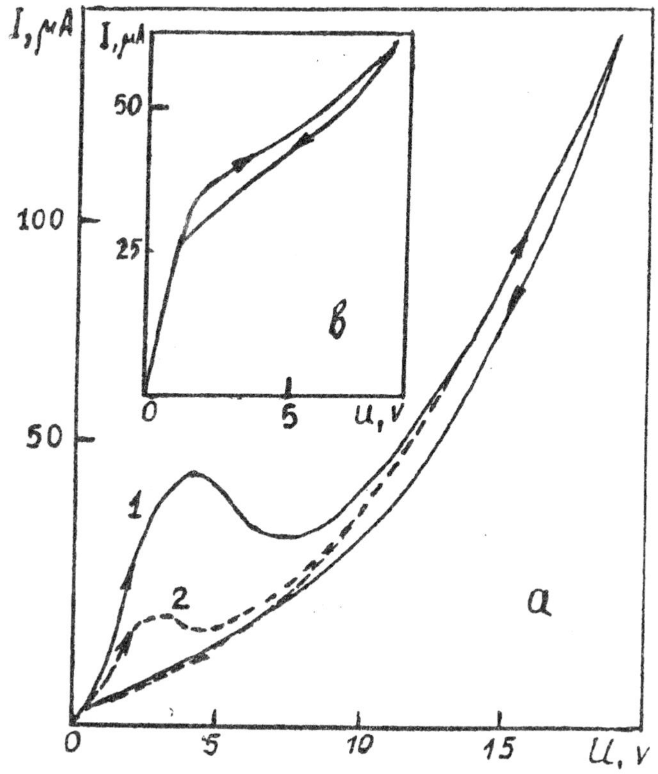

Fig. 1. IVC at the introduction and withdrawal of the voltage.
a) T=77 K, 1 -du/dt=0. 6 V/s; 2 -du/dt=0. 12 V/s;
b) T=100 K, du/dt=0. 6 V/s.

part of IVC curve at 77 K is observed only up to the voltage values U=0,25 V (E≃0,5 V/cm). With the temperature increase up to 100 K the ohmic part extends up to U≈ 0,8 V (E~1,5 V/cm) (Fig. 1b). At lower temperatures (T=67 K) the ohmic part of IVC curve practically disappears, however at the voltage more than 13 V on both ascending and descending curves of the IVC the current oscillations appear. Their amplitude increases with the voltage increasing and at U=25 V one is about 4-5% of the monotonous part of the current (Fig. 2).

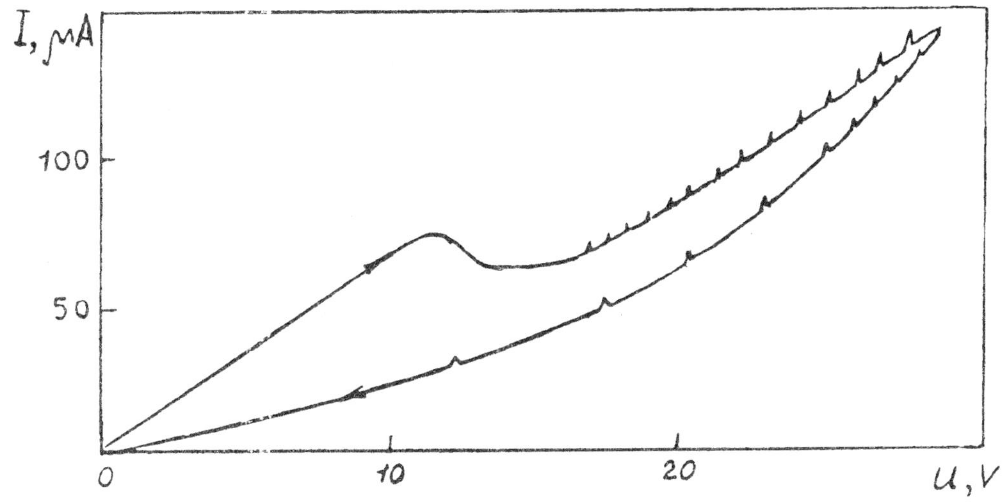

Fig. 2. The current oscillations on the IVC curve.
T=67 K; du/dt=0.3 V/s.

3. DISCUSSION

The complex form of the ascending branch of IVC may be caused by the influence of several mechanisms, one of which is the monopolar injection of electrons, trapping taken into account. The existence of the trapping processes is confirmed by the fact that at the fixed voltage the current through the sample decreases slowly in time. This mechanism determines mostly the dark conductivity at temperatures below 77 K when the ohmic section of IVC is practically absent. The ohmic contribution of free charge carriers to conductivity increases as the temperature rises.

The nature of the N-shaped part of IVC seems to be connected with the fact that in a certain range of electric fields the processes of carriers trapping overvail the processes of carriers generation in the sample, and the free carriers concentration decreases with the electric field increase. This is possible, firstly, if the trapping occurs in the result of tunneling through the Coulomb barrier. It is obvious that the tunneling probability increases with the increase of the electron energy. So, already in relatively weak fields, far from the breakdown range, the trapping coefficient will depend on the applied field and increases with increasing of the latter. In the region of rather strong fields the trapping coefficient increases slowly with the electric field because the considerable part of the carriers does not "feel" the Coulomb barrier action.

The current oscillation in a certain range of the electric field values we suppose to occur due to the formation and drift of electric domains along the sample resulting from the electrons redistribution between the band states and the trapping ones /1/. Really, if in some local region of the sample an instability appeared as a result of the increase of the electric field voltage, then such an increase leads to that of the trapping velocity, wherefore the free electron concentration in this region will decrease. This, in its turn, will lead to much larger growth of the field voltage, and as a result, to further decrease of the free electron concentration. At the same time near such a region, closer to the anode, there also forms a region with the bulk positive charge, as far as the electron influx to it from the cathode side out of the region with lowered concentration of electrons is less than the reflux to the anode side. As far as in the stationary state the electron influx from the cathode side corresponds to their reflux to the anode, the domain shifts to the side of the anode, that will result in the current spike at its teaching the anode.

CONCLUSION

IVC investigation in PbTe:Ga crystals shows that the formation of the latter is conditioned by the redistribution of electrons between the conduction band and the trapping centers, the electric field increase leading to the change of the trapping velocity.

REFERENCES

/1/. V. L. Bonch-Bruevich, I. P. Zvyagin, A. G. Mironov. Domain electric instability in semiconductors. Moscow, "Nauka", 1972. (in Russian).

Light and Heavy Charge Carrier Caused Breakdown of Silicon p-n Junction

T.Puritis

Laboratory of Semiconductor Physics, Riga Technical University,
1a Kalku str., Riga, LV-1658, LATVIA

SUMMARY - The initiation of impact ionisation is known to come mostly from light charge carriers. Transition from avalanche breakdown to thermoelectric breakdown, i.e., transition from light to heavy charge carrier caused breakdown is analysed in several microplasmas with increase in ambient temperature. Transition from light to heavy charge carrier caused secondary breakdown at increase current is analysed.

1. INTRODUCTION

We have previously [1-2] reported new aspects of the avalanche and secondary breakdown of the silicon p-n junction. An estimate was given for four models of energy accumulation, necessary for impact ionisation: Wolf's diffusion model [3], Shockley's model of "lucky" electrons [4], Ridley's model of "lucky-drift" [5] and Gribnikov's model of "light" electrons [6]. It was shown that impact ionisation is mainly realised in confirmity with the model of "light" charge carriers. According to this model, the energy that is necessary for impact ionisation may be accumulated by carriers located in the subbands of "light" carriers (the second conduction band for electrons and the spin-orbit split-off band for holes). This conclusion is based on three experimental evidences [1-2]. First, the strong effect of the magnetic field on the avalanche breakdown parameters [7]. Second, the coincidence of experimental results obtained for the impact ionisation threshold energies with the theoretical values of thresholds of "light" charge carrier subbands. Third, indications to the origin of inter-subband transitions of light emission in microplasma [8].

The first calculation recently performed by the Monte Carlo method using several bands showed that at field strength exceeding 500 kV/cm light charge carriers initiate more than 90 % of all the impact ionisation cases [9, 10].

Taking into account this new aspect we can analyse the experimental results of previous papers.

2. AVALANCHE AND THERMOELECTRIC BREAKDOWN

The avalanche breakdown (AB) is observed at lower ambient temperatures and is known to have a positive value of temperature coefficient of the breakdown voltage $\beta_B=(1/U_B)dU_B/dT$, where U_B is the breakdown voltage and T is the temperature. At higher ambient temperatures so called thermal breakdown related to a rapid increase in diffusion current is observed. It is characterised by large negative value of $\beta_B \approx -100\cdot10^{-4}$ deg^{-1}. A temperature region where β_B is negative but its magnitude is much lower than in the case of thermal breakdown has been observed by Vul and Shotov [11]. They called it the thermoelectric breakdown (TEB) assuming that both the impact ionisation and the thermal generation act simultaneously in this case.

When a saw-shaped voltage of a small amplitude is superimposed over the direct one, a breaking of the current-voltage characteristic (CVCH) caused by a separate microplasma may be observed [14]. In [15] a variation in the breaking caused by the first microplasma is studied with increase in the ambient temperature (Fig.1). At first the curvature of CVCH related to the unstable region of microplasma disappears (Fig. 1, a - d). Further temperature rise results in a rapid increase in premicroplasma current which, obviously, is the so called recombination-generation current (RGC).
The CVCH gradually straightens (Fig. 1, e - g) until the breakpoint completely disappears at 175°C (Fig. 1,h). Surprisingly, the slope of CVCH practically stays unchanged after the

0-7803-3223-7/96/$5,00

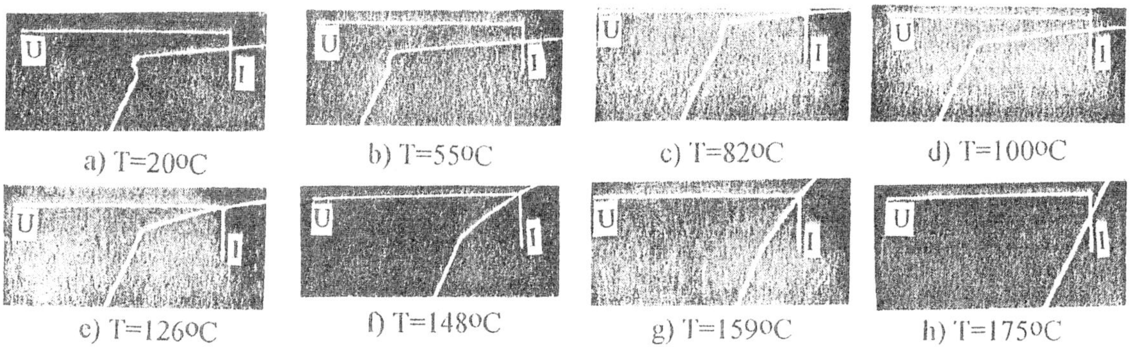

Fig.1.

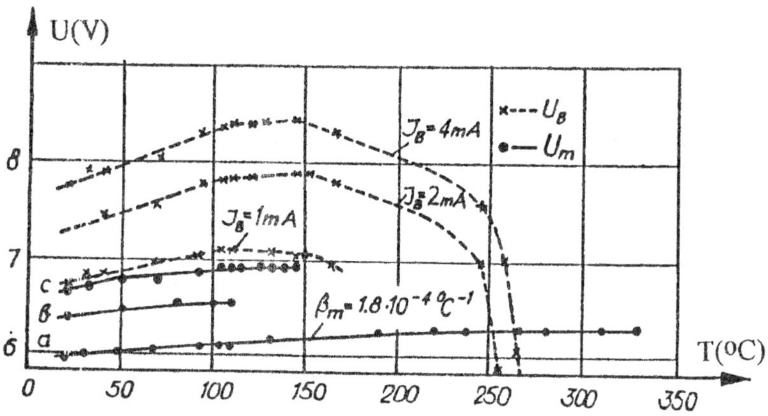

Fig.2.

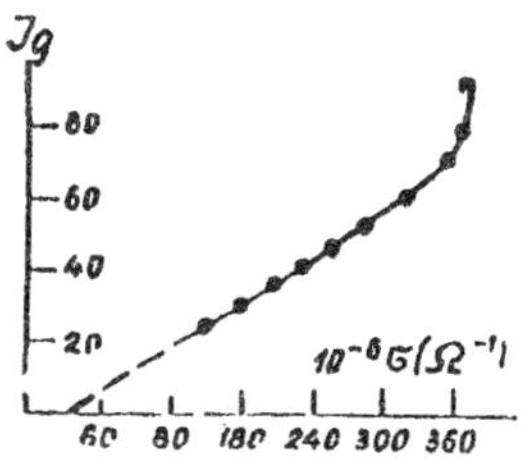

Fig.3.

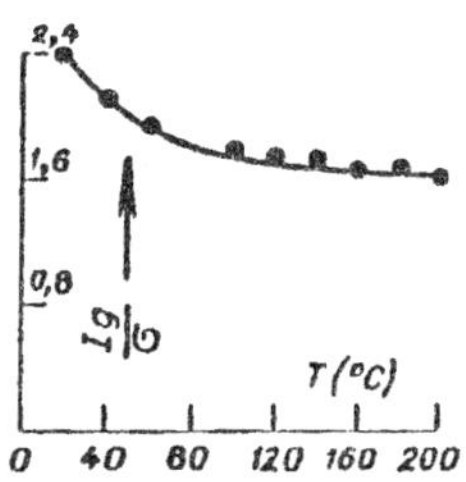

Fig.4.

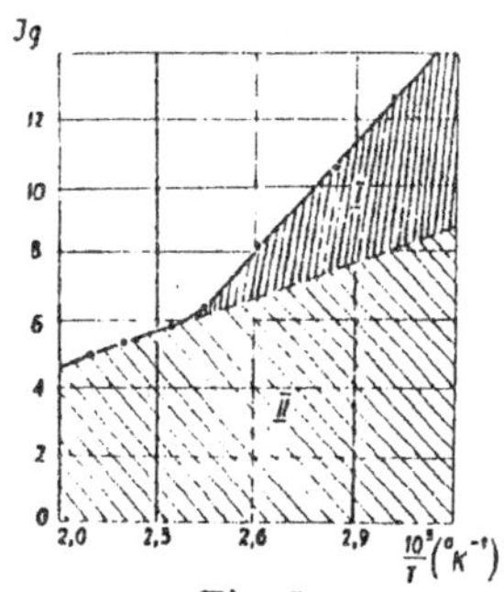

Fig.5.

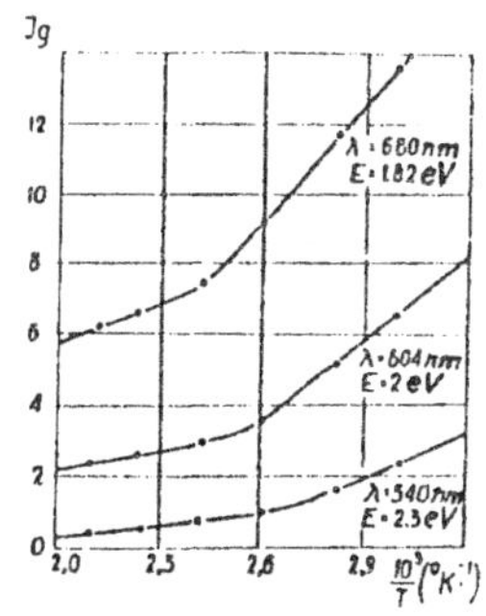

Fig.6.

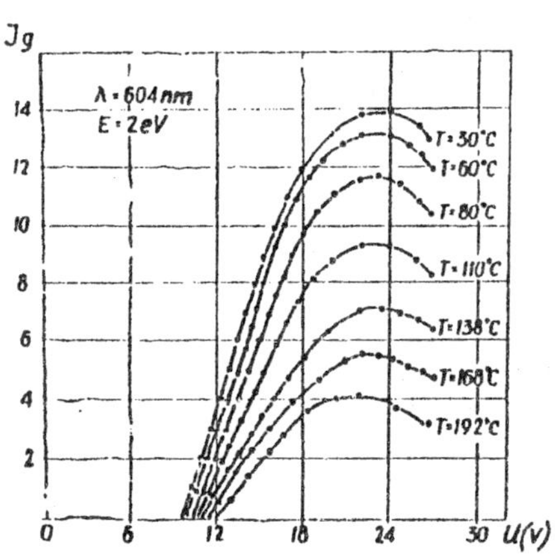

Fig.7.

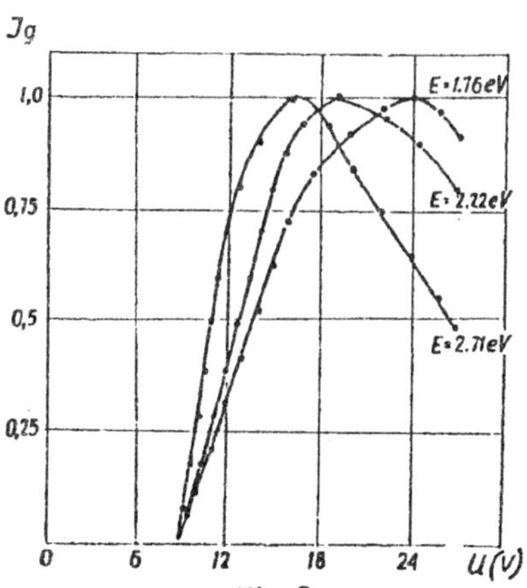

Fig.8.

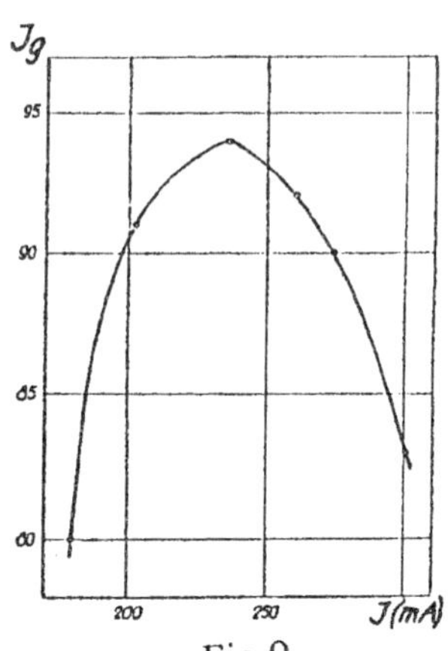

Fig.9.

microplasma lights up. An increase in total current through the p-n junction could be expected with temperature rise. However, the increase in RGC is accompanied by a simultaneous decrease in the avalanche current, the total current remaining unchanged. Such a result can be explained only if we assume that the RGC is localised and flows through the same microplasma channel.

Results of another experiment are depicted in Fig.2 where the temperature dependence of the breakdown voltage U_m for three microplasmas lighting up at much lower voltage than the rest is shown. The temperature dependence of the U_B of the whole p-n junction determined at three different current criteria: $I_m = 1$ mA, $I_m = 2$ mA and $I_m = 4$ mA, is also shown in this figure. The breakdown voltage curves a, b, c of separate microplasmas are depicted up to the temperature at which the breakpoint in the CVCH disappears. It may be seen that the temperature coefficient for separate microplasmas β_m is positive all the time, i.e., U_m grows with temperature. At the same time, a break in the U_m curves is seen at the disappearance of each microplasma. The rapid bending down of the U_m curves at 250°C is related to a rapid increase in diffusion current. From this figure we can see that the current flowing through the microplasma channel the value of β_m changes from positive to negative when the breakpoint of the microplasma disappears. Consequently, the term "thermoelectric breakdown" proposed by Vul and Shotov is correct, since the RGC in the spatial charge region of p-n junction is caused by thermal ionisation promoted by the electric field. However, it is not correct to think that impact ionisation takes place at thermoelectric breakdown.

In Fig.3 the light emission intensity I_g of the microplasma is shown in dependence of the additional conductivity σ [14] introduced by the avalanche current. σ decreases with temperature. It can be seen that the emitted light disappears even some before the complete disappearance of the breakpoint in the CVCH [14]. In Fig.4. the effectivity of light emission of the microplasma is shown, i.e., the ratio I_g/σ versus temperature. As it may be seen, I_g/σ decreases slightly with increasing temperature up to disappearance of the breakpoint in the CVCH.

The dependence of I_g on the inverse temperature 1/T (Fig.5 and Fig.6) is represented by a curve consisting of two straight lines [15]. The continuation of one of these lines/splits the whole region of radiation in to two parts. Each microplasma is known to consist of a nucleus and a shining cloud surrounding it. One may consider that the first region is related to the nucleus of the microplasma but the second one to the shining cloud.

In Fig.7 the variation intensity of a microplasma (at photon energy E = 2 eV) versus the applied voltage is shown at various ambient temperatures [15]. At the beginning the radiation increases, reaches a maximum and then decreases. The radiation intensity decreases with temperature. The higher the photon energy the lower is the voltage corresponding to the radiation maximum, as is shown on Fig.8.

Up to now it was not clear why the carriers of avalanche breakdown emit visible light, whereas the carriers of thermoelectric breakdown do not emit. It may be noted that in both cases charge carriers are generated at the same place of the microplasma, and they cross the same electric field, thus getting equal energy. This paradox becomes clear if we realize the breakdown can be caused by charge carriers of two types: light and heavy. The AB is caused by light charge carriers whereas the TEB by heavy ones. Light charge carriers can get higher energy from the electric field and emit radiation at transition to the heavy subband. The TEB is caused by heavy charge carriers which also get energy from the electric field. This energy, however, is transferred immediately to other charge carriers or, by means of phonons, to the lattice. Heavy charge carriers do not emit radiation since they are located in the lowest, i.e., the heavy charge carrier subband.

3. SECONDARY BREAKDOWN

The diffusion current (saturation current) is a current of heavy charge carriers and it does not emit radiation. When the heat sink from the p-n junction is bad, the diffusion current can increase unlimitedly and produce the secondary breakdown (with negative dynamic resistance), but without the light emission (it is the so called thermal breakdown). When the heat sink of the p-n junction is improved at sufficiently large avalanche current (high temperature) the shining clouds (encircled microplasmas) widen, move to the center of the p-n junction and meet there. The shining cloud gets the shape of a circle (pre-mesoplasma). Then it contracts quickly, and bright a bright circular-shaped localised plasma (called mesoplasma) lights up [16]. Radiative SB appears (with negative dynamic resistance).

Premesoplasma and mesoplasma are also caused by light charge carriers. Mesoplasma posseses similar light emission properties as the microplasma. The radiation intensity of the mesoplasma decreases with increase in ambient temperature. The intensity of emitted light first increases, reaches a maximum and then decreases (Fig.9) with increasing current through the mesoplasma, similarly as in the case of the microplasma (compare Fig.9 with Fig.7 and 8) [17]. When the light intensity of the mesoplasma falls to zero, the radiative SB is completely repleiced by the nonradiative SB (thermal breakdown). Consequently, the current of light charge carriers is exchanged by the current of heavy charge carriers.

4. CONCLUSIONS

1. The avalanche breakdown (caused by light charges) transforms into thermoelectric breakdown in separate microplasmas if the ambient temperature increases. Thermoelectric breakdown is due to the recombination- generation current and thus does not emit light.
2. If the mesoplasma current increases, light charge carrier flow changes to heavy charge flow (diffusion current) which does not emit light.

5. REFERENCES

[1]. T.Puritis. *Proc. of the 25th European Solid State Device Research Conference,* The Hague, The Netherlands 25the-27th September, Editions Frontiers, 635 - 638.

[2]. T.Puritis. *Lithuanian Journal of Physics,* N 5-6, 464-466 (1995).

[3]. P.A.Wolf. *Phys.Rev.,* **95,** 1415-1420.

[4]. W.Shockley. *Solid - State Electr.,* **2**, 35-67, (1961).

[5]. B.K. Ridley. *J.Phys.C: Solid - State Phys.,* **16**, 3373 - 3388 (1983).

[6]. Z.S.Gribnikov. *Zh. Eksp. Teor. Fiz.,* **78**, 2112 - 2122 (1978).

[7]. T.Puritis, J.Liberis, C. Maciulaitis and S.Lauza. *Lithuanian Journal of Physics*, **27**, 455-464, (1987).

[8]. W.Haecker. *phys.stat.sol.(a),* **25**, 301-310, (1974).

[9]. Y.Wang, K.F.Brennan. *J.Appl.Phys.,* **75**, 313-319 (1994).

[10]. J.H.Oguzman, Y.Wang, J.Kolnik and F.Brennan. *Institute of Physics, Conference Series,* **141**, 803-806, (1995).

[11]. B.M.Vul and A.P.Shotov. *Radiotehnika i elektronika,* **1**, 1080-1085, (1956).

[12]. A.G.Chynoweth and K.G.McKay. *J.Appl.Phys.,* **30**, 1811-1813, (1959).

[13]. R.Krike and T.Puritis. *Latvian J Phys.Techn.Sci.,* N 5, 20-27, (1966).

[14]. V.Eglitis, J.Balodis and T.Puritis. *Latvian J. Phys.Techn.Sci.,* N 2, 20-25 (1970).

[15]. V.Eglitis, J.Balodis, T.Puritis and E.Pentyush. *Latvian J. Phys.Techn.Sci.,* N 6, 58-66 (1970).

[16]. T.Puritis, D.Volks and J.Kaupuzs. *Solid-State Electron.,* **38**, 258-260 (1995).

[17]. J.Balodis, T.Puritis and A.Baduna. *Latvian J. Phys. Techn. Sci.* N1, 32-36 (1967).

THE INFLUENCE OF HIGHER MINIMA ON PHYSICAL PARAMETERS IN ULTRA HEAVILY DOPED n-TYPE GaAs

Miloš B. Živanov, Ljiljana D. Živanov and Jovan Dj. Jovović
Fakultet tehničkih nauka, Novi Sad, Yugoslavia

Summary: *In this paper was calculated the position of Fermi energy level and effective mass of electron in n-type GaAs as a function of donor concentrations in the range from 10^{17} to 10^{21} cm^{-3}, taking into account upper minima in conduction band, at 300 K. Calculations show that is necessary taken besides Γ and L minimum for concentrations greater than 10^{19} cm^{-3} and all three minima (Γ, L and X) for concentrations greater than 10^{20} cm^{-3}.*

Indexing terms: Ultra heavily doped semiconductors, Fermi level, density of states effective mass

1. INTRODUCTION

Semiconducting compounds such as GaAs used for the construction of lasers, FETs and similar devices are usually doped in the range up to 10^{19} cm^{-3}. Modern HBTs are fabricated with p-doping of 10^{20} cm^{-3} [1] and n-doping $3\cdot10^{19}$ cm^{-3} [2], and doping densities above 10^{21} cm^{-3} have been achieved [3]. n-type doping in thin layers about 10^{20} cm^{-3} was shown in [4]. Despite significant advances in GaAs processing many difficulties remain to be overcome. The transport properties of materials with such doping are therefore of great technical interest.

It was shown in [5] that heavy doping distorts the band structure radically. The relative insensitivity of the transport coefficients to the band distortion appears to be due to the fact that the contribution of the electrons occupying the distorted part of the band is very small compared to the contribution of electrons occupying the higher undistorted regions of conduction band.

For calculations the physical and transport parameters in GaAs the most autors[5,6,7] considered only Γ_6 minimum. The dependence of Fermi energy level and effective mass on concentration in n-type GaAs for concentrations lower than 10^{19} cm^{-3} were earlier observed [8]. In our previous paper[9] we also calculated Fermi level and average effective mass in n-type GaAs taking into account only Γ minimum of conduction-edge band. From obtained result it is seen that Fermi energy level enters the conduction band at donor concentrations about 3 10^{17} cm^{-3}. It was shown that characteristics (for example effective mass) would be changed from this concentration. Fermi energy level begins to increase rapidly for concentrations that are greater than concentration at which Fermi level enters the conduction band and to attain upper minima. Because, for higher concentrations it is necessary to include in the calculation and simulation, and other minima, particularly in ultra heavily doped (UHD) GaAs. Γ_6, L_6 and X_6 minima have energy gaps of 1.42 eV, 1.71 eV and 1.90 eV above the top of valence band, respectively[6].

2. THEORETICAL MODEL AND RESULTS

The total number of conduction electrons (assuming that on 300 K all impurity atoms are ionized) should be written as a sum of three contributions

$$n = n_\Gamma + n_L + n_X = \int_0^{1eV} \rho_\Gamma(E) f(E) dE + \int_{\Delta_{\Gamma L}}^{1eV} \rho_L(E-\Delta_{\Gamma L}) f(E) dE + $$
$$+ \int_{\Delta_{\Gamma X}}^{1eV} \rho_X(E-\Delta_{\Gamma X}) f(E) dE \quad , \tag{1}$$

0-7803-3223-7/96/$5,00

where n_Γ, n_L and n_X are number of electrons in Γ, L, X minima, respectively, $\rho_\Gamma(E)$ is nonparabolic density of states for Γ minimum taken as in Altschul *et al.*[10], $\rho_L(E)$ and $\rho_X(E)$ are parabolic density of states for L and X minima, $\Delta_{\Gamma L}$ is energy elevation (E_L-E_Γ)=0.284 eV, $\Delta_{\Gamma X}$ is energy elevation (E_X-E_Γ)=0,476 eV, $f(E)$ - Fermi-Dirac distribution function of electrons.

Nonparabolicity factor, α, and effective mass of electrons at the conduction band minimum Γ, m_{co}, are functions of energy gap at Γ minimum , $E_{g\Gamma}$. These connections are obtained from the formula given in [6]. In heavily doped semiconductors it was observed the band gap narrowing as a function of impurity concentrations. In the present paper we took published results for the band gap narrowing given in [11] and we substituted it into relations for α and m_{co}. Band-gap narrowing of GaAs as a function of doping concentration has been measured using n-doped samples with concentrations from $3\cdot10^{17}$ cm^{-3} to $3\cdot10^{18}$ cm^{-3}. Assuming that same relation is valid above this range, we calculated that ΔE_g = 158 meV for $N_D=10^{19}$ cm^{-3} , ΔE_g = 340 meV for $N_D=10^{20}$ cm^{-3} and ΔE_g = 734 meV for $N_D=10^{21}$ cm^{-3}. These values are more applicable than the values ΔE_g = 1006 meV for $N_D=10^{20}$ cm^{-3} in [12] and have better agreement with experiment at $N_D=10^{19}$ cm^{-3}. In ρ_L and ρ_X, density of states masses are m_L=0.56 m_o and m_X=0.85 m_o [6]. m_o is mass of free electron. Using relation (1), we calculated the concentration dependence of the three-way split of conduction electrons among the Γ_6, L_6 and X_6 bands. That is shown in Fig. 1.

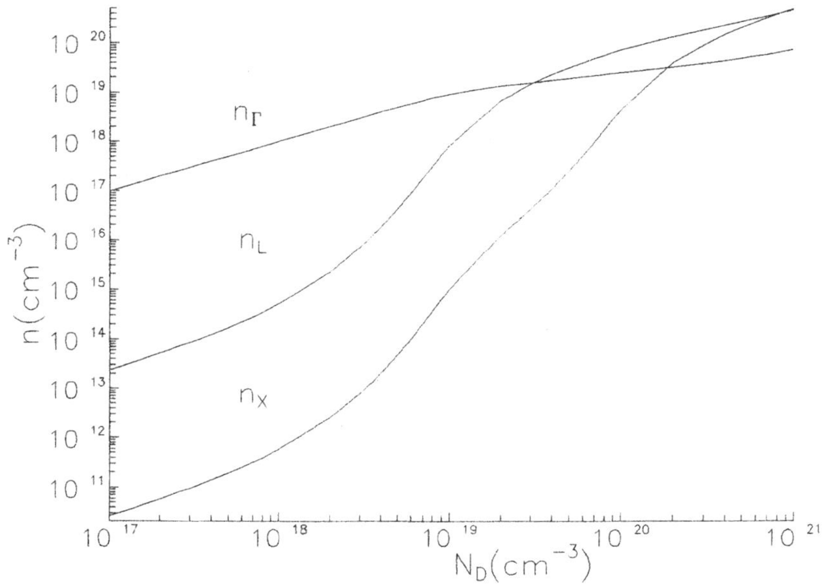

Fig. 1. *Concentration dependence of the three-way split of conduction electrons among the Γ_6 (n_Γ), L_6(n_L), and X_6(n_X) bands.*

The presented results in Fig. 1 show that the number of electrons at the L_6 minimum attained number of electrons in Γ_6 minimum at donor concentrations (N_D) of 10^{19} cm^{-3} and that for greater concentrations this number is several times greater. Similarly it is happened with number of electrons at X_6 minimum for N_D >10^{20} cm^{-3}. Numbers of electrons in L_6 and X_6 minima become same for N_D $\approx10^{21}$ cm^{-3} .

Assuming that all donors, N_D , are ionized at room temperature, the total number of majority electrons $n= N_D$. From this condition, using self-consistent method we obtained the dependence of Fermi energy (E_F) 'on donor concentration. Integrating limits are taken from 0 to 1 eV, because all significant influence on results is in this limit. The result for E_F as a function of N_D is given in Fig.2, for case labeled (b) with only Γ minimum and (a) all three minima and this results we are using late in this work. Taking into account only Γ_6 minimum for calculations E_F, it reaches 1 eV at concentrations about 10^{20} cm^{-3} and begins to increase rapidly. Including upper minima (L_6 and X_6)

in calculations it is seen that E_F changed much slower and reaches 1 eV at concentrations about 10^{21} cm^{-3}.

For parabolic isotropic bands electron concentration, n, is given with

$$n = N_c F_{1/2}(\eta) = 2\left(\frac{2m^* \pi kT}{h^2}\right)^{2/3} F_{1/2}(\eta) \quad , \tag{2}$$

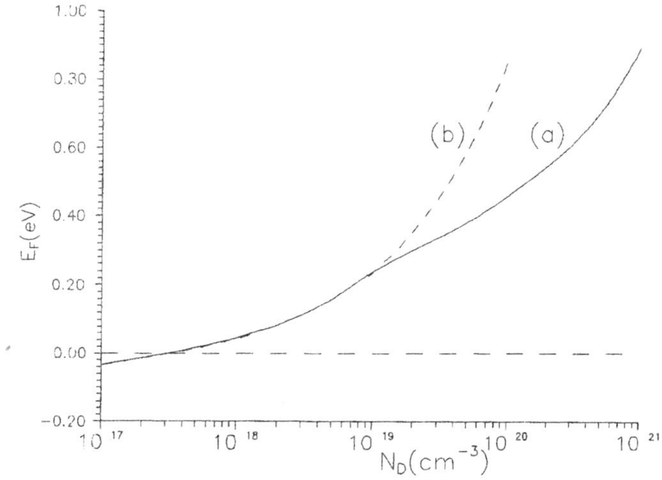

Fig.2. *Variation of the Fermi energy level with donor concentrations taking into account (a) all three minima (Γ_6, L_6 and X_6) and (b) only Γ_6 minimum.*

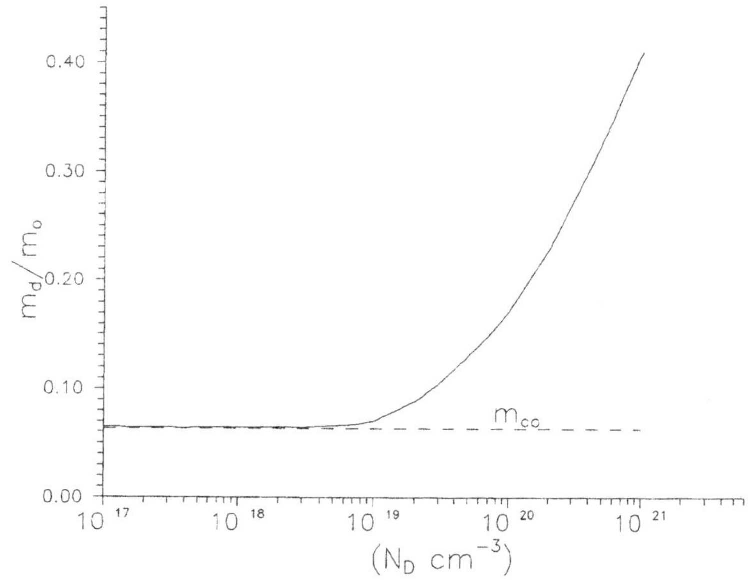

Fig.3 *Density of states effective mass, m_d in the n-type GaAs as a function of the donor concentration N_D*

where $F_{1/2}(\eta)$ is Fermi-Dirac integral and $\eta = E_F/kT$. m^* is effective mass at the bottom of the parabolic band. For a simplified view of the distribution among the band minima at high concentrations, we may rewrite (1) in the form of (2) by defining a density of states effective mass given by

$$m_d = \frac{h^2 N_D^{2/3}}{2^{5/3} \pi kT \left[F_{1/2}(\eta)\right]^{2/3}} \tag{3}$$

Then the three quantities on the right of eqn. 1 ($n = N_D$), we can describe by model for one parabolic, spherical, conduction band with effective mass m_d. This approach is similar to that given by Nag[5] and Blakmore[6] and it is convenient for some kind of device simulations. The calculated results from relation (3) for m_d as a function of donor concentrations are shown in Fig.3. Our results of calculations show that is justifiable the approximation of constant effective mass up to 10^{19} cm^{-3}, which, for example, was used by Lowney and Bennett [7].

For the practical using obtained results it is convenient to introduce the approximate relation, which fits the results in Fig.3,

$$m_d = \frac{m_{d\min} - m_{d\max}}{1 + (N_D / N_0)^a} + m_{d\max} \quad (4)$$

Where, m_{dmin} =0.063 m_o is effective mass in bottom of Γ minimum, m_{dmax}=0.85 m_o is density of states effective mass at the X minimum and the fitting parameters are N_D =1.2 10^{21} cm^{-3} and a=0.8.

3. CONCLUSION

The proposed model describes the dependence of E_F and m_d in UHD n-type GaAs. This approach can be used to analyze the mobility and other transport parameters in UHD semiconductors. It is evident, from obtained results, that the modified distribution of electrons can have great influence on the characteristics of UHD semiconductors. The structures of UHD n-type GaAs at this moment is not possible to realize due the technological problems. For that these results are interesting for prediction the properties of such structures.

References:

[1] P.M.ASBECK et al.: “Heterojunction bipolar transistors for microwave and milimetar-wave integrated circuits”, IEEE Trans. on Electron Devices, Vol.**ED-34**(12),1987, pp.2571-2577

[2] MAU-CHUNG F.CHANG et al.: “AlGaAs/GaAs heterojunction bipolar transtors fabricated using a self-aligned dual-lift-off process”, IEEE Elect. Dev. Lett., Vol.**EDL-8**(7), 1987, pp.303-305

[3] M.S. LUNDSTROM, et al: “Device related material properties of hevily doped gallium arsenide”, Solid-St Electron., Vol. **33**(6), 1990, pp.693-704

[4] J.P. de SOUZA and D.K.SADANA: Ion implantation in galium arsenide MESFET technology”, IEEE Trans. on Electron Devices, Vol.**ED-39**(1), 1992, pp.166-175

[5] B.R.NAG:"Electron Transport in Compound Semiconductors", Springer Verlag, 1980

[6] J.S.BLAKEMORE: "Semiconducting and other major properties of GaAs", J. Appl. Phys., Vol.**53**, (10), 1982, pp.R123-R181.

[7] J.R.LOWNEY AND H.S. BENNETT, "Majority and minority electron and hole mobilities in heavily doped GaAs", J. Appl. Phys., Vol.**69**, 1991, pp.7102-7109

[8] M.CARDONA, “Electron effective masses of InAs and GaAs as a functin of temperature and doping”, Phys.Rew. Vol.**121**(3), 1960, pp.752-758

[9] LJ.D.ŽIVANOV, M.B.ŽIVANOV, "Determination of average effective masses of majority carriers as a function of impurity concentrations for heavily doped GaAs", Proc. Conf, CAS'95, 1995, pp.103-106

[10] V.ARIEL ALTSCHUL, A.FRAENKEL, E.FINKMAN: "Effects of band nonparabolicity on two-dimensional electron gas", J.Appl.Phys., Vol.71, 1992, pp.4382-4384

[11] G. BORGHS, , et al: "Band-gap narrowing in highly doped n- and p-type GaAs studied by photoluminescence spectroscopy", J. Appl. Phys., Vol. 66, 1989, pp.4381-4386

[12] S.C. JAINT and D.J.ROULSTON: “A simple expression for band gap narrowing (BGN) in heavily doped Si, Ge, GaAs and Ge_xSie_{1-x} strained layers”.Solid-St Electron., Vol. **34**(5), 1990, pp.453-465

NONLINEAR FREE NUTATION OF POLARITONS IN DIRECT-GAP SEMICONDUCTORS

Igor V. Beloussov and Vladimir V. Frolov
Institute of Applied Physics, Academy of Sciences of Moldova,
Academiei Street 5, 277028 Kishinev, Moldova

SUMMARY

Time evolution of a nonequilibrium polariton condensate has been studied in the framework of a microscopic approach. It has been shown that due to polariton-polariton scattering a significant condensate depletion takes place in a comparatively short time interval. The condensate decay occurs in the form of multiple echo signals. Distribution-function dynamics of noncondensate polaritons have been investigated.

1. INTRODUCTION

It is known [1] that coherent electromagnetic radiation resonant to the isolated exciton energy level excites in the crystal the coherent polariton wave with the wave vector $\mathbf{k}_0 \neq 0$ — the nonequilibrium polariton condensate. Different scattering processes accompanying its propagation lead to the loss of initial coherence of the polariton wave, complete or partial depletion of condensate, excitation of polaritons characterized by some statistics with wave vector $\mathbf{k} \neq \mathbf{k}_0$, and other phenomena.

At sufficiently high excitation energies the processes of polariton scattering due to exciton-exciton collisions prevail. This scattering mechanism is of considerable interest due to the recent experimental investigations [2] and many interesting physical results obtained in theoretical study of dynamic and kinetic processes in the system of interacting polaritons (see e.g. [3,4]).

According to [4] exciton-exciton scattering is of significant importance in the situation where coherent polaritons are excited in a certain spectral region where energy and momentum conservation laws allow real processes of two-quantum excitation of polaritons from the condensate. These processes lead to instability of the completely condensed state of the polariton system. Existence of this spectral region situated around the isolated exciton resonance is due to the peculiarities of the polariton dispersion law.

Because of essential nonstationarity of the processes in the system methods of nonequilibrium mechanics should be used to describe it adequately. Derivation of the equations that describe kinetics of the polariton condensate decay and excitation of quantum fluctuations has some specific features owing to degeneracy in the system. As the energy and resulting wave vector of two non-condensate polaritons can be equal to the energy and wave vector of two condensate polaritons, respectively, there is degeneracy of two-particle states. Moreover, the presence of the condensate in the system also leads to degeneracy due to its macroscopic amplitude [5]. The correct description of the system with degeneracy requires the introduction of abnormal distribution functions $\Psi_{\mathbf{k}_0} = \langle \Phi_{\mathbf{k}_0} \rangle$ and $F_{\mathbf{k}} = \langle \Phi_{\mathbf{k}} \Phi_{2\mathbf{k}_0-\mathbf{k}} \rangle$ together with the usual (normal) ones $N_{\mathbf{k}} = \langle \Phi^+_{\mathbf{k}} \Phi_{\mathbf{k}} \rangle$. Here $\Phi^+_{\mathbf{k}}(\Phi_{\mathbf{k}})$

0-7803-3223-7/96/$5,00

are Bose operators of creation (annihilation) of a polariton on the lower branch with the wave vector $\mathbf{k}$. The appearance of abnormal averages indicates a breaking of the selection rules connected with the gauge invariance of the system [6]. In this problem the breaking of gauge symmetry arises as a result of the noninvariant initial condition due to the assumption that there is a condensate in the system at the initial moment $t = 0$.

Kinetic equations describing evolution of partially coherent polaritons that take into account the degeneracy were obtained in [7,8]. The right-hand sides of kinetic equations obtained in [7,8] include terms linear in the constant of exciton-exciton interaction $\nu > 0$ and the ones $\sim \nu^2$.

Terms $\sim \nu$ correspond to the self-consistent field approximation, which neglects the higher-order correlation functions. In this approximation the processes of two-particle excitations of polaritons from the condensate, backward processes, and transformation of a created pair of noncondensate polaritons into another pair with the same value of total momentum are taken into account. It describes the fastest processes in the system and is sufficient for study of the early (before-kinetic) stage of the system's evolution.

Terms $\sim \nu^2$ take into account scattering processes in which only one polariton can belongs to the condensate. Therefore, they differ from zero only if noncondensate polaritons exist in the system. Terms $\sim \nu^2$ in comparison with terms $\sim \nu$ describe slower changes of the system characteristics and are significant only at the kinetic stage of evolution.

2. NUMERICAL RESULTS

In self-consisted field approximation time evolution of the polariton system is described by the closed set of nonlinear integrodifferential equations for the coherent part of polariton field $\Psi_{\mathbf{k}_0}$ and the normal $n_{\mathbf{k}} = N_{\mathbf{k}} - \delta_{\mathbf{k},\mathbf{k}_0}|\Psi_{\mathbf{k}_0}|^2$ and abnormal $f_{\mathbf{k}} = F_{\mathbf{k}} - \delta_{\mathbf{k},\mathbf{k}_0}\Psi^2_{\mathbf{k}_0}$ distribution functions respectively. The numerical solution to these equations is shown in Figs. 1–3. We use the notation: $w_{\mathbf{k}} = \lambda^{-1}(\Omega_{\mathbf{k}} + \Omega_{2\mathbf{k}_0-\mathbf{k}} - 2\Omega_{\mathbf{k}_0})$, where $\lambda = (\nu/2\hbar)\,(|\Psi_{\mathbf{k}_0}(0)|^2/V)$, and $\hbar\Omega_{\mathbf{k}}$ is the energy of polariton with wave vector $\mathbf{k}$ on the lower polariton branch.

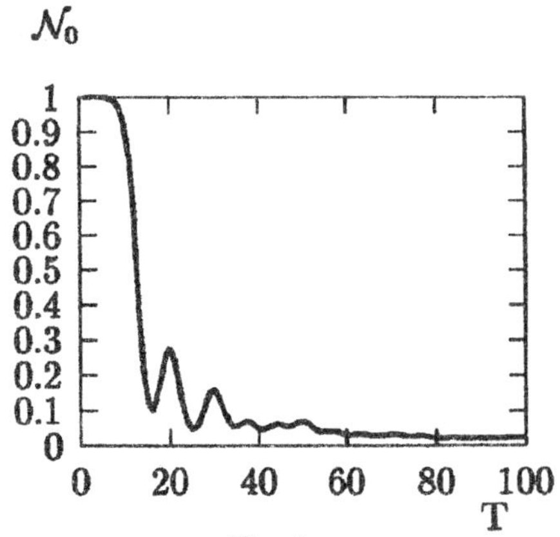

Fig. 1.

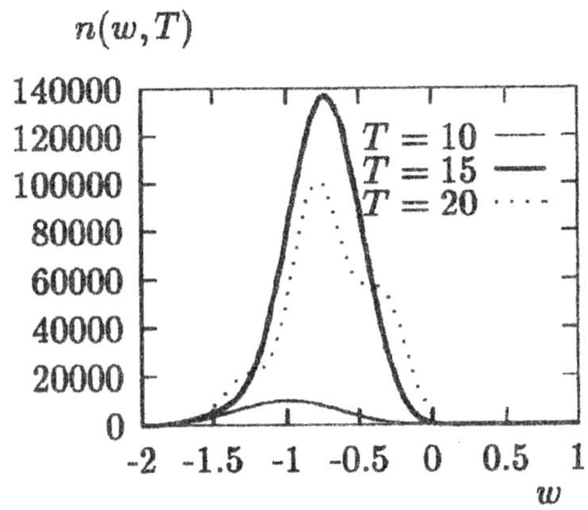

Fig. 2.

The decay of the nonequilibrium polariton condensate is shown in Fig. 1. Before the moment of dimensionless time $T \approx 7$ the condensate depletion proceeds rather slowly. Then a drastic fall takes place. As a result, by the moment $T \approx 15$ only 10% of the initial polariton number remain in the condensate. The further condensate decay is accompanied by nutation oscillations. This leads to the partial restoration of the condensate (up to 30% at $T \approx 20$). The nutation oscillations do not occur if the integral influence of all pairs of noncondensate polaritons with the same total momentum $2\hbar\mathbf{k}_0$ on each individual pair is neglected. Thus, the occurrence of oscillations is due to correlation of the states of individual pairs of noncondensate polaritons. By the moment $T \approx 50$ the oscillations disappear and further condensate evolution has the form of slow monotonous decay.

Fig. 2 depicts the distribution function of noncondensate polaritons at time moments in the interval where the polariton condensate undergoes the most rapid and significant changes. The curve corresponding to the moment of time $T = 10$ has the form of a bell with the maximum at $w = -1$. By the moment $T = 15$ the distribution function without changing its symmetrical form increases significantly its value and its maximum shifts to greater energies of non-condensate polaritons. An increase of the number of noncondensate polaritons by more than an order of magnitude corresponds to the presence of the local minimum of $\mathcal{N}_0(T)$ function at $T \approx 15$. When $T = 20$ the distortion of the distribution function takes place and the area under curve diminishes. This is in accordance with the local maximum of the $\mathcal{N}_0(T)$ at $T \approx 20$. Thus, the distribution function is "breathing" in conformity with the variations of condensate density.

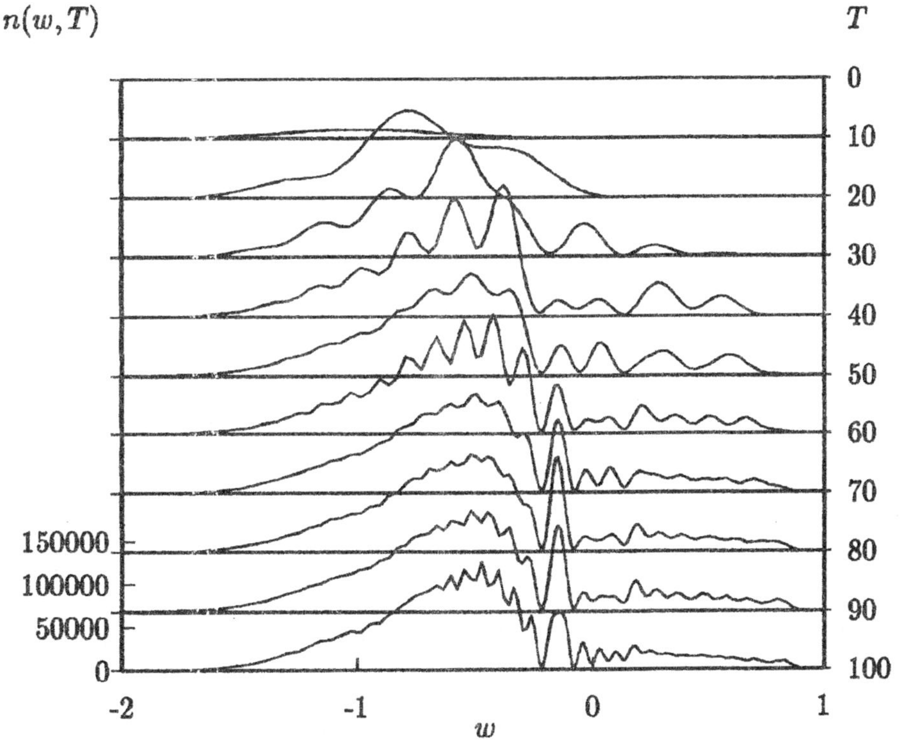

Fig. 3.

Further evolution of the distribution function is shown in Fig. 3 (the function scale is shown by the marks in the left vertical axis). At the moment of time $T = 30$ its contour gets the oscillating character. The oscillations arise not only in the dependence of distribution function on parameter w but in its dependence on dimensionless time

T (unfortunately, we could not demonstrate this fact in the picture). This means that if, e.g., the distribution function at the moment $T \approx 40$ possesses a local maximum at $w = -1$ it can possess at this point a local minimum at another moment of time. Later on, the frequency oscillations become denser, smaller, and transform into chaotic ripples on the smooth contour of the distribution function. From the moment $T \approx 30$ a short-wave wing begins to arise in the distribution function curve. By the moment of time $T = 100$ the function localizes in the interval $-2 < w < 1$. Very important is the enhancement and then the partial extinction of the sharp extra peak localized within the region $-0.22 < w < -0.08$ with a maximum at $w = -0.15$. The time interval where the extra peak arises and exists is characterized by small values of condensate density and its relatively slow change (see Figs. 1,3). The extra peak does not change its position with time and reaches its maximum at $T \approx 80$.

3. CONCLUSIONS

Therefore, the significant condensate depletion takes place already at the before-kinetic stage of the system evolution. Within a short time $t_* \sim 15\lambda^{-1}$ the condensate density diminishes 10 times compared to the initial density. For CdS single crystals and initial condensate density $(10^{16} \div 10^{18})\,\mathrm{cm}^{-3}$ the time t_* equals to $(120 \div 1.2)$ ps. The short time of the polariton condensate decay makes problematic the possibility of observing such coherent nonlinear phenomena as, e.g., the phenomenon of self-induced transparency and creation of solitons in exciton spectral range (see e.g. [9-12]).

Note the irreversible-in-time character of the described solution of the time-reversible equations [13].

4. REFERENCES

[1] A. L. Ivanov and L. V. Keldysh, Zh. Eksp. Teor. Fiz. **84**, 404 (1983).

[2] M. V. Lebedev, Zh. Eksp. Teor. Fiz. **101**, 957 (1992).

[3] S. A. Moskalenko, *Introduction into Theory of High Density Excitons* (Ştiinţa, Kishinev, 1983).

[4] M. I. Shmigliuk and V. N. Pitei, *Coherent Polaritons in Semiconductors* (Ştiinţa, Kishinev, 1989).

[5] J. R. Schrieffer, *Theory of Superconductivity* (Addison-Wesley, Menlo Park, CA, 1964).

[6] N. N. Bogoliubov, *Quasi-Averages in the Problems of Statistical Mechanics* (Preprint of JINR R-1415, Dubna, 1963).

[7] I. V. Beloussov and Yu. M. Shvera, *Z. Phys.* **B90** (1993) 51.

[8] I. V. Beloussov and Yu. M. Shvera, Theor. Math. Phys. **85**, 1177 (1991).

[9] J. Goll and H. Haken, Opt. Commun. **24**, 1 (1978).

[10] J. Goll and H. Haken, Phys. Rev. A **18**, 2241 (1978).

[11] S. N. Belkin, P. I. Khadzhi, S. A. Moskalenko and A. H. Rotaru, J. Phys. C **14**, 4109 (1981).

[12] P. I. Khadzhi, S. A. Moskalenko, A. H. Rotaru and G. D. Shibarshina, phys. stat. sol. (b) **114**, K25 (1982).

[13] G. M. Zaslavskii, *Stochasticity of Dynamic Systems* (Nauka, Moscow, 1984).

NONLINEAR INTERACTION BETWEEN ULTRASHORT LASER PULSES AND A THIN SEMICONDUCTOR FILM DUE TO TWO-PHOTON EXCITATION OF BIEXCITONS

P. I. Khadzhi and S. L. Gaivan
Institute of Applied Physics, Academy of Sciences of Moldova
Academiei str. 5, Kishinev 2028, Moldova

SUMMARY

The transmission and reflection of ultrashort laser pulses by a thin semiconductor film are studied theoretically under the conditions corresponding to two-photon, two-pulse creation of biexcitons from the ground state. Specific examples are given to demonstrate various possibilities of controlling the transmission of both incident pulses by changing the amplitude of one of them.

1. INTRODUCTION

In recent years several theoretical studies [1-5] were dedicated to the resonant nonlinear optics of thin films. Like the authors of these papers, we define a thin film as an object whose thickness, L, is much smaller than the wavelength of incident light. Within this approach, the set of partial differential equations for the field and medium can be reduced to a relatively simple set of ordinary differential and algebraic equations, and in many cases exact analytic solutions can be derived. Note than until now, the effects of nonlinear transient transmission and reflection of ultrashort laser pulses have been studied largely in thin films of two-level atoms under the conditions of one-photon [1,2] or two-photon [3,4] resonance.

We shall report a theoretical study of transient nonlinear transmission and reflection of ultrashort laser pulses by a thin semiconductor film (TSF) under the conditions of resonant two-photon excitation of biexcitons from the ground state. It is well known that two-photon biexciton generation is characterized by a giant oscillator strength [6]. Since typical times of interaction between optical pulses and media are determined by oscillator strength of specific optical transitions, it is feasible that two-photon excitation of biexcitons from the ground state may facilitate our understanding of transient nonlinear transmission of ultrashort pulses through the TSF.

2. BASIC EQUATIONS

Suppose that two monochromatic laser pulses with electric field envelopes $E_{01}(t)$ and $E_{02}(t)$ and frequencies ω_1 and ω_2, respectively, are incident on the TSF in vacuum. The durations of the pulses are much smaller than the biexciton relaxation time in the film, and the envelopes are slow functions of time. Suppose that the frequency of each pulse is resonant with neither the excitonic nor M-band exciton–biexciton transitions [7], but that the sum of the photon energies equals to that of the ground state to biexciton transition.

The Hamiltonian for a definite photon–biexciton mode includes the sum of the free biexcitons and field Hamiltonian and the interaction Hamiltonian [7,8]:

$$H_{int} = -\hbar\mu(b^+ E_1^+ E_2^+ + bE_1^- E_2^-), \qquad (1)$$

0-7803-3223-7/96/$5,00

where μ is the constant of the two-photon generation of biexcitons from the ground state [6], b is the amplitude of the biexciton wave, $E_i^+(E_i^-)$ is the component of the transmitted pulse amplitude ($i = 1, 2$) with the positive (negative) frequency. From Eq. (1) we derive the Heisenberg equation for the biexciton wave amplitude

$$i\dot{b} = \Omega_{bi}b - \mu E_1^+ E_2^+, \tag{2}$$

where Ω_{bi} is the frequency of the ground state to biexciton transition. Assuming that the resonance condition $\omega_1 + \omega_2 = \Omega_{bi}$ is fulfilled and taking the wave amplitudes in the form $E_1^+ = E_1 \exp(-i\omega_1 t)$, $E_2^+ = E_2 \exp(-i\omega_2 t)$, $b = (u + iv)\exp(-i\omega_1 t - i\omega_2 t)$, one can easily prove that if the crystal is initially in the ground state, then the real component u of the biexciton wave is exactly zero, and its imaginary component is described by the equation

$$\dot{v} = \mu E_1 E_2. \tag{3}$$

As a consequence, the envelopes of the transmitted and reflected pulses are not phase modulated.

Using the approach described in Refs. [1-4], we derive from the continuity of the tangential field components at the vacuum–film interface the following electrodynamic relations for amplitudes of incident, (E_{01}, E_{02}), transmitted, (E_1, E_2), and reflected, (E_{r1}, E_{r2}), pulses:

$$E_1 = E_{01} - \alpha_1 v E_2, \qquad E_2 = E_{02} - \alpha_2 v E_1, \tag{4}$$

$$E_{r1} = E_1 - E_{01}, \qquad E_{r2} = E_2 - E_{02}, \tag{5}$$

where $\alpha_1 = 2\pi\hbar\omega_1\mu L/c$, $\alpha_2 = 2\pi\hbar\omega_2\mu L/c$. From Eqs. (4) and (5) we obtain

$$\frac{dv}{dt} = \mu \frac{(E_{01} - \alpha_1 v E_{02})(E_{02} - \alpha_2 v E_{01})}{(1 - \alpha_1\alpha_2 v^2)^2}. \tag{6}$$

Equations (4)–(6) yield a complete solution to the problem in terms of the shapes of transmitted and reflected pulses derived from the envelopes of incident pulses $E_{01}(t)$ and $E_{02}(t)$ under the conditions of two-photon excitation of biexcitons.

3. INTERACTION WITH TWO IDENTICAL PULSES

Consider the degenerate case, in which the two incident pulses have identical shapes, $E_{01}(t) = E_{02}(t) = E_0(t)$, and frequencies, $\alpha_1 = \alpha_2 = \alpha$. Then Eqs. (4)–(6) can be substantially simplified:

$$E(t) = \frac{E_0(t)}{1 + \alpha v}, \quad E_r(t) = E(t) - E_0(t), \quad \frac{dv}{dt} = \frac{\mu E_0^2(t)}{(1 + \alpha v)^2}. \tag{7}$$

Given an arbitrary shape of the incident pulse $E_0(t)$, we can derive the general solution for the fields of the transmitted and reflected pulses

$$E(t) = E_0(t)\left(1 + 3\alpha\mu \int_{-\infty}^{t} E_0^2(t')dt'\right)^{-1/3}, \quad E_r(t) = E(t) - E_0(t). \tag{8}$$

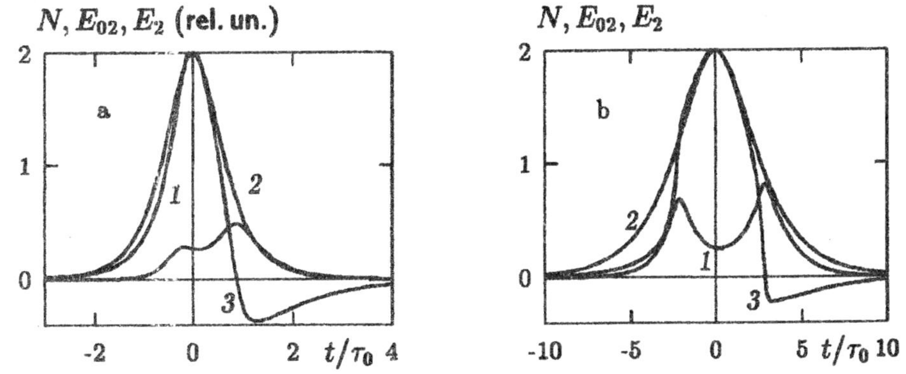

Fig. 1. (*1*) Biexciton concentration N, and amplitudes of (*2*) incident, E_{02}, and (*3*) transmitted, E_2, pulses at $E_{01} = E_0 = \text{const}$ and $E_{02}(t) = 2E_0/\cosh(t/T)$. The biexciton concentration is normalized to $\alpha_1\alpha_2$, light to E_0. The half-width of the incident pulse $E_{02}(t)$ equals (a) $0.5\tau_0$ and (b) $2.0\tau_0$.

Consider as an example a solution for a soliton-like incident pulse, $E_0(t) = E_0/\cosh(t/T)$ with amplitude E_0 and half-width T. We obtain from Eqs. (8)

$$E(t) = E_0 \frac{1}{\cosh(t/T)} \left[1 + 3\frac{T}{\tau_0} e^{t/T} \frac{1}{\cosh(t/T)}\right]^{-1/3}, \quad E_r(t) = E(t) - E_0 \frac{1}{\cosh(t/T)}, \tag{9}$$

where $\tau_0^{-1} = \alpha\mu E_0^2$. The typical time τ_0 over which the amplitude of the transmitted (reflected) pulse changes is determined by the square of the incident pulse amplitude, i.e. its intensity.

It follows from Eq. (9) that when the field of the incident pulse increases, the amplitude of the transmitted pulse also rises, ahead of the incident pulse maximum ($t = 0$), and then drops rapidly. The reflected pulse is generated at a considerable delay and peaks when most of the incident pulse has traversed the film. The shape of the transmitted pulse and position of its maximum with respect to that of the incident pulse is controlled by the half-width of the incident pulse, T and the parameter τ_0. Thus, under the conditions of single-pulse, two-photon excitation of biexcitons from the ground state, the TSF efficiently discriminates pulses with respect to their intensity and duration, transmitting short pulses of low intensity without distorsion and reflecticting longer and more intense ones.

4. INTERACTION WITH TWO DIFFERENT PULSES

We now consider a nondegenerate case. It follows from Eq. (6) that pump E_{01} and E_{02} can transfer the system to a steady state. Taking $dv/dt = 0$ in Eq. (6), we find the steady state values of v: $\alpha_1 v = E_{01}/E_{02}$, $\alpha_2 v = E_{02}/E_{01}$. Assuming that one pump field, for example E_{02}, is constant, and the other is adiabatically slow, one can derive v, E_1, and E_2 as functions of E_{01} at $E_{02} = \text{const}$. For $\sqrt{\alpha_2}E_{01} < \sqrt{\alpha_1}E_{02}$, the stable solutions are $v = E_{01}/\alpha_1 E_{02}$, $E_1 = 0$, $E_2 = E_{02}$, and for $\sqrt{\alpha_2}E_{01} > \sqrt{\alpha_1}E_{02}$: $v = E_{02}/\alpha_2 E_{01}$, $E_1 = E_{01}$, $E_2 = 0$.

These equations indicate that with increasing E_{01} the function v first grows linearly to $\sqrt{\alpha_2}E_{01} = \sqrt{\alpha_1}E_{02}$, at which point $\sqrt{\alpha_1\alpha_2}v = 1$, and then drops monotonically as a hyperbolic function. Hence the steady-state amplitudes of the transmitted pulses depend on those of the incident pulses, so that the film transmits a pulse with a high amplitude, whereas a less intense pulse is totally reflected.

We next consider time-dependent transmission when incident are different. If only their amplitudes are different, but their shapes are similar, i.e $E_{01}(t) = E_{01}f(t)$, $E_{02}(t) = E_{02}f(t)$, then the solution of Eq. (6) in terms of $v(t)$ is given by the transcendental equation

$$Av + Bv^2 + Cv^3 + D\ln\left|1 - s^{-1}\sqrt{\alpha_1\alpha_2}v\right| - F\ln\left|1 - s\sqrt{\alpha_1\alpha_2}v\right| = \\ = \sqrt{\alpha_1\alpha_2}\mu E_{01}E_{02}\int_{-\infty}^{t} f^2(t')dt', \tag{10}$$

where $A = (s^2 + s^{-2} - 1)\sqrt{\alpha_1\alpha_2}$, $B = (s + s^{-1})\alpha_1\alpha_2/2$, $C = (\alpha_1\alpha_2)^{3/2}/3$, $D = s^3 - s$, $F = s^{-3} - s^{-1}$, $s = \sqrt{\alpha_2}E_{01}/\sqrt{\alpha_1}E_{02}$.

Equations of (4)–(6) can be solved in the case of arbitrary pulse shapes only numerically. Below we consider a fairly simple case in which one of the pulses, e.g. E_{01}, is rectangular, and the amplitude of other pulse varies with time as $E_{02}(t) = E_{02}/\cosh(t/T)$. If, $\sqrt{\alpha_1}E_{02}/(\sqrt{\alpha_2}E_{01}) > 1$, as follows from Eq. (6) the time derivative of $v(t)$ vanishes twice, and its sign changes at the leading and trailing edges of the incident pulse. As a result, the density of biexcitons (curve *1* in Fig. 1) has two maxima and one minimum. If $T < \tau_0$ (Fig. 1a), the film easily transmits its leading edge and maximum, and at the trailing edge of the incident pulse, where the density of biexcitons in the film is a maximum, the transmission drops to zero. At a lower amplitude of $E_{02}(t)$, radiative recombination of biexcitons is induced, and a second subpulse with negative amplitude is generated. For $T > \tau_0$ (Fig. 1b), the transmitted pulse is shorter than the incident one. Thus, the TSF easily transmits short pulses and shortens longer ones. Since $\tau_0 \sim E_0^{-2}$, the film also easily transmits weak pulses and shortens those of higher intensity.

5. CONCLUSIONS

We have obtained basic equations and studied features of transient transmission and reflection of ultrashort resonant laser pulses by a TSF under the conditions of two-pulse, two-photon generation of biexcitons from the ground state. We predict that this technique can be widely used to control the transmission of both pulses by varying the amplitude of one pulse, and therefore can be employed in fast optical integrated circuits.

6. REFERENCES

[1] V. I. Rupasov and V. I. Yudson, Sov. J. Quant. Electr. **9**, 1415 (1982).

[2] V. I. Rupasov and V. I. Yudson, JETP **66**, 282 (1987).

[3] S. M. Zakharov and E. A. Manykin, JETP **68**, 457 (1989).

[4] S. M. Zakharov and E. A. Manykin, JETP **78**, 566 (1994).

[5] M. G. Benedict and E. D. Trifonov, Phys. Rev. **A38**, 2854 (1988).

[6] E. Hanamura, J. Phys. Soc. Japan **39**, 1506 (1975).

[7] P. I. Khadzhi, *Nonlinear Optical Processes in a System of Excitons and Biexcitons in Semiconductors* [in Russian], Ştiinţa, Kishinev (1985).

[8] P. I. Khadzhi, *Kinetics of Recombination Radiation from Excitons and Biexcitons in Semiconductors* [in Russian], Ştiinţa, Kishinev (1977).

ENERGY BAND STRUCTURE OF $CuInS_2$ CRYSTALS

N.N. Syrbu [a], *V.E. Tazlavanu* [b], *R. Cretu* [b]

[a] *Technical University of Moldova, MD-2012, 168 Stefan cel Mare Bul., Chisinau, Moldova*

[b] *Inst. of Applied Physics, Moldavian Academy of Sciences, MD-2028, 5 Academiei str., Chisinau, Moldova*

SUMMARY

In this work, the wave-length derivative reflection (WDR) spectra of $CuInS_2$ (CIS) crystals have been investigated in the region of A, B and C series. The n=1, n=2 and n=3 states are determined and contours of exciton lines (n=1) are calculated. The parameters of excitons and bands have been determined for the region of band gap minimum.

1. INTRODUCTION

Belonging to the I-III-VI_2 group the CIS compound crystallises into the chalcopyrite structure with the space group I42d-D_{2d}^{12} and have direct band gap (E_g~1.6eV). There is a great of research interest to materials of this group, as a result of this, an expected application in optoelectronic devices. On the basis of these materials, converters of the solar energy are being developed. The energy band structure was calculated as nearest analogue of zinc-blend one [1-4]. The S states of exciton series A and B(C) were revealed in the region of band gap minimum. Was shown, that splitting because of crystal field are comes to naught. At the same time, the investigation of one-phonon IR reflection have revealed the some anisotropy of these crystals [5]. It is essential to determine experimentally values of the band gap among three upper valence bands (V_1, V_2, V_3) and lowest conduction band (C_1). The exciton binding energies are not known. The wave-length derivative reflection (WDR) spectra have been investigated for these purposes. The n=1, n=2 and n=3 states of the A, B, C exciton series are determined. The exciton binding energies and parameters of bands are defined.

2. EXPERIMENTAL

The crystals under study were grown by the chemical transport method. The specimens had the shape of ~5x10 mm^2 plates with mirror surface and a thickness of 300-400 μm. WDR spectra were measured by using MDR-2 spectrometer. MDR-2 has a strength light 1:2,5. The size of diffraction gratings of 600 rulings per mm is 140 x 140 mm, focal distance - 400 mm and provided resolution - 40 Å/mm. The mirror vibrated at ~73 Hz near the exit slit and produced a ~6-8 Å wavelength modulation. A modulated signal was registered by photomultiplier FEU-62 with a phase-sensitive detector. During the WDR

0-7803-3223-7/96/$5,00

measure, the samples were immersed in liquid nitrogen. An orientation error of the polarisation vector of a light wave and of crystal axes did not exceed 5°.

3. RESULTS AND DISCUSSION

For the E||c and E⊥c polarisations the WDR spectra of CIS have been investigated in the region of n=1, n=2 and n=3 exciton lines at 77K (Fig.1). The energy positions of A excitons (n=1) coincide in both polarisations and differ on 0.6±0.3 meV for B(C) excitons. The bands at 1.59 - 1.63 eV we attribute to n=2 and n=3 of A and B(C) exciton states. These bands are fitting into hydrogen-like dependence of exciton series. The continuum of A exciton series is observed at 1.6139 eV and are formed by electronic transitions $\Gamma_6(V_1)$-$\Gamma_6(C_1)$. For B(C) exciton series this continuum is observed at 1.6293 eV and is formed by transitions from $\Gamma_6(V_4)$, $\Gamma_7(V_3)$ bands to $\Gamma_6(C_1)$ band.

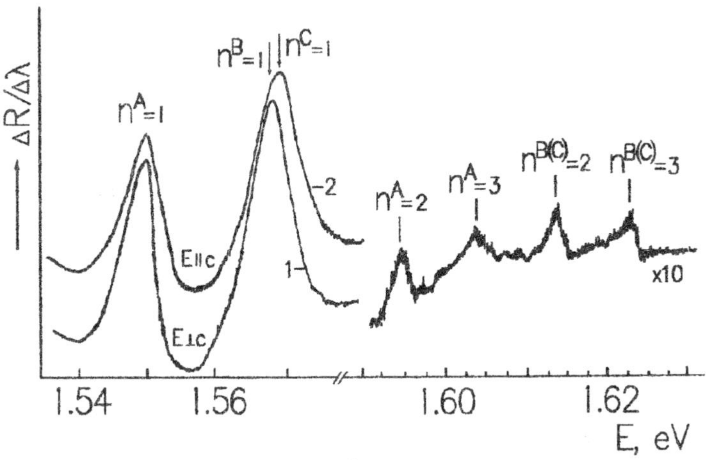

Fig. 1

The contours of exciton lines n=1 are calculated using dispersion relations taking into account spacial dispersion (M) and dead layer (t). Calculations shown, that longitudinal-transverse splittings (ω_{LT}) were 0.33 and 1.0 eV for A and B(C) excitons, respectively and dampings - 6.5 and 10.0, respectively (see table). These data correspond to the intermediate case (between strong and weak polariton effect). The contours of spectra are considerable affected by variation of spacial dispersion and dead layer values. It is necessary to note, that the exciton translation mass was calculated with the error of about 20%.

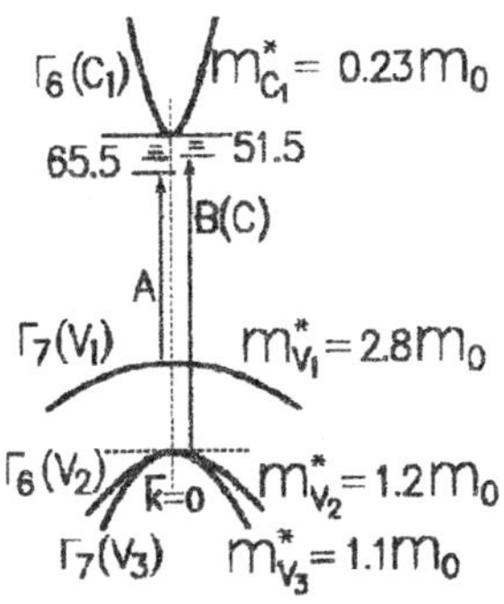

Fig. 2

Since in CIS the splitting because of crystal field comes to naught, we suppose that the small anisotropy of S-state exciton contours is caused by weak difference of effective masses of two degenerated valence bands. The effective masses of conduction band (m^*_{C1}) and valence bands (m^*_{V1}, m^*_{V2}, and m^*_{V3}) were calculated using values of M and exciton effective masses. The energy band structure of CIS near the center of Brillouin zone is shown in Fig.2. The binding energy magnitudes (R) of A and B(C) exciton series are presented in the table. The values of band gaps are 1,6139 eV and 1.6293 eV for $\Gamma_7(V_1)$ - $\Gamma_6(C_1)$ and $\Gamma_6(V_2)$, $\Gamma_7(V_3)$ - $\Gamma_6(C_1)$, respectively. Thus the splitting because of spin-orbital interaction of $\Gamma_7(V_1)$ - $\Gamma_6(V_2)$, $\Gamma_1(V_2)$ bands constitutes 15.4 meV in k=0 (Fig.2). Considering these data the contributions of p-states in the upper valence band (β) are 0.38. In that case the d-state's contributions in the upper valence band are 0.62 in exchange for 0.45 [2].

Exciton and bands parameters of $CuInS_2$ crystals

parameters	$\Delta R/\Delta\lambda$		
	A	B	C
ω_0	1.5536	1.5723	1.5730
ω_{LT}, meV	0.5	0.7	0.7
γ	9	11	10
ε_0	6.9 - 7.0	6.9 - 7.0	6.9 - 7.0
t, Å	50	-	-
M	3	1.3	1.2
R, eV	0.0655	0.0515	0.0515
μ	0.320	0.182	0.182
m^*_V	$2.75m_0$	$1.08m_0$	$0.98m_0$
m^*_C	$0.25m_0$	$0.22m_0$	$0.23m_0$

REFERENCES

1. B. Tell, J.L. Shay, H.M. Kasper: Phys.Rev. B, 4 (1971), 2463
2. J.L. Shay, B. Tell, H.M. Kasper, L.M. Schiavone: Phys.Rev. B, 5 (1972), 5003
3. B. Tell, J.L. Shay, H.M. Kasper, R.L. Barus: Phys.Rev. B, 10 (1974), 1748
4. J.E. Jaffe and A. Zunger: Phys.Rev. B, 28 (1983), 5822
5. H. Neumann, W. Kissinger, H. Sobotta, V. Riede, R.D. Tomlinson, N. Argerinos: Czech. J. Phys. B34 (1984), 370

Microhardness and Elastic Properties of Bulk Glasses and Thin Films of the Ge_x-Sb_{40-x}-S_{60} family

V. Pamukchieva, E. Savova, E. Skordeva
Institute of Solid State Physics of the Bulgarian Academy of Science
Blvd. "Tzarigradsko Chausse" 72, Sofia 1784, BULGARIA
M. Popescu, F. Sava, A. Lörinczi
Institute of Physics and Technology of Materials
Bucharest,-Magurele, P.O. Box MG.7 ROMANIA

Abstract

The microhardnesses (H) of bulk glasses and thin films in the Ge_x-Sb_{40-x}-S_{60} family has been measured. The volume of the micro-voids (V_h), their formation energy (E_h) and the module of elasticity (E) have been calculated. It is assumed that the bulk glasses and thin films have different elasticity levels due to the deformation of specific chemical bonds.

1. Introduction

The molecular-kinetical processes in glasses are, to a large extent, defined by the local fluctuation disorder in the structure at the expense of the deformation of the glass network, without breaking the valence bonds. These processes can be considered within the free-volume theory as formation and migration of the fluctuation micro-volumes [1]. This gives reasons to correlate the microhardness, glass transition temperature (T_g) and the elastic constants (E).

Based on the free-volume concept, the microhardness (H) can be expressed by the following formula:

$$H = E_h / V_h \qquad (1)$$

in which E_h is equal to the micro-void formation energy with a volume V_h, which is defined by the relation:

$$E_h \cong 3k_B\, T_g \qquad (2)$$

where k_B is Boltzmann's constant, and T_g the glass transition temperature.
From equation (1) and (2) one gets

$$V_h \cong 3k_BT_g / H \qquad (3).$$

Based on the equation for the energy of molecular interaction and on the microhardness definition as pressure necessary to overcome the intermolecular forces, Sanditov [1] wrote the following relationship:

$$H = (1-2\mu)\cdot E / 6\,(1+\mu) \qquad (4),$$

where E is the module of the elasticity at single axis expansion/shrinking (kg f/ mm^2) and μ is the Poisson's coefficient, which can be defined by the following relation:

$$\Delta\alpha\cdot T_g = (1-2\mu)^2 / 2\,(1+\mu) \qquad (5).$$

$\Delta\alpha\cdot T_g$ is the free volume part, which is maintained below T_g and according to Simha's and Boyer's rule [2] for a number of glasses with nearly equal μ, is around 0.1.

The purpose of the present article is to reveal the relation between H, measured in bulk glasses and thin films from the Ge-Sb-S system and the average coordination number Z. An attempt was made to assess the relationship between V_h, E_h, E and the medium-range order of glasses, on the basis of the correlation between H and E.

2. Experimental

The glasses were prepared by the melt quenching method generally adopted for chalcogenide glasses. Appropriate quantities (8 g total per batch) of 99,999 % Ge, Sb and S were sealed into fused silica ampoules (16 mm in diameter), after evacuating to a vacuum of 10^{-5} Torr. They were heated to 950^0 C in a rotary furnace; after 24 h the melt was quenched in air.

The T_g of the bulk glasses was determined using a Perkin-Elmer DSC-4 apparatus with the maximum sensitivity of 8 μW$\cdot$cm^{-1}. The DSC thermogram was measured by using ~20 mg of sample in powdered form sealed in Pt-pans. The values of T_g were determined at heating rate 5 K/min with accuracy ±0.1K. The glass transition temperature T_g of the thin films was evaluated on the basis of de Neufville's formula [3], using the optical band gap [4].

0-7803-3223-7/96/$5,00

The films with a thickness of ~3μm were prepared on the aluminium substrate by vacuum thermal evaporation with high deposition rate (~70 Ås $^{-1}$). Their composition analyzed with a microprobe, is close to that of the parent glass.

For the Vickers hardness measurements, specimens of about 1 cm^2 and 1 mm thick were cut and polished to obtain smooth and flat parallel surfaces before indentation testing. Vickers indentations were obtained using a PMT-3 microhardness tester. The hardness was measured at loadings of 5, 10, 20, 30, 40 and 50 g for bulk glasses and 2, 3, 4, 5 and 6 g for films. The hardness value corresponding to the largest load that did not form cracks in the samples was selected. At least ten indentations were performed and measured for each sample for the same indentation loading. The indenter was manipulated in order to penetrate slowly the glass samples and thin films and loading time was fixed for all tested samples (5 s). The measurements were carried out under normal atmosphere conditions at ~ 20 °C. The accuracy of the average values was ±5%. The Vickers hardness H (kgf / mm^2) or (× 10^7 N / m^2) was calculated from:

$$H_V = A \cdot (P / d^2) , \qquad (6)$$

where A is a constant (1854.4) that takes into account the geometry of the square based diamond indenter with an angle of 136° between opposing faces and P is the applied load (gf) while d is the diagonal length (μm).

3. Results

The results obtained for microhardness of bulk glasses and fresh films vs. the average coordination number (Z), calculated using the formula (1) are presented in Fig. 1 a, b.

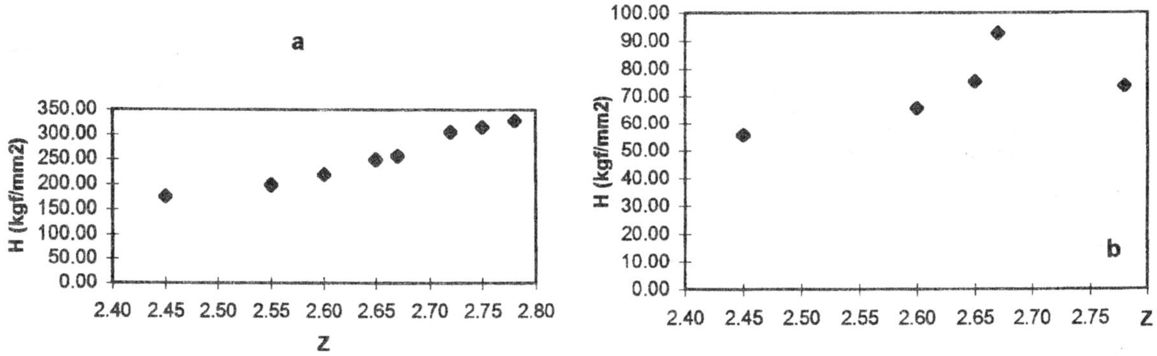

Fig. 1 Microhardness H dependence on the average coordination number Z of Ge_x-Sb_{40-x}-S_{60} bulk glasses (a) and thin films (b).

On the basis of correlation (1) ÷ (5) and on the experimentally obtained H and T_g, the minimal micro-void volume, energy of formation of the micro-voids and module of elasticity were evaluated for bulk glasses and thin films of the Ge_x-Sb_{40-x}-S_{60} family. The V_h-Z correlation for bulk glasses and thin films is shown in Fig. 2 a, b.

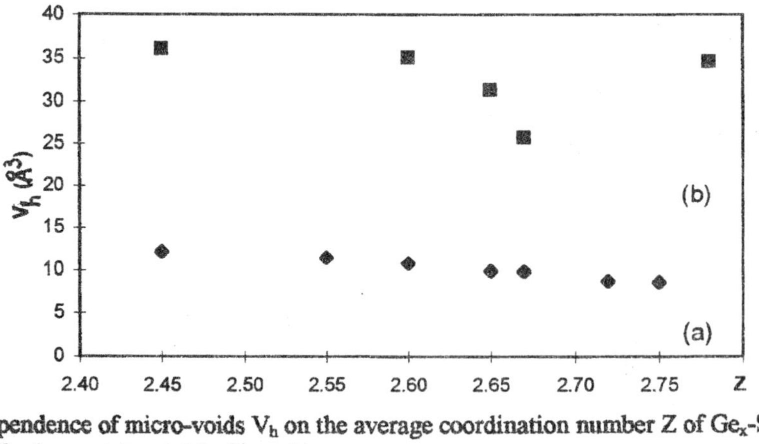

Fig.2 Dependence of micro-voids V_h on the average coordination number Z of Ge_x-Sb_{40-x}-S_{60} bulk glasses (a) and thin films (b).

Changes of the micro-void formation energy E_h vs. Z is shown in Figure 3. Figure 4 a, b shows the evolution of the module of elasticity (E) of bulk glasses and thin films vs. Z.

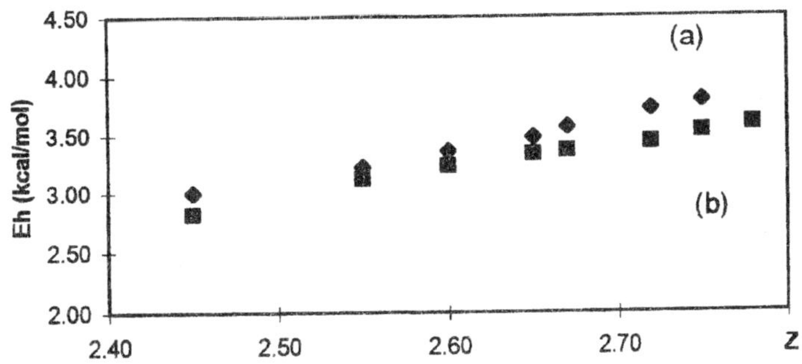

Fig. 3 The micro-void formation energy E_h dependence on the average coordination number Z of Ge_x-Sb_{40-x}-S_{60} bulk glasses (a) and thin films (b).

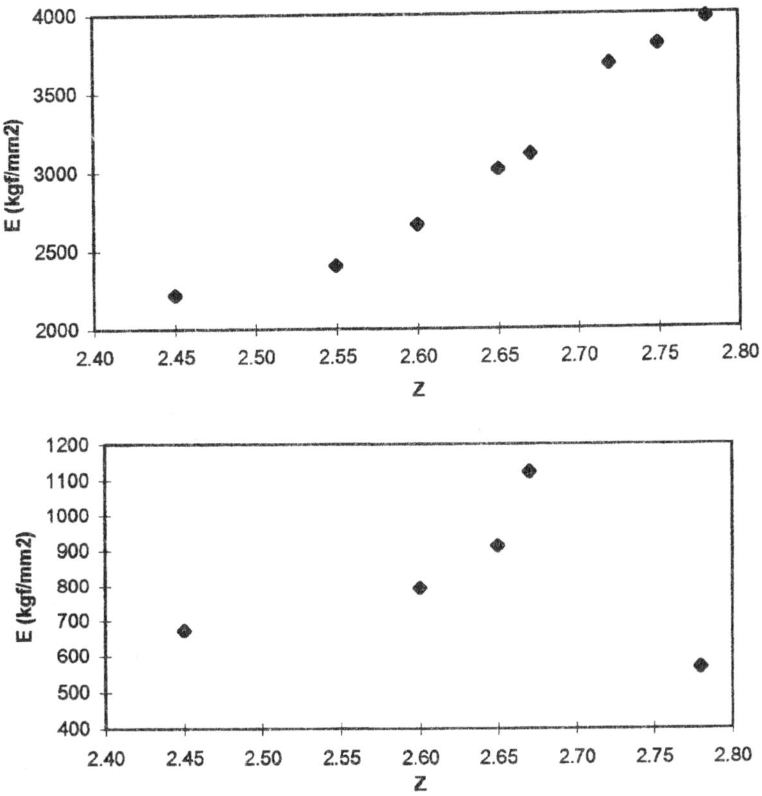

a.

b.

Fig. 4 Dependence of the module of elasticity E on Z of bulk glasses (a) and thin films (b).

4. Discussions

A glance at Fig. 1 reveals a good correlation between the average coordination number (Z) and the microhardness. Microhardness increasing with Z has also been observed in the Ge-As-Se system by Tille et al. [5] and in a limited number of compositions in Ge-Sb-Se system by Borisova and Pazin [6]. Our measurements of H for bulk glasses in Ge-Sb-Se system fit well in the line which describes the hardness of the isostructural glasses as that of several other chalcogenide glasses plotted as a function of Z by D.R. Swiler et al. [7]. The microhardness of the chalcogenide glasses proves a gradually progression from 1-D to 2-D and 3-D substructures.

To support the idea that microhardness can be considered as a typical property related to the glass strength, many investigations on the relationship between H and viscosity or T_g of glasses have been made. H is also indicative of the plastic properties of the glasses. It has been emphasized [8] that the glass immediatly beneath the indentation is densified. According to Frenkel's theory [9], glass packing is related to the formation/dissapearance of voids, bearing information about the structure, frozen at T_g. V_h dependency on Z for bulk glasses (Fig. 2a) shows that when the content of Ge increases the micro-void volume decreases, which is indicative of the cross-linking of the structure. The formation and disappearance of the holes (micro-voids) defined by the order level of the structure, can be reffered to its dimensionality. The activation energy of these processes coincides with the glass packing processes (looseness of the structure). The activation energy of the micro-voids E_h

(Fig. 3a) is about 3 kcal/mol, which means that this energy is not connected with breaking of chemical bonds but is defined by the inter-molecular interaction energy at the expense of the Van der Waals interaction energy. When Z increases, the weak intermolecular forces decrease, and the covalent forces increase; the temperature of glass transition T_g, activation energy of micro-voids E_h and microhardness H increase, and micro-void volume V_h decreases.

The calculated module of elasticity for bulk glasses show a hump around $Z \sim 2.67$, which can be connected with elasticity phase transition of the structure. Many authors who have measured the elastic properties of a number of chalcogenide glasses [7, 10, 11, 12, 13] remarked an increase in all elastic constants when cross-linking increases. It has been demonstrated by experimental ways that these chalcogenide glasses have layer-like type structure for $2.4<Z<2.7$ [13, 14]. The continuous network model is considered to be applicable to samples with $Z>2.7$ in which the layer-like structure is cross-linked. This phase transition is also confirmed by a number of physico-chemical, optical and diffraction measurements on bulk glasses and thin films of the Ge_x-Sb_{40-x}-S_{60} family done by us [4, 15, 16].

The existence of topological phase transition at $Z \sim 2.67$ is well expressed in films too (Fig. 1, 2, 4b) which confirms the change of the structure from layer-like to cross-linking layers, both for bulk glasses and for thin films. Compared to bulk glasses, the thin films have bigger micro-void volumes V_h at closer E_h values. Therefore the layers are looser and the volumes in which the fluctuations of the particles take place, are much bigger.

On the basis of the simple solid state model, it was shown that the micro-void volume V_h is proportional to r_0^3, where r_0 represents the equilibrum distance between the particles. The dimensions of the micro-voids in bulk glasses is about $V_h^{1/3}$ = 2.29 - 2.04 Å, very close to the bond length of Ge-S and S-S [17]. Since the chalcogenide atoms in glasses form a backbone chain with cross-linking provided by 3 or 4 coordinated group V and group IV atoms, the deformation is at the expense of the bending of these chains, linked by weak Van der Waals forces.

The micro-void linear dimensions ($V_h^{1/3}$) in layers are ~ 3.3-2.95 Å, and correspond to the short Ge-Ge interactions of the edge sharing tetraedra (2.92 Å) and long Ge-Ge interactions of corner sharing tetraedra (3.405 Å). Therefore, it can be assumed that in layers, the elastic deformations are associated with the network deformations at the linking places of cluster-like substructures.

5. Conclusions

It was shown that a strong correlation exists between H and Z in Ge-Sb-S system, which can be associated with the progressively growing dimensionality in chalcogenide glasses (1D → 2D → 3D). A topological phase transition of bulk glasses and thin films of the Ge-Sb-S system at Z ~ 2.67 has been established. The thin layers have longer equilibrum distances than the corresponding bulk glasses. The deformation is associated with specific chemical bonds both for films and bulk glasses.

References:

[1]. D. S. Sanditov - Physics and Chemistry of Glasses (in Russian), 3, 1, (1977), 14 -19.

[2]. R. Simha, R.Boyer, J. Chem. Phys., 37, (1962), 1003;

[3]. J. DeNeufville, H. Rockstad, in: Proc. 5th Int. Conf. on Amorphous and Liquid Semiconductors (Taylor and Francis, London, 1974), 419.

[4]. E. Savova, E. Skordeva, E. Vateva, J. Phys. Chem. Solids, 7, (1994), 575-578;

[5]. U. Tille, G. H. Frischat, K. J. Leers in: Non-Cryst. Solids(Trans. Tech. Publications, Amsterdam)(1977), 631.

[6]. Z.U. Borisova, A.V.Pazin in: Solid State Chemistry, ed. Z.U. Borisova (Consultants Bureau, NY) (1966), 63.

[7]. D. R. Swiler, A. K. Varshneya and R. M. Callahan, J. Non-Cryst. Solids 125, (1990), 250-257.

[8]. F. M. Ernsberger, J. Am. Ceram. Soc., 51, (1968), 545.

[9]. J.I. Frenkel - Kinetic theory of fluids (in Russian) - ed. AS USSR (Moscow) (1945).

[10]. K. Tanaka, Sol. St. Commun. 60, 3 (1986), 295-297.

[11]. S. Mahadevan, A. Giridhar, A.K. Singh, J. Non-Cryst. Solids, 57, (1983), 423-430.

[12]. A.N. Sreeram, A.K. Varshneya, D.R. Swiler, J. Non-Cryst. Solids, 128, (1991), 294-309.

[13]. C. Zin, L.E. Busse & S.R. Nagel, Phys. Rev., B29, (1984), 5060.

[14]. K. Tanaka, Sol. St. Commun., 58, (1986), 469.

[15]. E. Vateva, E. Savova, J. Non-Cryst. Solids, 192-193, (1995), 145-148.

[16]. E. Savova, B. Pashmakov, V. Pamukchieva - J. Phys. Chem. Solids, in press.

[17]. D. Armand, A. Ibanez, E. Philippot, Q. Ma, D. Raoux, J. Non-Cryst. Solids, 150, (1992), 371-375.

[18]. D. S. Sanditov, G. M. Bartenev - Physics and Chemisry of Glasses (in Russian), 1, 5, (1975), 414-420.

Session 14

SUPERLATTICES
(Poster)

OPTICAL ABSORPTION IN A QUANTUM WELL WITH AN APPLIED ELECTRIC FIELD

V. A. Sinyak
Institute of Applied Physics, Academy of Sciences of Moldova,
Academiei Street 5, 277028 Kishinev, Moldova

Summary

The light absorption by the semiconductor quantum well with infinite walls in the constant electric field is studied. Optical transitions between the levels of the electron size quantization in the conduction band are considered. On the basis of exact orthonormalized wave functions of the electron in the constant electric field in the quantum well analytical expressions for the oscillator strength and absorption coefficient are obtained.

Optical properties of semiconductors with lowered dimensions are an object of special attention recently. It is connected with the unique possibilities of their use at the development of infrared lasers, photodetectors, high-frequency electrooptical modulators. A large quantity of works are devoted to interband optical transitions, exciton effects, Stark optical effects and many other linear and non-linear optical properties of semiconductor quantum wells (see review [1]).

It is known, that constant electric field changes optical properties of semiconductors considerably (Franz-Keldysh effect). In a number of papers some authors they study the constant electric field influence on size quantization levels and spectra of interbands and intraband light absorption of semiconductor quantum wells with infinite barriers [2, 3].

In the present work in contrast to [2, 3] the analytical expression for intraband, absorption coefficients in constant electric fields is obtained on the basis of the exact wave functions of the charge carries. The calculation of the absorption coefficient was carried out by the density matrix method in the relaxation time approximation. The optical transitions between two levels of size quantization of the electron in the conduction band are considered. The peak of the intraband absorption coefficient is shown to shift in the direction of higher energies, its amplitude increasing with the increase of the constant electric field intensity. This is connected with the fact that the distance between the size quantization levels increases with field intensity increase.

Let us consider the semiconductor quantum well of width d with infinite walls to which the constant electric field of intensity F is applied being, directed perpendicularly to the well plane. The orthonormalized wave function of the electron in the effective mass approximation has the form:

$$\Psi_n(\mathbf{r},t) = \frac{U(\mathbf{r})}{\sqrt{L_x L_y}} \exp\left[\frac{i}{\hbar}(p_x x + p_y y)\right] \exp\left[-\frac{i}{\hbar}(E+E_n)t\right] \varphi_n(z), \tag{1}$$

where p_x, p_y are the electron impulses in the $x-y$ plane of thewell, $E=(p_x^2+p_y^2)/2m^*$, m^* is the electron effective mass, $U(\mathbf{r})$ is the periodic Bloch function. The $\varphi_n(z)$ function satisfies the Schrödinger equation:

$$\left(-\frac{\hbar^2}{2m}\frac{d^2}{dz^2} + |e|Fz - E_n\right)\varphi_n = 0, \qquad 0 < z < d. \tag{2}$$

and has the following form [4]:

$$\varphi_n(z) = \pi\sqrt{\frac{b}{a}}\frac{\mathrm{Ai}(\rho_n)\mathrm{Bi}(\beta_n) - \mathrm{Ai}(\beta_n)\mathrm{Bi}(\rho_n)}{(1-\gamma_n^2)^{1/2}} \tag{3}$$

Here

$$b = \pi\sqrt{\frac{\hbar\omega}{E_0}}; \quad \omega = \frac{(|e|\,F)^{2/3}}{(2m^*\hbar)^{1/3}}; \quad E_0 = \frac{\hbar^2\pi^2}{2m^*d^2};$$

$$\beta_n = -\frac{E_n}{\hbar\omega}; \quad \alpha_n = b - \beta_n; \quad \rho_n = \frac{z}{d}b - \beta_n; \quad \gamma_n = \frac{Ai(\beta_n)}{\mathrm{Bi}(\alpha_n)};$$

E_n is the energy corresponding to the size quantization level n ($n = 1, 2, 3, ...$) which is determined by the dispersion equation

$$\mathrm{Ai}(\beta_n)\mathrm{Bi}(\alpha_n) - \mathrm{Bi}(\beta_n)\mathrm{Ai}(\alpha_n) = 0 \tag{4}$$

$\mathrm{Ai}(x)$ and $\mathrm{Bi}(x)$ are Airy functions having the integral presentations,

$$\mathrm{Ai}(x) = \frac{1}{\pi}\int \cos\left(\frac{u^3}{3} + xu\right) du;$$

$$\mathrm{Bi}(x) = \frac{1}{\pi}\int \left[\sin\left(\frac{u^3}{3} + xu\right) + \exp\left(-\frac{u^3}{3} + xu\right)\right] du.$$

If one uses asymptotes $\mathrm{Ai}(x)$ and $\mathrm{Bi}(x)$ at large negative values of the argument,

$$\mathrm{Ai}(x) \approx \frac{1}{\sqrt{\pi}}\frac{1}{|x|^{1/4}}\sin\left(\frac{2}{3}|x|^{3/2} + \frac{\pi}{4}\right), \quad x \to -\infty$$

$$\mathrm{Bi}(x) \approx \frac{1}{\sqrt{\pi}}\frac{1}{|x|^{1/4}}\cos\left(\frac{2}{3}|x|^{3/2} + \frac{\pi}{4}\right), \quad x \to -\infty$$

then it is possible to show that at $F > 0$ function (3) transform into the wave function of the free electron in the quantum well

$$\varphi_n(z)\,|_{F=0} = \sqrt{\frac{2}{d}}\sin\left(\frac{\pi n}{d}z\right)$$

and from (4), for values of the size quantization levels energy, we obtain

$$E_n = \frac{\hbar^2\pi^2n^2}{2m^*d^2}, \quad n = 1, 2, 3\ldots\,.$$

Let us consider the light absorption connected with the optical transitions between size quantization levels n and m inside the conduction band of the quantum well in the presence of the constant electric field F. The electromagnetic wave of the frequency Ω and intensity $\varepsilon(t) = \mathrm{Re}\left(\varepsilon_0 e^{-i\Omega t}\right)$ is perpendicular to the well plane. The absorption coefficient $K(\Omega)$ is bound the imaginary part of susceptibility $\mathrm{Im}\chi(\Omega)$ in the following way:

$$K(\Omega) = \frac{4\pi\Omega}{cn_r}\mathrm{Im}\chi(\Omega) \tag{5}$$

Using the wave function (1) and the density matrix formalism in the relaxation time approximation, we obtain for (5) the expression [2]

$$\begin{aligned} K(\Omega) &= \frac{\Omega}{n_r}|M_{mn}|^2\left(\frac{m^*kT}{d\pi\hbar^2}\right)\ln\left\{\frac{1+\exp\left[(E_F-E_n)/kT\right]}{1+\exp\left[(E_F-E_m)/kT\right]}\right\} \\ &\times \frac{\hbar/\tau}{(E_m-E_n-\hbar\Omega)^2+(\hbar/\tau)^2} \end{aligned} \tag{6}$$

Where n_r is the medium index of refraction, k is the Boltzmann constant, τ is the relaxation time, T is the absolute temperature, E_n and E_m are the energies corresponding to the size quantization levels n and m, E_F is the Fermi energy. The transition matrix element M_{mn} is equal to

$$M_{mn} = |e|\int_0^d z\varphi_m^*(z)\,\varphi_n(z)\,dz \tag{7}$$

where $\varphi_m(z)$ and $\varphi_n(z)$ are the wave functions of the electron located on levels m and n, and satisfying the Schrödinger equation (2).

In all the works on the light electroabsorption the matrix element (7) was found numerically, or the experimental values of the oscillator strength were used for its determination.

On the basis of the wave functions (3) we have found the analytical expression for M_{mn}. Here we used the equation, to which Airy functions

$$\left(\frac{d^2}{dx^2}-x\right)\mathrm{Ai}(x) = \left(\frac{d^2}{dx^2}-x\right)\mathrm{Bi}(x) = 0. \tag{8}$$

Wronskian

$$\mathrm{Ai}(x)'\mathrm{Bi}(x)-\mathrm{Bi}(x)'\mathrm{Ai}(x) = \frac{1}{\pi}$$

and the following integrals

$$\int f^2(x)\,dx = xf^2(x)-f'^2(x), \quad \int f_1(x)\,f_2(x)\,dx = xf_1(x)\,f_2(x)-f_1'(x)\,f_2'(x),$$

$$\int f_1(x)\,f_2(x+\alpha)\,dx = \frac{1}{\alpha}\left[f_1(x)\,f_2'(x+\alpha)-f_1'(x)\,f_2(x+\alpha)\right], \ (\alpha\neq 0) \tag{9}$$

satisfy, where f, f_1, f_2 are any of Airy function $\mathrm{Ai}(x)$ and $\mathrm{Bi}(x)$. The primes mean the functions derivativs.

Taking into account (8),(9) we obtain for the matrix element (7)

$$M_{mn} = |e|\int_0^d z\varphi_m^*(z)\,\varphi_n dz = \frac{2|e|d}{b}\frac{1}{(\beta_n-\beta_m)^2}\frac{\gamma_n\gamma_m-1}{\left[(1-\gamma_n^2)(1-\gamma_m^2)\right]^{1/2}}. \tag{10}$$

At $F>0$ from (10) results the transition matrix element for the free electron is obtained

$$M_{mn} = \frac{4d|e|}{\pi^2}\frac{nm\left[(-1)^{m+n}-1\right]}{(n^2-m^2)^2}. \tag{11}$$

Let us also give the expressions for the oscillator strength $f_{mn}(F)$ in the presence of the constant electric field

$$f_{mn}(F) = 4\frac{m_0}{m^*}\frac{1}{(\beta_n - \beta_m)^2}\frac{(\gamma_n\gamma_m - 1)^2}{(1-\gamma_n^2)(1-\gamma_m^2)} \quad (12)$$

At $F = 0$ from (12) we find the known expression

$$f_{mn}(0) = \frac{64}{\pi^2}\frac{m_0}{m^*}\frac{n^2m^2}{(n^2 - m^2)^3}$$

where m_0 is the free electron mass.

Let us estimate the absorption coefficient value of the quantum well $GaAs - AlGaAs$ for the transition between the levels with $n = 1$ and $m = 2$ at the following values of the parameter [3]:

$$T = 77K,\ m^* = 0.0665m_0,\ n_r = 3.2,\ \tau = 0.14ps,\ E_F = 25.38meV,\ d = 100\text{Å}$$

1. $F = 0$
   The absorption maximum corresponds to the quantum energy $\hbar\Omega_0 \approx 105meV$ and is equal to $\alpha(\Omega_0) \approx 990cm^{-1}$.

2. $F = 100kV/cm$
   The absorption coefficient maximum corresponds to the light quantum energy $\hbar\Omega_F \approx 111meV$, $\alpha(\Omega_F) \approx 1750cm^{-1}$. The absorption peak shifts in the constant electric field by $\Delta\hbar\Omega = \hbar\Omega_F - \hbar\Omega_0 \approx 6meV$. The absorption peak increase $\Delta\alpha(\Omega) = \alpha(\Omega_F) - \alpha(\Omega_0) \approx 760cm^{-1}$.

Let us discuss the obtained results. In thc fields absence the light absorption coefficient differs from zero only under the condition that the difference $m - n$ is an odd number. This follows from the expression for the matrix element (11). At $F \neq 0$ the quantum transitions can happen between only two levels of the size quantization. The calculations show that with the increase of the constant electric field intensity the distance between the size quantization levels increases. It leads to the absorption coefficient peaks shift in the direction of higher energies of the photon and to its growth.

The above given results show that with the help of the constant electric field one can rather effectively modulate the light absorption by semiconductor quantum wells.

References

[1] S. Schmitt-Rink, D. S. Chemla and A. B. Miller, *Advances in Physics*, **38**, No 4, 89 (1989).

[2] D. Ahn and S. L. Chuang, *Phys.Rev.*, **35**, No 8, 4149 (1987).

[3] D. Ahn and S. L. Chuang, *IEEE Journal of Quantum Electronics*, **QE-23**, No 12, 2196 (1987).

[4] C. Trallero Giner and J. Lopez Gondar, *Physica B+C*, **138**, No 3, 287 (1986).

OBSERVATION OF SURFACE AND INTERFACE PHONON-POLARITONS IN SUPERLATTICES

N.L.Dmitruk*, A.V.Goncharenko*, O.S.Gorea, A.S.Keyanu,
V.R.Romaniuk*, O.M.Tatarinskaya, E.F.Venger*
State University of Moldova, A.Mateevich str.,60, Kishinau, MD2009, Moldova
*Institute of Semiconductor Physics, av.Nauki, 45, Kiev, 252650, Ukraine

Abstract

The dispersion curves and electric fields spatial distribution of phonon-polaritons in finite superlattices were calculated. The experimental investigation of GaAs/GaAsP SLS by ATR-spectroscopy has given the possibility to interpret adequately the observed peculiarities.

1. Introduction

There has been a very active interest recently in the far-infrared (FIR) properties of the superlattices (SL) and at semiconducting heterostructure interfaces. FIR experiments have already been used to determine relative layer thicknesses, interface roughness, superlattice strain, electron effective masses and free-electron distributions [1]. Still in many cases a reasonable agreement between the results of calculations and the experimental data has not been achieved.

A popular outline of the classical approach leads to two model: the "bulk-slab" model [2] in wich each component of the infinite SL is treated as a slab of the corresponding bulk material and a description of the SL as an anisotropic effective medium (AEM). It is worth noting that the first model being applied to optical phonons investigation gives solutions only for interface modes forming phonon zones [3]. As concerns AEM-model, one can obtain a satisfactory description of experimental spectra of the dispersion Fourier-transform spectroscopy (DFTS) [1] and attenuated total reflection (ATR) [4], but a problem in the interpretation of the peculiarities in observed spectra arises.

Here we present our concepts concerning some problems that have arised during investigation of SLs by FIR spectroscopy.

2. Theoretical consideration

The problem of the polaritonic spectrum in planar systems with up of an arbitrary number of isotropic, homogeneous layers $(0|1|2|3|.....|N|0)$, each of which could be described by a dielectric function $\varepsilon_n(q_{||},\omega)$ is examined in [5]. Its solution for the determination of vibrational modes of collective excitation (CE) in the case of p-polarization reduces to zero the determinant of the 3-diagonal block-matrix $\mathbf{D}$:

$$det\mathbf{D} = \begin{vmatrix} [1] & [2^0] & & & & \\ [1_0] & [2] & [3^0] & & & \\ & [2_0] & [3] & [4^0] & & \\ & & [3_0] & [4] & \cdots & \\ & & & & \cdots & [N^0] \\ & & & & [(N-1)_0] & [N] \end{vmatrix} = 0, \qquad (1)$$

0-7803-3223-7/96/$5.00 © 1996 IEEE

where

$$[n]=\begin{bmatrix}\delta_n^- e^{-iq_nd_n} & \delta_n^+\\ \delta_n^+ & \delta_n^- e^{-iq_nd_n}\end{bmatrix},\qquad [n^0]=\begin{bmatrix}0 & 0\\ -\delta_n^+ e^{-iq_nd_n} & -\delta_n^-\end{bmatrix},$$

$$[n_0]=\begin{bmatrix}-\delta_n^- & -\delta_n^+ e^{-iq_nd_n}\\ 0 & 0\end{bmatrix},\qquad q_n=[\varepsilon_n\left(\frac{\omega}{c}\right)^2-q_{\|}^2]^{1/2},\qquad \delta_n^{\pm}=i(\varepsilon_n q_0\pm q_n),$$

d_n - is the width of n-layer. $q_{\|}$ - wave vector in parallel to the layers, $i=\sqrt{-1}$, $\varepsilon_0=1$. Matrices of the second order $[n]$ describe the eigenfrequencies of CE in the separately taken layer in vacuum, matrices $[n^0]$, $[n_0]$ determine the mutual influence of these oscillations when the spectrum of eigenfrequencies in a multilayer structure is formed.

The solution of the Maxwell equations for Fourier components of the electric field ($\mathbf{E}=\{E_x,0,E_z\}$) is obtained by using the matrix equation $\mathbf{DC}=0$, where $\mathbf{D}$ is the 2Nx2N matrix (1) and $\mathbf{C}$ is the column 2Nx1 matrix with C_1'; C_1''; C_2'; C_2'';...; C_N'; C_N''. In the layer n the field components will be written as:

$$E_{nz}(q_{\|},\omega,z)=C_n' e^{iq_n(z-(d_1+d_2+...+d_n))}+C_n'' e^{-iq_n(z-(d_1+d_2+...+d_{n-1}))} \tag{2}$$

$$E_{nx}(q_{\|},\omega,z)=\frac{-q_n}{q_{\|}}\{C_n' e^{iq_{\|}(z-(d_1+d_2+...d_n))}-C_n'' e^{-iq_n(z-(d_1+d_2+...+d_{n-1}))}\} \tag{3}$$

where $z\in[z_{n-1},z_n], z_0=0, d_0=0$. Axis z is perpendicular to the layers and z_n is the coordinate of the boundary between n and $n+1$ layers. According to the dielectric function $\varepsilon_n(q_n,\omega)$, equation (1) determines the spectrum of vibrational modes of polaritons, plasmons and a combined CE for a given layer structure. For the case of infinite superlattice (SL) composed of such heterolayers (after applying the Floquet theorem $(E(z+d)=e^{iK_Bd}E(z))$), we obtain from (1):

$$cos(K_Bd)=cos(q_1d_1)cos(q_2d_2)-\frac{1}{2}\left(\frac{\varepsilon_1q_2}{\varepsilon_2q_1}+\frac{\varepsilon_2q_1}{\varepsilon_1q_2}\right)sin(q_1d_1)sin(q_2d_2), \tag{4}$$

where period $d=d_1+d_2$, K_B - Bloch wave vector. The type of the concrete mode (surface and interface, bulk, wave-guided and confined) is determined from z wavevector components q_n. In the Fig.1, the vibrational modes (a) and the corresponding spatial distributions of E_x (b) for surface (SP) and interface (IP) optical phonons in a GaAs/GaP SL composed from five layers ($|GaAs|GaP|GaAs|GaP|GaAs|$) are shown. The dielectric functions $\varepsilon_1(\omega)$, $\varepsilon_2(\omega)$ were taken in the standard form for bulk samples in approximation of harmonic oscillators [8]: for GaAs - $\varepsilon_\infty=11.1$; $\omega_{TO}=268cm^{-1}$; $\omega_{LO}=292cm^{-1}$; for GaP - $\varepsilon_\infty=10.06$; $\omega_{TO}=348cm^{-1}$; $\omega_{LO}=357cm^{-1}$.

It is noteworthy that the spatial distribution of vibrational amplitudes is essentially different for the IP and SP. This allows one to suggest an experimental way to distinguish them.

3. ATR-investigation

It can be seen from Fig.1b that SP (curves 5,6) and IP (curves 1-4; 7-10) can be identified by ATR-spectroscopy. Really, the depth of field penetration in the vacuum gap

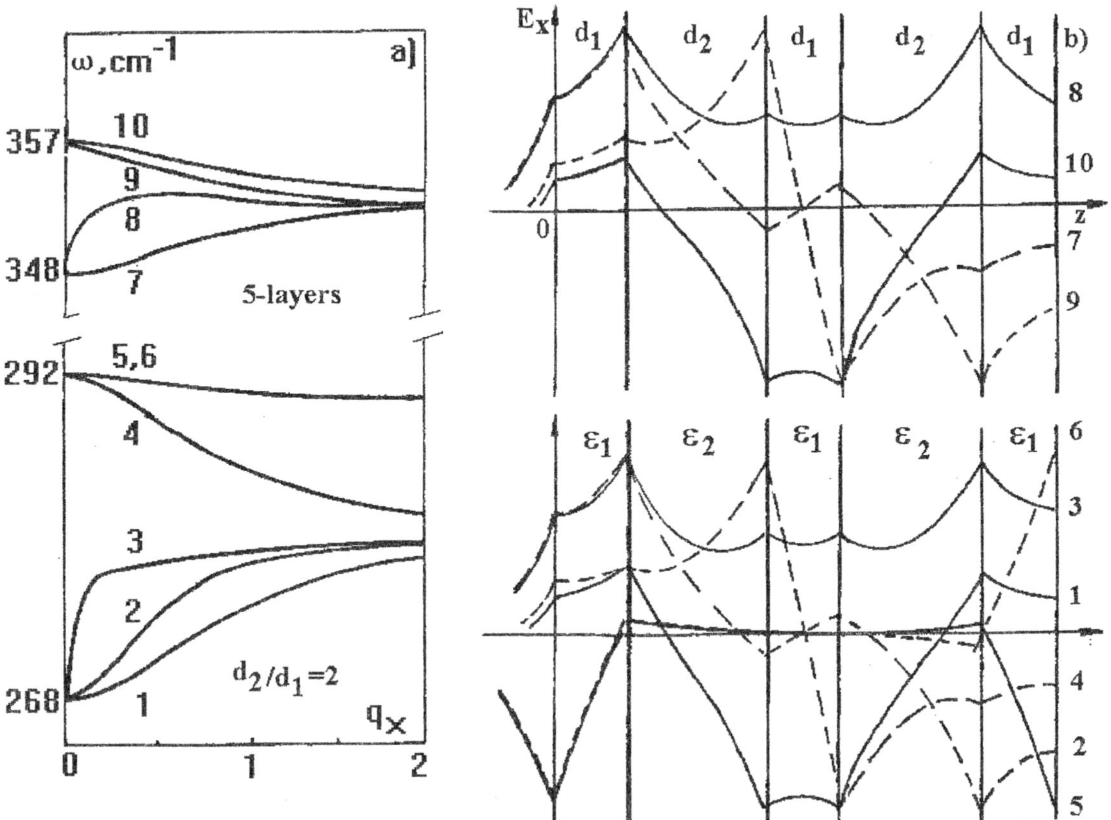

Fig.1 Vibrational modes (a) and corresponding spatial distributions of E_x (b) for SP (5,6) and IP (1-4,7-10) in SL (ε_1, d_1-GaAs, ε_2, d_2-GaP, $q_x d_1=1$)

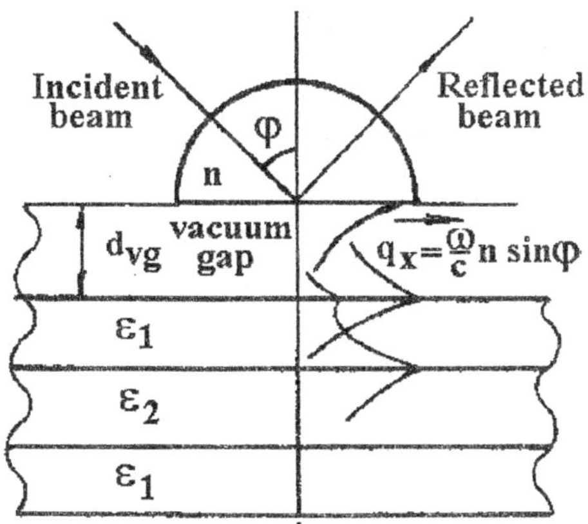

Fig.2 Illustration of the principle of ATR experiment

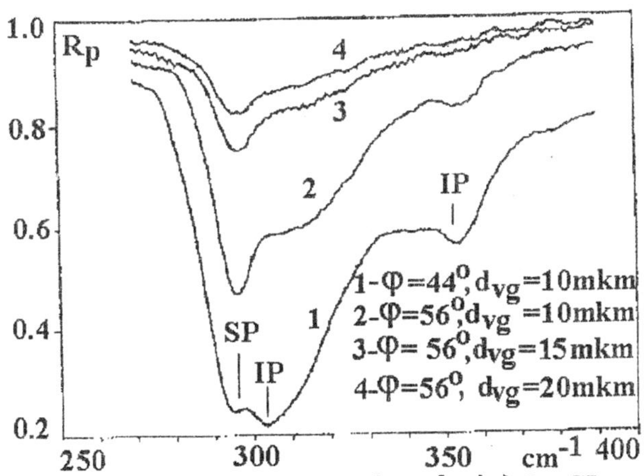

Fig.3 ATR spectra (p-polariz.) for SL $GaAs_{0.76}P_{0.24}$(80Å)/GaAs (80Å)

between the semicylinder (prism) and the specimen (fig.2) decreases when the angle φ is increased [6]. Consequently, the depth of minimum caused by SP will be less sensitive to the angle change. The identical effect can be achieved by enlarging of the vacuum gap when the angle φ is fixed. In the fig.3, ATR-spectra measured at p-polarized light for the SL samples GaAsP/GaAs consisting of 20 periods on the GaAs substrate are presented. The sample growth details are described in [7].

Our experimental arrangement for ATR in the far-infrared is made to work with a fixed incidence angle φ determined by the geometry of ATR attachement to spectrometer. The incident radiation couples to surface polaritons and guided waves of the sample by means of the semicylinder of CsI (n=1.72). Otto configuration for the experimental setup has been used and frequency scan has been carried out.

As one can see from fig.3 (curves 1, 2), there are two minima in the spectral dependence of reflectance coefficient (R_p)in the frequency range of 290-310cm^{-1} and one minimum at 356cm^{-1}. This minima are absent in spectra, measured at s-polarization. The depth of the minimum at about 296cm^{-1} does not change as the angle of incidence increases. This indicates on its origin from SP, whereas two other minima became approximately two times weaker, what allows one to attribute them to IP. Only SP is revealed in the spectra when the gap is enlarged (curves 3, 4). Shift of all observed frequencies to the larger values is caused by the strains arising in the layers.

References

1. T.Dumelov, T.I Parker, S.R.P.Smith, D.R.Tilley, Surf.Sci.Rep.**17**(1993) 151

2. S.M.Rytov, Zh.Eksp.Teor.Fiz. **29**(1955) 605

3. M.Cardona, Superlatt.& Microstr.,**5**.N1(1989) 27

4. N.L.Dmitruk, A.V.Goncharenko, O.S.Gorea et.al. Proc.SPIE,v.2648 (1995)

5. O.S.Gorea, E.P.Pokatilov, N541M-D85, MoldNIINTI, Kishinau (1985),21p.

6. N.L.Dmitruk, V.G.Litovchenco, V.L.Strijevski. Surface polaritons in semiconductors and dielectrics (in russian), Kiev,1989,375p.

7. O.S.Gorea, V.V.Zelenin, A.V.Koval, A.V.Simashkevich, J.Cryst.Prop.and prep.,**32-34** (1991) 503-507

8. P.Baranski, V.Klotchkov, I.Potykevitch, Electronique des semiconductors. Ed. Mir, Moscou, (1978), 672p.

Photoelectric Emission from Quantum Wells with Various Potential Profiles

C. Bose, C. Chakraborty and C. K. Sarkar*, *Senior Member, IEEE*

Department of Electronics & Tele-Communication Engineering, Jadavpur University, Calcutta-700 032, INDIA

**Department of Physics, B. E. College (D.U.), Howrah -711 103, INDIA*

SUMMARY— The subband energies in a quantum well (QW) with rectangular, triangular and parabolic potential profiles have been calculated. The subband energies for rectangular QW is found to be the lowest among all the three cases, while the energies for triangular well is higher compared to the parabolic potential profile case. The photoelectric (PE) current density as a function of the well width and the incident photon energy have been investigated for all the cases. The effect of size quantization is quite evident in all cases from the oscillatory variation of the PE current density.

I. Introduction

THE modern crystal growth techniques have reached such a level of sophistication that the band structure may be tailored to produce any desired variation of the potential profile. The molecular-beam epitaxy (MBE) method is well suited not only to the fabrication of rectangular quantum well with smooth and sharp boundaries, but also to that of quasi-two-dimensional (Q2D) structures with parabolic as well as triangular profiles [1,2]. All these structures possess electronic subbands, quite different from each other. Therefore their optical properties are also expected to differ significantly. In the present communication, a comparative study of the photoelectric emission (PE) from quasi-two-dimensional electron gas (Q2DEG) formed in quantum wells (QW) with three different potential profiles have been made theoretically.

II. Theoretical background

Without going into the fabricational details of how a QW with specific potential profile can be generated, we consider here, three QWs - rectangular, parabolic and triangular. In all cases, the potential variations appear in the z-direction, the direction along which the well dimension is kept restricted. We further assume that as indicated in **Fig.1**, all three wells have the same dimension d (~ nm) at the vacuum level, i.e. at energy W, the electron affinity of the material considered.

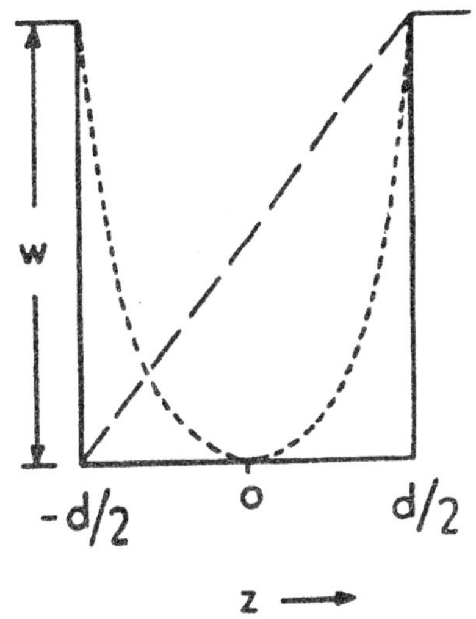

Fig. 1. Schematic diagram showing the height and width of three QWs –Rectangular (solid), Parabolic (dotted) and Triangular (dashed).

Let us first investigate the energy expressions for the three cases assuming infinite well. Energy levels for electrons within a rectangular QW are given as [3]

$$E_n = \frac{\hbar^2}{2m^*}\left(\frac{n\pi}{d}\right)^2 \tag{1}$$

where, the size quantization number n possesses only integral values starting from 1.

In the case of parabolic QW, a potential energy term kz^2 appears in the Schrödinger equation, which is then identical to that for an harmonic oscillator. The energy levels for electrons confined within such an well can therefore be written as [1]

$$E_n = \left(n + \frac{1}{2}\right)\hbar\omega \tag{2}$$

where, n = 0, 1, 2,..., $\omega = \sqrt{k/m^*}$ and k is the curvature of the potential well. The curvature can be related with height and width of the well through the relation $k/2(\pm d/2)^2 = W$.

The width of the potential well at the quantized energy level can similarly be obtained as

0-7803-3223-7/96/$5.00

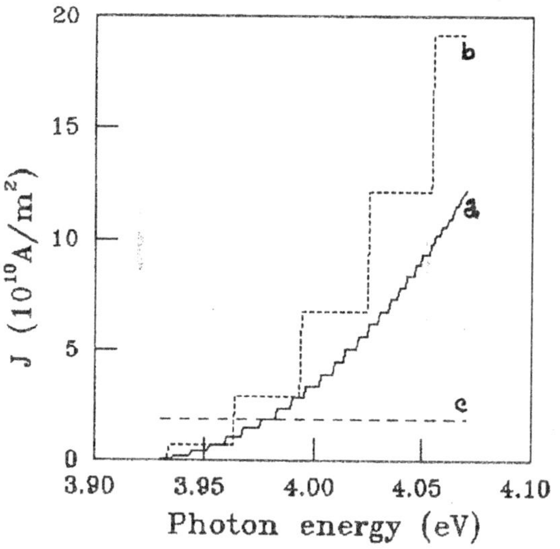

Fig. 3. Photoelectric current density vs. incident photon energy with d = 200 nm for various QWs. **a** : rectangular, **b** : parabolic, and **c** : triangular.

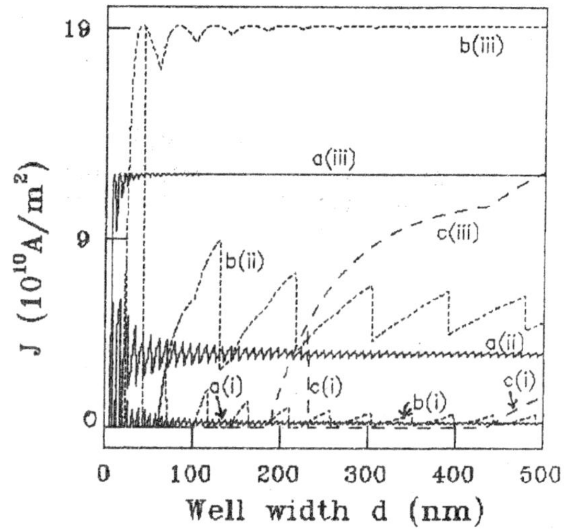

Fig. 4. Photoelectric current density vs. width of the **QW** for three different photon energies [**i:** 3.94 eV, **ii:** 4.0 eV, and **iii:** 4.07 eV], for three different wells. **a, b,** and **c** are as in Fig. 3.

The curve **a** exhibits a gradual increase in step length with decrease of photon energy. This indicates that for a rectangular **QW**, inter-subband separation increases towards higher energy. The opposite variation in step length is noted for the triangular well (*Fig. 5*), where the higher energy levels are more closely spaced. The existence of equispaced energy levels in parabolic well is clear from equal step length appeared in curve **b**.

However, only the lowest level falls below the Fermi level in the case of a 200 nm triangular **QW** and therefore, no step appears in the corresponding emission characteristics (curve **c**). In triangular well, a number of steps appear in the current density for an well width as large as 1200 nm (*Fig. 5*). For such a high value of d, more than one subband fall below the Fermi level and thereby participate in the emission process.

Fig. 4 illustrates the dependence of photoemission on well width for three potential profiles and three different values of photon energy. For a particular value of incident photon energy, contributions come from various subbands down to that with quantum number n_{min}. With increasing d, the value of n_{min} increases, as well as energy levels above the Fermi level gradually fall below it thereby taking part in emission. Thus, at photon energies significantly lower than the electron affinity, each subband contributes discretely, as evident from the set of curves (**i**), more distinctly from curve **b(i)** for parabolic well. This discreteness in the **PE** current is found to diminish gradually with increasing d, since the **QSE** becomes insignificant for higher d. In the case

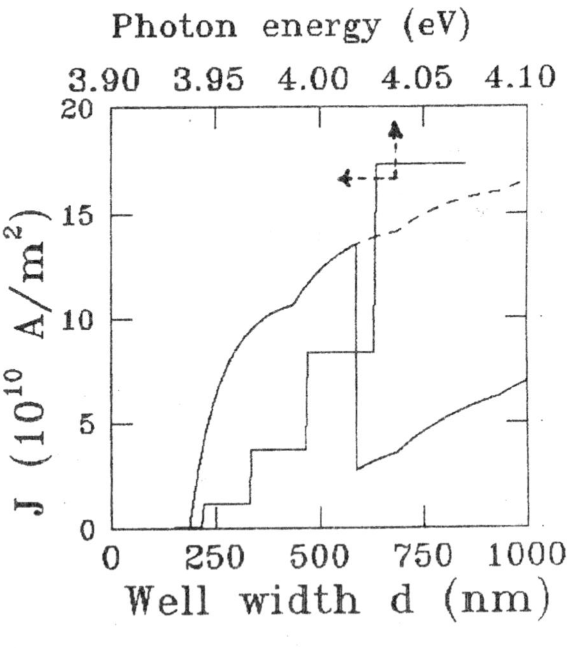

Fig. 5. Variation of **PE** current density for triangular well : with incident photon energy for d=1200 nm, and with well width for photon energies 4.0 eV (solid) and 4.07 eV (dashed).

$$E_n = (n + \frac{1}{2})\hbar\omega \quad (3)$$

where, n = 0, 1, 2,..., $\omega = \sqrt{k/m^*}$ and k is the curvature of the potential well. The curvature can be related with height and width of the well through the relation $k/2(\pm d/2)^2 = W$.

The width of the potential well at the quantized energy level can similarly be obtained as

$$d_n = 2[\frac{(n + 1/2)\hbar d}{\sqrt{2m^* W}}]^{1/2} \quad (4)$$

In the case of a triangular **QW**, due to the linearly varying potential energy term, the Schrödinger equation yields the solution to be an Airy function. The energy eigen values can then be expressed in terms of well height and width as

$$E_n = (\frac{\hbar^2}{2m^*})^{1/3}[\frac{3\pi W}{2d}(n + 3/4)]^{2/3} \quad (5)$$

with n = 0, 1, 2,....

At the energy level E_n, the width of the potential well is simply $d_n = E_n\, d/W$.

The above wells are now assumed to be subjected to a shower of monochromatic photons of energy $h\nu$, incident along the z–direction. The resulting emission of photoelectrons is considered to take place along the same direction. The incident photons are assumed to have energy sufficient to emit conduction electrons only. The phenomenological expression for current density $J = n_0 ev$ is used here to estimate contribution from individual electric subbands.

The volume density of carriers in the quantized energy level E_n can be written as

$$(\Delta n_0)_n = \frac{m^*}{\pi\hbar^2 d_n}(E_F - E_n) \quad (6)$$

assuming very low temperature, so that the Fermi distribution function f(E) = 1 can be retained. The velocity v_n of electrons in a particular energy level E_n is calculated from its kinetic energy, to get $v_n = \sqrt{2E_n/m^*}$. Thus obtaining all the terms to be used, the **PE** current density for the above three cases can be estimated from the final expression [4]

$$J = \sum_{n_{min}}^{n_{max}} \frac{1}{2} a e v_n (\Delta n_0)_n \quad (7)$$

Since only the subbands lying below the Fermi level will participate in emission, n_{max} will be limited as $E_{n_{max}} \leq E_F$, other terms in equation (6) are as defined elsewhere [4].

III. Results and discussion

The material parameters for GaAs [5] are taken in our computation. Here, the well depth has been taken to be W (=4.07 eV), although the energy expressions, obtained from infinite well treatment, are considered. For the sake of simplicity, we have computed the **PE** current density keeping Fermi energy E_F fixed at 0.15 eV (which corresponds to a degenerate carrier concentration of about 10^{16} m^{-2}). Again, the photon absorption co-efficient 'a' is taken to be unity, as normally done for a direct band gap material.

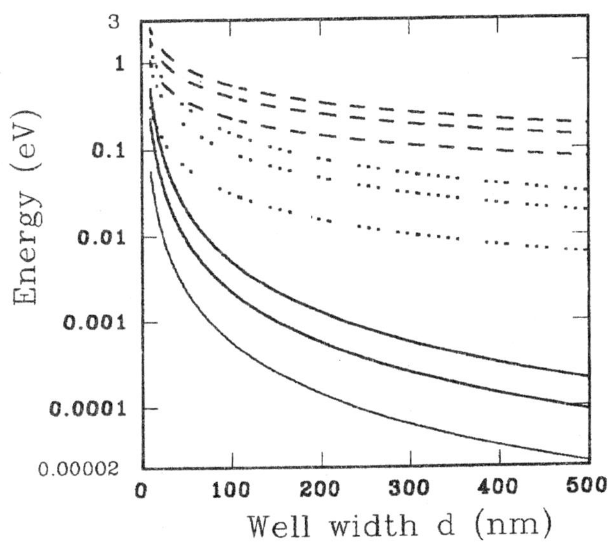

Fig. 2. The lowest three energy levels as a function of well width in case of rectangular **QW** (solid), parabolic **QW** (dotted) and triangular **QW** (dashed).

Fig. 2 shows the lowest three energy levels plotted against well depth for all the three **QW**s. The energy levels are the highest for triangular **QW**, and the lowest for rectangular **QW**, while levels for parabolic **QW** fall in between.

Fig.3 shows the variation of **PE** current vs. incident photon energy for all the three wells, each being 200 nm wide. As incident photon energy is increased, inner subbands gradually contribute to yield the net current density as ladder-like steps. For a parabolic **QW**, due to large intersubband separation, the effect of quantization is very prominent, giving rise to sharp stair-case **PE** current (curve **b**). The few number of steps appearing in curve **b** signify that due to large subband energies, only a few subbands fall below the Fermi level. For a rectangular **QW**, contributions from a large number of closely spaced energy levels add together. Consequently the **PE** current is characterized with a large number of small steps, as shown in curve **a**. The quantum size effect (**QSE**) in this case, becomes less significant for the large well width.

of rectangular well, subband energies and their separations are quite small. As a result, contributions from various subbands are less discrete for rectangular well [curve **a(i)**] compared to that for a parabolic well [curve **b(i)**]. Similar variation of **PE** current was reported elsewhere for narrower rectangular **QW** [4]. The curve **c(i)** shows that for the same range of incident photon energy, only two subbands emit discretely in the case of triangular **QW**.

On the other hand, for photon energy equal to or very close to the electron affinity, all subbands up to the Fermi level will contribute. Their contributions will overlap to give the net current density, which is found to oscillate with changing well width in the set of curves **(iii)**. The curve **a(iii)** for rectangular **QW** shows that the oscillation dies out completely for about 100 nm wide well and a steady current density for bulk specimen is achieved beyond this value. For parabolic **QW** [curve **b(iii)**], similar oscillation occurs with larger period, due to higher value of inter-subband separation. Again, the oscillation retains up to a large width of 250 nm (approx.) due to higher value of energy levels compared to those in the rectangular case. Also the steady value of **PE** current in parabolic well is 1.5 times of that in the rectangular well. The curve **c(iii)** for a triangular **QW** shows that with 4.07 eV photon energy, only one subband comes below the Fermi level for well width 190 nm (approx.). Only this single level contributes unless the width is increased to a value of about 435 nm, when the second subband also starts emitting.

For an intermediate value of incident photon energy, resulting PE current density is due to contributions from few subbands, where discreteness or overlapping of various contributions occurs to a degree intermediate of those appeared in case **(i)** and **(iii)**. The nature of the curves **a(ii)** and **b(ii)** reveals this fact. However within the range of d taken in *Fig. 4*, the **PE** current remains unaltered for photon energies 4.0 eV and 4.07 eV in the triangular **QW** [curve **c(iii)**]. The observation implies that, the lowest subband lies much above the bottom of the well, and hence causes no additional contribution due to that increment of photon energy. *Fig. 5* shows that the current density for photon energy of 4.07 eV (dashed) can be distinguished from that of 4.0 eV (solid), while investigated for larger range of well width.

ACKNOWLEDGMENT

The work is financially supported by UGC (India). The authors are grateful to Prof. M. N. Roy and K. K. Mallik for providing computational facility.

References

[1] R. C. Miller, A. C. Gossard, D. A. Kleinman and O. Munteanu, *Phys. Rev. B*, vol. 29, p. 3740, 1984.

[2] A. C. Gossard, W. Brown, C. L. Allyn and W. Wiegmann, *J. Vac. Sci. Technol.*, vol. 20, p. 694, 1982.

[3] R. Dingle, *Festkörperprobleme*, vol. 15, p. 21, 1975.

[4] C. Majumdar, A. B. Maity and A. N. Chakravarti, *phys. stat. sol.(b)*, vol. 140, p. K7, 1987.

[5] S. Adachi, *J. Appl. Phys.*, vol. 58, p. R-1, 1985.

Session 15

PROCESS CHARACTERIZATION (Poster)

IMPLANTATION BEAM ANGLE STUDY FOR AL IMPLANTED IN 6H-SIC

E. Morvan, J. Montserrat, P. Godignon, J. Fernandez, D. Flores, M.L. Locatelli**
J. Millán and G. Brezeanu*
Centro Nacional de Microelectronica (CNM-CSIC), Campus UAB, 08193 Bellaterra, SPAIN
*Faculty of Electronics, University Politehnica Bucharest, 7000, S6, Bucharest, ROMANIA
**CEGELY-ECPA, INSA Lyon, 69621 Villeurbanne, FRANCE

SUMMARY

In this paper a simulation study of Al implantation into 6H-SiC single crystal is presented. Beam orientation with respect to crystal axis has been investigated. this aspect is of crucial importance to achieve good implantation process control. A first step analysis is given and values for optimum ion beam angles are proposed.

1. INTRODUCTION

Silicon carbide 6H-SiC single crystal is known as an exceptional semiconductor material for specialized devices fabrication in the fields of power and high frequency electronics, optoelectronic and sensors /1/. Its possible industrial applications are automotive and aerospace electronics, power devices for converters, switches or microwave technology. Ion implantation technology, which has been shown to be a very interesting tool to incorporate doping impurities in Silicon in a controlled manner, begins to be transposed to SiC for device fabrication. Large differences between the two materials give rise to many problems for SiC technology which do not appear in the case of Silicon technology. Different implantation conditions have to be used for SiC in order to achieve semiconductor doping. In the same way, ion implant simulators for crystalline silicon have to be modified in order to account for the different crystal structure and composition.

In order to study the different aspects of ion implantation technology for 6H-SiC, a specific simulator has been developed. It uses Montecarlo method together with semiempirical and physically based models to calculate the implanted ion trajectory through the SiC single crystal. The slowing down of an incoming ion is evaluated within the framework of binary collision approximation for nuclear stopping /2,3/ and with advanced models for electronic stopping, following the recent work of other research groups /4-6/. In this paper, we present the simulation of one of the numerous problems of ion implantation technology: the tilt and rotation angle optimization for maximum dopant depth profile uniformity across the implanted wafer which is of crucial importance for process reproducibility.

2. STATEMENT OF PROBLEM

A first aspect to be considered is channelling which occurs when a swift ion travels in a special crystallographic direction inside a crystal. A channelled ion can be viewed as a particle guided between the walls of a cylinder which generatrices are made by rows of crystal atoms. This case is named "axial channelling" along a given crystallographic direction. An other configuration exists when the ion travels between two crystal planes and is called "planar channelling" parallel to a given crystal plane. In the first case, which is predominant, the moving ion has an oscillatory trajectory inside the cylinder. The walls of this channel are somewhat "porous" for the ion, due to the nonuniformity of the repulsive potential generated by the dispersed positive nucleus of the lattice. Generally speaking, the channelling behaviour of the ion depends on many parameters such as the *atomic structure of the cylinder walls* related to the crystalline structure of the target and the direction of motion of the ion, the *ion energy* which determines the sensitivity of the ion trajectory on the target nucleus potential, the *initial angle of incidence* of the velocity vector of the ion with respect to the channel direction, the *initial*

0-7803-3223-7/96/$5,00 © 1996 IEEE

position of the ion in the plane perpendicular to the channel axis. A schematic representation of some parameters effects is given in Fig 1. Further explanation will be found in part 4.

The second aspect to be considered is the optimization of the tilt/rotation angles (tilt is the angle of the ion beam with respect to <0001> crystal axis, rotation is measured from <11-20> axis). When implanting a wafer for device fabrication, uniformity of the depth doping profile of the implanted layer has to be taken into account. With electrostatical beam sweep implanters this problem is critical with respect to channelling since the orientation of the ion beam (tilt and rotation) with respect to the <0001> axis of the crystalline wafer changes during the sweep. Then, channelling being tilt and rotation sensitive, both angles have to be chosen such as to minimize channelling and/or to keep constant the channelling conditions during the whole implantation process. Determination of optimum tilt/rotation angles requires the evaluation of the channelling yield with respect to both angles for a given implantation energy. In practice, large tilt angles must be avoided because of wafer surface erosion, strong backscattering of incoming ions and mask shadowing effect. For (100) oriented Silicon crystals, 7º tilt and 30º rotation are used in standard implantation processes, giving about random implantation (i.e worst channelling conditions) with low depth doping profile variations across the wafer.

3. CHANNELLING SIMULATION IN 6H-SIC

The channelling behaviour of ions in the 6H-SiC crystal being a consequence of the crystal structure and periodicity, we have simulated the trajectory of implanted ions in a perfect crystal. This means no thermal vibration of lattice atoms and no defects generation during the implantation process simulation. The defect generation would tend to randomize the crystal lattice but the minimization of channelling by crystal amorphisation is a different aspect of ion implantation and will not be treated through this paper. Moreover, zero degree beam divergence will be assumed as well as oxide free crystal surface.

We have chosen a simple procedure to evaluate the channelling yield relative to a given implantation process: ions are assumed to be channelled and participate to the tail of the impurity depth profile if they cross a depth limit in the crystal defined as:

$$Z_{LIMIT}(E) = R_p(E) + \text{sigma}\ (E)$$

where R_p and sigma(E) are the projected range and standard deviation for random implantation conditions at energy E. This criterion is somewhat crude but represents quite well the atom depth profile nonuniformity problem since it is a measurement of the dose which penetrates too deeply into the crystal and thus modify the configuration of a junction formed by implantation.

4. RESULTS AND DISCUSSION

The sensitivity of the channelling with respect to the tilt and rotation angles can be visualized by plotting, for a given implantation energy, the percentage of channelled ions (following the criterion described above) versus the tilt and rotation angles on a 3D surface plot. Due to the crystal structure symmetries, we only need to represent the plot for rotation angles ranging from 0º to 30º. For practical purpose, the tilt angle is varied between 0º and 30º.

A 3D plot for 30 keV $^{27}Al^+$ ions implanted into 6H-SiC is proposed in Fig. 2 and shows some interesting aspects of the implantation process. First, we can verify that for a 0º tilt angle, the channelling yield (CY) is the same for any rotation angle used in the simulation. Next, it can be seen that increasing the tilt leads to a decrease in CY. This is a consequence of incident ions having a tilted velocity vector with respect to the <0001> crystallographic axis, which are more likely to leave the channel at the beginning of their trajectory and suffer more random collisions with target atoms. The main feature of the plot is that appreciable reduction of CY appears only when tilt angle reaches about 5º, which is a surprising high value. This means that when simulating ion implantation around the <0001> direction, the implanted atom depth profile is not affected until about 5º tilted implantation is used.

For tilt angles above 5º, the CY decreases rapidly from 90% to about 50% whichever rotation angle is used. However, an increase of the tilt angle above 15º give rise to a special behaviour of The CY: for rotation angles around 30º, the CY strongly increases to about 80%. This sharp pic for CY around 20º tilt and 30º rotation can be explained by the hexagonal structure of 6H-SiC crystal which presents open channels in this direction with respect to the <0001> axis. Even if those channels are discontinuous due to the relatively complex stacking sequence of 6H-SiC, they are effective in channelling implanted ions. An other consequence of this geometrical configuration is the small peak for 20º tilt/0º rotation implantation.

For rotation angles around 10º this channelling effect is greatly reduced because it is more difficult for implanted ions to enter a channel. They penetrate inside the crystal within a more random direction. As a consequence, one can use tilt angles superior to 10º together with about 10º rotation angle as defined previously, in order to minimize atom depth profile nonuniformity for a given implantation process. Increasing the implantation energy will not affect much this optimum conditions since higher energy ions have even lower probability to be channelled. The energy of the moving ion affects its channelling probability in two ways: increasing the energy will cause the critical angle of channelling to decrease. In other words, for a given incidence angle of implantation with respect to a given channel axis, a high energy ion will have a higher transverse energy with respect to the same axis and will have a higher probability to cross the potential barrier formed by the atoms of the channel walls. Even if this ion can be rechannelled in a parallel channel in the crystal, the probability of random scattering after the crossing of the wall is high. On the other hand, for the same incidence angle, if the ion has a low enough energy, it will suffer a "progressive" reflection by the wall repulsive potential and will then have an oscillatory motion between the walls of the channel (Fig. 3). Analytical formulae for critical angle evaluation usually assume an energy dependence in $E^{-0.25}$. Our simulations for <0001> channels give about $E^{-0.66}$ behaviour (fig.4). Further investigations have to be carried out.

CONCLUSIONS

We have investigated the tilt/rotation dependence of channelled dose during Al implantation into 6H-SiC. The first step of this work leads to optimum tilt of 10º or more with respect to <0001> crystal axis and rotation of 10º with respect to <11-20> crystal axis. The simulator for 6H-SiC developed at the CNM has been used here and will serve for better understanding and optimisation of the implantation process.

REFERENCES

/1/ H.Morkoç & al. Journal of Applied Physics. 76 (3), August 1994.

/2/ M.T Robinson & al. Physical Rev. B9 (1974) p.5008.

/3/ J.F Ziegler & al. The stopping and range of ions in solids (Pergamon, New York 1985)

/4/ W.Brandt & al. Physical Rev. B25 nº9 (1982).

/5/ P.M Echenique & al. Physical Rev. A33 nº2 (1986)

/6/ W. Neng-ping Physical rev. A52 nº5 (1995)

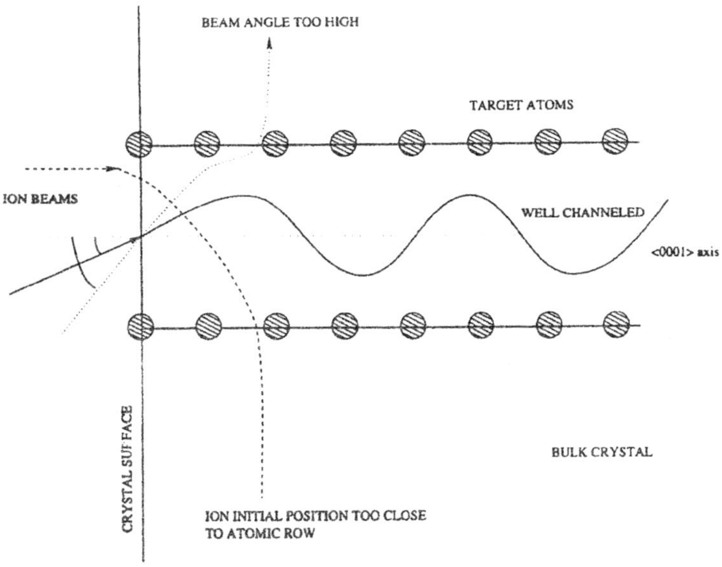

Fig.- 1: Different channelling configurations. A 2D channel is represented for simplicity. The oscillating trajectory correspond to a well channeled ion (solid line). The large angle tilted implantation causes the ion to cross the channel wall (dotted line) and the ion implanted too close to an atomic row is directly scattered out of the channel (dashed line).

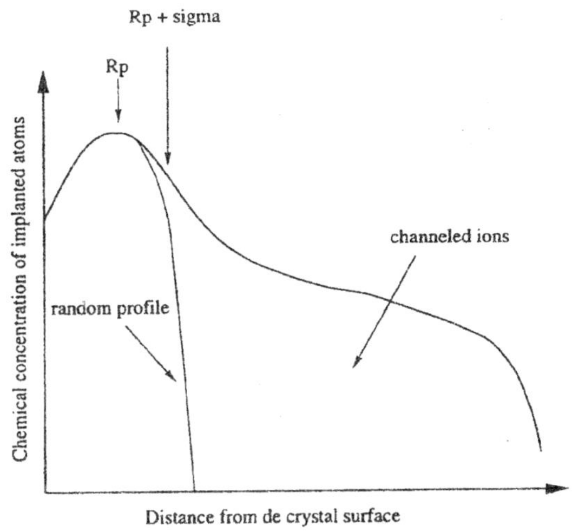

Fig.- 2: definition of the criterion used for the calculation of the channelling yield (CY). Ions that go beyond the (R_p + sigma) depth limit are sumed to give the CY.

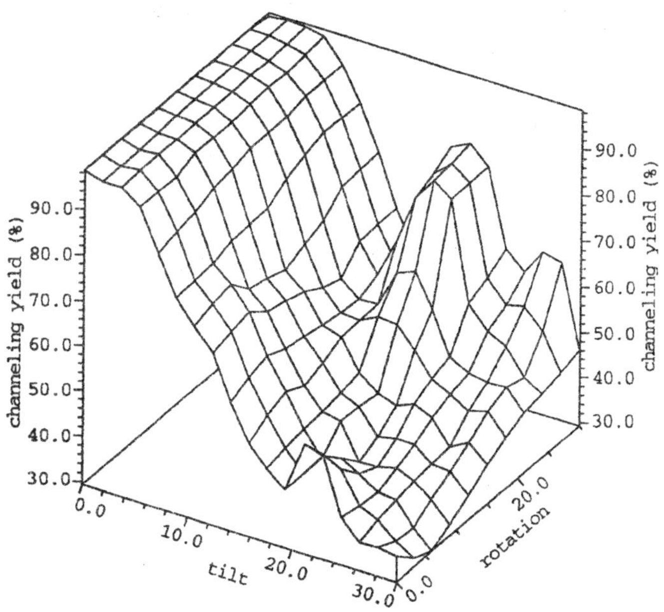

Fig.- 3: Channelling yield versus tilt and rotation angles for $^{27}Al^{+}$ ions implanted into 6H-SiC.

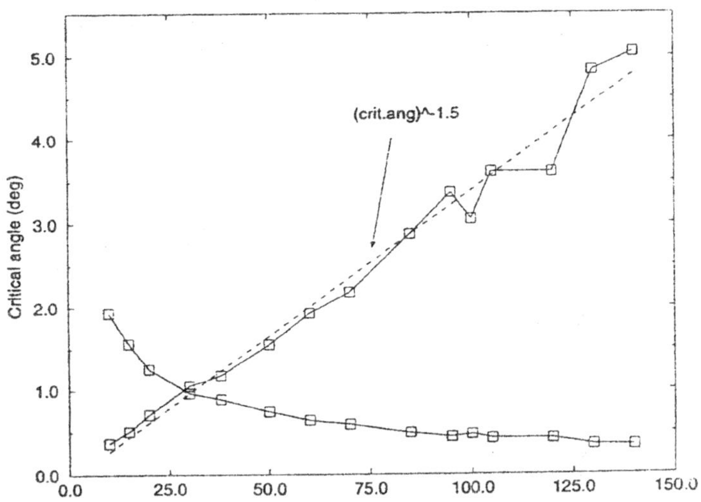

Fig.- 4: Critical angle versus energy of implanted ions. The curve labeled "crit.ang$^{-1.5}$" is the best fit for a linear function of the energy. Resulting simulated energy dependence of critical angle has the form crit.ang = k. E^{-066} where k is a constant.

Mechanism of Strain Relaxation and Defect Introduction in Ge–GaAs Heterostructure:Influence of Ultrasound Treatment

T.G.Kryshtab, M.A.Mazin, I.V.Prokopenko, G.N. Semenova.

Institute of Semiconductor Physics NASU, pr.Nauki 45, Kyiv 252650, UKRAINE.
Tel. (38044)2658317, Fax.(38044)2658342
E-mail: class@class.semicond.kiev.ua

Summary

For the MBE grown Ge films on the semiinsulating (001) GaAs substrates the effect of the substrate processing and the subthreshold ultrasound treatment on the mechanisms of strain relaxation and the structural transformations in the heterosystem is reported.

X-ray topography and X-ray diffraction measurements have shown that the native oxide influences the strain relaxation. This is evidenced by appearing the lattice tilting and mosaic structure in the films. The ultrasound treatment leads to structural relaxation by the misfit dislocation nucleation, insignificant redistribution of strain deformation fields and improvement of structural perfection due to interaction between the point defects and dislocations.

1. Introduction

The "Ge on GaAs" heterostructure (HS) is of practical interest and has actual industrial applications in different types of devices and in nonalloyed Ohmic contacts to GaAs. Molecular-beam epitaxy (MBE) is the nonalternative nonequilibrium growth technique to control the heterojunction properties by optimization of technology conditions and substrate surface preparation [1,2].One of the main problems in MBE growth of Ge–GaAs HS for industrial applications is connected with interdiffusion and interface broadening. It was partly solved by use of stoichiometric substrates or by forming native oxide films [1].

The practical importance of the ultrasound (US) wave interaction with different types of defects in semiconductors is determined by the possibility of improving the material parameters by ultrasound treatment(UST). Scientific interest in US is due to its efficiency for investigation of the mechanisms of defect diffusion and transformation [3]. The influence of US on the material properties depends critically on the US wave power. Even the use of low-power (subthreshold) US waves leads to important changes in the properties of bulk materials3 or epilayers [4,5].

In this paper the mechanisms of both strain relaxation and formation of structural properties of Ge films grown by MBE, as well as the effect of subthreshold UST on structural transformation in strained HS, were studied using X-ray diffraction.

0-7803-3223-7/96/$5,00

2.Experimental

The Ge films were MBE grown on (001) GaAs semiinsulating substrate. The growth temperature was 400 ± 20^0 C, growth rate ~ 1 monolayer per second and base pressure ~ 10^{-8} Pa [1,2]. There was a native oxide layer (~15Å) on the substrate surface after chemical polishing in $H_2SO_4 : H_2O_2 : H_2O$ (10 : 1 : 1) solution at 30^0 C for 1 min. This layer prevented As or Ga diffusion into Ge films and its contamination by carbon. The films were of p-type with charge carrier concentration $p = 5 \div 7 \cdot 10^{15}$ cm^{-3} and mobility $\mu = 400 \div 600$ $cm^2/ V s$.

The structure of Ge films and heterosystems has been investigated by X-ray diffraction methods (sources - CuK_{α}): a) topography (transmission - in the Borrmann mode and reflection - double-crystals for Bragg asymmetrical diffraction); b) analysis of rocking curves from the double-crystal diffractometer [6,7]; c) measurements of integrated intensity (II) of X–ray Bragg diffraction for structural and quasiforbidden reflections; d) the II of X-ray anomalous transmission as a function of crystal thickness.

US treatment was realized by piezoceramic transducer at the frequency f = 160 kHz with amplitude of acoustic deformation $\tilde{a}_{us} \sim 10^{-5}$.

3.Results

The Ge films were grown on the commercial GaAs substrates. As was shown using Borrmann topography, they had inhomogeneous W-shape distribution of dislocation density ($N_d = 10^4 \div 10^5$ cm^{-2}) with a cell structure. All attempts to detect the misfit dislocations (MD) for Ge–GaAs heterostructures ($d_{Ge} \leq 2$ μm) by using Bragg topography for optimum asymmetrical reflections have failed.

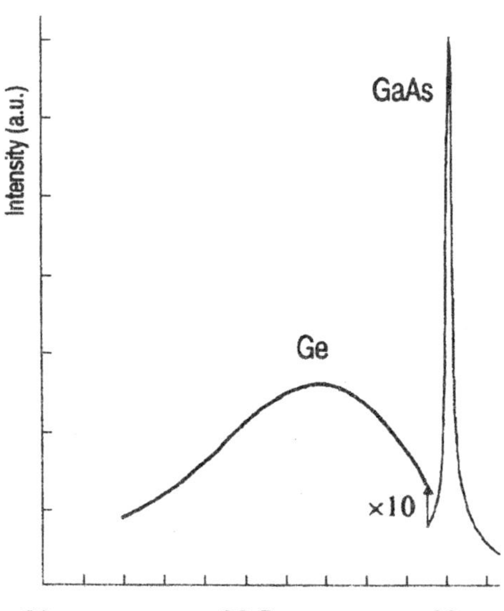

Fig. 1. (004) X-ray diffraction curve for Ge-GaAs (d_{Ge}=2,2 μm) heterostructure

The rocking curves exhibit a very broad peak from Ge film, and a full width at half maximum for Ge film 2,2 μm thick was $\Delta\theta = 26'$ for the symmetrical 004 reflection (Fig. 1). The peak position changes with the incident X-ray beam direction up to arising on the opposite (right) side from the GaAs peak. The symmetrical and asymmetrical Bragg reflections (the directions of incidental and reflected X–ray beams being changed by 180^0) were used to determine the discrepancy angle β between the substrate and epilayer (Fig.2) and the film deformation, respectively. The lattice deformations in films were defined by measuring the lattice constant ratio $a^{\parallel} / a^{\perp}$ [6]:

$$a^{\parallel} / a^{\perp} = 1 - [(k^2 + l^2)^{1/2} / h] \operatorname{cosec}^2\{\arctan[(k^2 + l^2)^{1/2} / h]\} \Delta\alpha / 2,$$

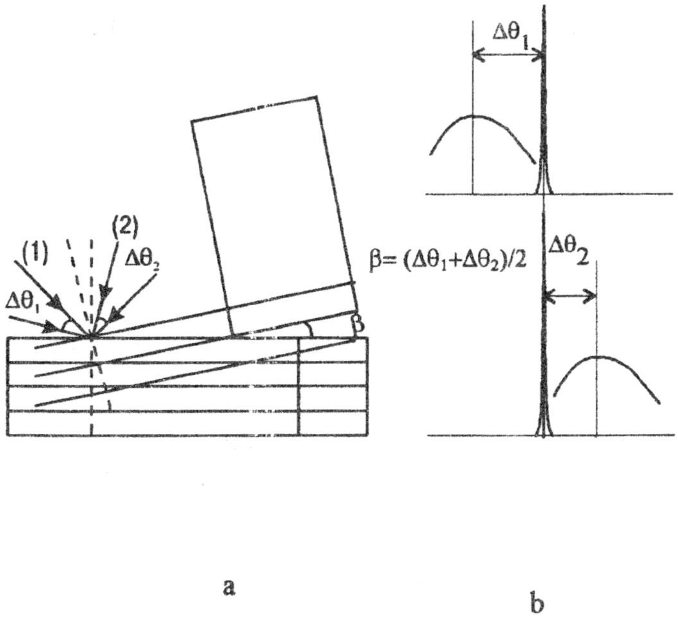

Fig. 2. Principle of measurement: a) the lattice plane and X-ray direction; b) the rocking curves for (400) reflection from Ge-GaAs.

where h, k and l are the asymmetrical reflection plane indices, $\Delta\alpha$ is an angle between two diffraction peaks from the obtained film by interchanging the incidence and reflection directions of the X-ray beam.

A tilt angle appeared between the lattice plane of the film and substrate up to $\beta > 30'$ in $\langle 100 \rangle$ direction. The rocking curve linewidth $\Delta\theta$ may be due to different structural defects: lattice strain, bending, dislocations and accompanying mosaic structure. The peak broadening for our HS is due rather to mosaic structure then to other defects, and change in its position can be explained only by misorientation between the film and the substrate. The microstructure consists of crystalline mosaic blocks which are slightly misoriented with respect to one another [7]. The thickness of each block, which gives the particle size perpendicular to the diffraction plane, was estimated from $\Delta\theta$ and equalled ~ 250Å. The crystal lattices in these mismatched films are deformed from cubic to tetragonal once with the ratio $(a^{\parallel} - a^{\perp}) / a^{\perp} = 2\times10^{-3}$.

After UST no changes in dislocation structure were detected by Borrmann topography. The essential transformation of rocking curves was not observed. The macrobend of HS was unchanged, it was the microcurvature redistribution along a slab diameter.

On Bragg asymmetrical topogramms we could see the MD nucleation. The measurements of II for Bragg diffraction (400 - reflection) indirectly confirm the nucleation of MD by increasing the intensity of reflection from the substrate by ~ 10%. At the same time the II for quasiforbidden 200 reflection from GaAs – substrate, which was higher than for stoichiometric samples, decreased and approached the value for stoichiometric GaAs after UST. This indicates that UST results in the stress relaxation not only by MD but also by redistribution of point defects.

When investigating the II of X-ray anomalous transformation for hkl and $\overline{hkl}$ we observed changes in parameter $Y = I_i^{hkl} / I_i^{\overline{hkl}}$ along the coordinate of sample, which gave an information about bending of atomic planes and increasing of $W=0.5(I_i^{hkl}+I_i^{\bar{h}\bar{k}\bar{l}})$. This indicated the improvement of structural perfection [8] after UST. We studied also the ln W dependence on crystal thickness t in the areas, where Y ≅ 1. These measurements were performed by inclination of sample around the axis perpendicular to diffraction planes. The effective absorption coefficient μ^* was determined from the dependence ln $W = f(t)$. After UST the μ^* value decreased, thus indicating that the structural perfection of the heterosystem improved. Such results may be explained only by the redistribution of point defects (decreasing of their concentrations) when taking into account that we could only observe the nucleation of MD (introduction of additional defects) and not modification of the substrate dislocation structure. The results obtained show that US treatment causes significant changes in dislocation surrounding by enhancing the gettering of point defects by dislocations.

Conclusions

Using the X-ray diffraction methods, which permit to investigate the changes in structural imperfection, it was shown that the subthreshold US treatment of the Ge-GaAs heterosystem leads to strain structural relaxation by the misfit dislocation nucleation, slight redistribution of strain deformation fields and improvement of structural perfection due to the point defect - dislocation interaction.

Reference

1. Yu.G.Sadof'ev, Pis'ma v Zhurn. Tekhn. Fiz. **19,** No.10 ,1993, P. 5.
2. T.V.Belousova, T.I.Kitayeva, Yu.G.Sadof'ev and A.B. Tolstoguzov, Poverkhnost No.6 1991, P.60.
3. V.P.Gromashevski, V.V.Dyakin, E.A.Sal'kov, S.V.Sklyarov and N.S.Halimova, Ukr.Fiz. Zhurn. **29,** 1984 , P.550
4. A.P.Zdebskii, V.L.Korchnaya, T.V.Torchinskaya and M.K.Sheinkman, Pis'ma v Zhurn. Tekhn. Fiz.**12,** 1986, P.76.
5. V.F.Britun, V.L.Korchnaya, G.N,Semenova, M.Ya.Skorohod, Yu.A.Thorik, L.S.Hazan and M.K.Sheinkman, Fiz.Tverdogo Tela **33,** 1991, P.2340.
6. Y.Kawamura and H.Okamoto, J.Appl.Phys. **50,** 1979, P. 4457.
7. B.E.Warren, X-ray Diffraction (Addison - Wiley, Reading), 1969, p.315.
8. I.V.Prokopenko, SPIE Proceedings "Time-Resolved Electron and X-Ray Diffraction".-San-Diego, USA, **2521,** 1995, P.320.

Zn^+/P^+ and Zn^+/As^+ co-implantation in InP single crystals

V.V. Ursaki, I.M. Tiginyanu, V.M. Ichizli, A.I. Terletsky, N.B. Pyshnaya, S.I. Radautsan
Institute of Applied Physics, 277028 Kishinev, Moldova

ABSTRACT

The activation efficiency of zinc impurity co-implanted with P^+ and As^+ ions in InP was studied by Hall-effect measurements. Both P^+ and As^+ co-implantations followed by post-implantation annealing at 400 to 600 °C in InP single crystals have been found to result in a decrease of impurity activation. At the same time an improvement of activation efficiency was observed at annealing temperatures $T_{ann} > 600$ °C.

1. INTRODUCTION

Dual implantation has been established to be a powerful tool for improving the activation efficiency of *p*-type dopants in GaAs /1, 2/. The co-implantation of As^+ host-component ions as well as of P^+ isoelectronic-impurity ones leads to nearly the same enhancement of the activation efficiency of implanted Zn-acceptor impurity /3, 4/. Such behaviour seems to be related with the formation of gallium vacancies in the As^+ and P^+ implanted layers, which facilitates the incorporation of impurity atoms into the cationic sublattice. As concerns InP, comparatively little attention was paid to the effect of co-implantations on the activation efficiency of p-type dopants. Moreover, the analysis of the data published on this matter /5, 6/ shows that they are rather contradictory. The goal of this work was to study the influence of P^+ and As^+ co-implantations upon activation efficiency of Zn^+ impurity in InP single crystals. For the purpose of comparison the effect of ordinary P^+ implantation on electrical properties of p-type InP has been studied as well.

2. EXPERIMENTAL

(100)-oriented liquid encapsulated Czochralski grown n-type crystals have been used for dual implantation. The carrier concentration and mobility in as-grown samples at 300 K were 2×10^{16} cm^{-3} and 3550 cm^2/(V×s) respectively. Dual implantation at doses 5×10^{13}, 5×10^{14} and 5×10^{15} cm^{-2} was performed at room temperature. The ion energies used (150, 75 and 163 keV for Zn^+, P^+ and As^+ correspondingly) provided an overlap of the implant-depth profiles. Ordinary P^+ implantation was performed at 80 keV energy in p-type wafers possessing different hole concentrations. The post-implantation annealing was carried out in a H_2 atmosphere for 15 min at different temperatures from 400 to 750 °C. A face-to-face proximity was used to avoid the surface decomposition. Free-carrier concentration and mobility were determined by Hall-effect measurements using Van der Pauw method. When measuring p-type layers on p-InP substrates the hole concentration was determined by C-V techniques.

3. RESULTS

First we shall consider the results on P^+ implantation in p-InP.

0-7803-3223-7/96/$5.00

Fig. 1 illustrates the effect of ordinary P^+ implantation and subsequent annealing on free-carrier concentration in InP:Zn single crystals. The drawing is divided into three parts. The middle part corresponds to semi-insulating material while the upper and lower parts represent *n*- and *p*-types regions respectively. One can see that P^+ implantation leads to *n*-type layer formation in InP:Zn. The higher the P^+ ion dose, the higher the electron concentration. On the other hand, the critical dose necessary to reach a conductive-type conversion depends upon the hole density in the as-grown samples, it being higher in heavily doped substrates. A rather weak dependence of electron concentration upon annealing temperature was observed in the range 400 °C < T_{ann} < 600 °C. At the same time further increase of T_{ann} leads to a fast diminution of free electron concentration and to a partial recovery of the hole conductivity.

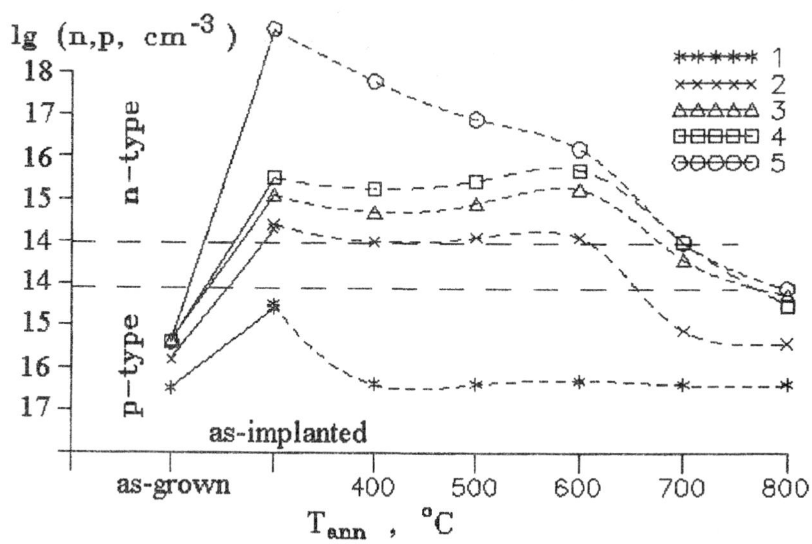

Fig. 1. Annealing curves for the InP:Zn crystals P^+-implanted at the doses: 1, 2 - $4 \cdot 10^{13}$ cm^{-2}; 3 - $7 \cdot 10^{13}$ cm^{-2}; 4 - $2 \cdot 10^{14}$ cm^{-2}; 5 - $7 \cdot 10^{12}$ cm^{-2}. Hole concentration in the substrates: 1 - $3 \cdot 10^{17}$ cm^{-3}; 2 - $6 \cdot 10^{16}$ cm^{-3}; 3, 4, 5 - $2 \cdot 10^{16}$ cm^{-3}.

The observed behaviour of conductivity may be explained by the predominant formation of donor like centers as a result of P^+ implantation. Activation of these centers turns out to take place in the interval of annealing temperatures from 400 to 600 °C. One can note, in this regard, that high electron conductivity was observed recently in low-temperature molecular-beam-epitaxy-grown InP, it being attributed to the autoionization of the P_{In} antisite defects /7/. Taking into account our results, the defects responsible for n-type conductivity may be supposed to anneal at temperatures T_{ann} > 600 °C.

Let us pass now to the discussion of results on dual implantation in n-InP single crystals. Table 1 summarizes the electrical parameters for Zn^+, Zn^+/P^+ and Zn^+/As^+ implanted layers. One can see that the implantation of P^+ or As^+ with subsequent annealing at 600 °C does not improve the activation efficiency. Moreover, the impurity activation is diminished and semi-insulating or n-type layers are formed. In case of high doses of implantation the behaviour involved was observed at T_{ann} > 600°C as well. Only moderate doses of P^+ and As^+

Table 1. **Parameters of InP implanted layers**

Annealing temperature, °C	Impl. dose, cm^{-3}	Implantation schedule	Sheet hole concentration, cm^{-2}	Mobility, $cm^2/(V \times s)$	Activation, %
		Zn^+	$1.7 \cdot 10^{13}$	51	34
	$5 \cdot 10^{13}$	Zn^+/P^+	*i*-type		
		Zn^+/As^+	*i*-type		
		Zn^+	$1.4 \cdot 10^{14}$	44	28
600 °C	$5 \cdot 10^{14}$	Zn^+/P^+	*i*-type		
		Zn^+/As^+	*i*-type		
		Zn^+	$2.1 \cdot 10^{14}$	45	4.2
	$5 \cdot 10^{15}$	Zn^+/P^+	*n*-type		
		Zn^+/As^+	*n*-type		
		Zn^+	$2.3 \cdot 10^{13}$	72	46
	$5 \cdot 10^{13}$	Zn^+/P^+	$1.9 \cdot 10^{13}$	66	38
		Zn^+/As^+	$2.0 \cdot 10^{13}$	91	40
		Zn^+	$2.0 \cdot 10^{14}$	51	40
700 °C	$5 \cdot 10^{14}$	Zn^+/P^+	$1.5 \cdot 10^{14}$	68	30
		Zn^+/As^+	$1.7 \cdot 10^{14}$	60	34
		Zn^+	$2.9 \cdot 10^{14}$	48	5.8
	$5 \cdot 10^{15}$	Zn^+/P^+	*n*-type		
		Zn^+/As^+	*n*-type		
		Zn^+	$2.3 \cdot 10^{13}$	71	46
	$5 \cdot 10^{13}$	Zn^+/P^+	$2.7 \cdot 10^{13}$	61	54
		Zn^+/As^+	$3.1 \cdot 10^{13}$	84	62
		Zn^+	$2.0 \cdot 10^{14}$	55	40
750 °C	$5 \cdot 10^{14}$	Zn^+/P^+	$2.4 \cdot 10^{14}$	46	48
		Zn^+/As^+	$2.5 \cdot 10^{14}$	81	50
		Zn^+	$2.5 \cdot 10^{14}$	50	5
	$5 \cdot 10^{15}$	Zn^+/P^+	*i*-type		
		Zn^+/As^+	*i*-type		

co-implantation and annealing temperatures higher than 600 °C lead to an increase of zinc activation efficiency. This behavior is consistent with the results obtained in the case of ordinary P^+ implantation and can be explained as follows. Unlike GaAs, there are two competitive mechanisms of defect formation in P^+ or As^+ coimplanted InP crystals. One of them is the formation of donor-like centers as in the case of ordinary P^+ implantation. The other is the formation of indium vacancies in the crystal lattice which are easily filled in by impurity atoms during the post-implantation annealing, like gallium vacancies in GaAs crystals. In the range of annealing temperatures up to 600 °C the formation and activation of donor-type defects is predominant giving rise to the n-type conductivity observed. At $T_{ann} >$ 600 °C the second mechanism prevails which is responsible for the improvement of the activation efficiency.

Although the free carrier concentrations in P^+ or As^+ coimplanted layers were nearly the same, the hole mobility was found to be higher in the case of As^+ coimplantation.

4. CONCLUSIONS

The activation efficiency of Zn impurity coimplanted in n-type InP with P^+ or As^+ ions was established to decrease in the interval of annealing temperatures from 400 °C to 600 °C. At the same time an improvement of activation efficiency was observed after dual implantation and annealing at $T_{ann} > 600$ °C. The values of the hole mobility were higher in the case of Zn^+/As^+ coimplantation.

REFERENCES

/1/ A.N.M. Masum Choudhury et al., "Improved Activation of Mg^+ and As^+ Dual Implants in GaAs by Capless Rapid Thermal Annealing", Appl. Phys. Lett., vol.50, p.488, 1987.

/2/ G. Landgren et al., "Extremely High Hole Concentration in GaAs by Dual Mg/As Implants", J. Appl. Phys., vol.63, p.2783, 1988.

/3/ S.I. Radautsan et al. "Zn^+/As^+ and Zn^+/Ar^+ Co-implantations in GaAs Single Crystals", CAS'93 Proceedings, p.465.

/4/ V.V. Ursaki et al., "Raman Scattering Study of Zn^+/P^+ Co-implanted GaAs Single Crystals", CAS'95 Proceedings, p.99.

/5/ Kon-Wei-Wang, "Formation of a Shallow p^+-layer on InP Using a P/Be Co-implantation", Appl. Phys. Lett., vol.51, p.2127, 1987.

/6/ Mulpuri V. Rao and Ravi K. Nadella, "Be^+/P^+, Be^+/Ar^+ and Be^+/N^+ Co-implantation into InP:Fe", J. Appl. Phys., vol.67, p.1761, 1990.

/7/ W.M. Chen et al., "Origin of *n*-type Conductivity in Low-temperature Grown InP", J. Appl. Phys. vol.76, p.600, 1994.

RBS ANALYSIS OF Au/Cr CONTACT ON GaAs

Dan Pantelica, George Drafta, Cristina Borcan
Institute of Physics and Nuclear Engineering (IFIN),
P.O.Box MG-6, Magurele-Bucharest

Dorina Diaconescu, Manuela Buda, Rodica Ghita
Institute of Physics and Technology of Materials (IFTM),
P.O.Box MG-7, Magurele-Bucharest

Abstract

This paper investigates the mass transport in Au/Cr contact deposited on GaAs. Such contacts are commonly used as ohmic contacts on p type GaAs. For characterisation we used RBS technique, replacing the classical α particles with ^{7}Li ions. This allows a great improvement in depth and mass resolution.

1. INTRODUCTION

The ohmic contacts in semiconductor devices must be homogenous and uniform over the entire device area. Our interest concerns ohmic contacts on p-type GaAs for laser diode applications.

Au based contacts show a poor thermal stability because during annealing the volatile group V-components (As) outdiffuse easily and Au penetrates into the bulk/1,2/. Rapid thermal annealing offers an effective alloying process without a strong interdiffusion. For GaAs contacts an annealing temperature in the range 400-520°C /3/ for 20 sec is optimal to ensure the components interdifussion at the metal-semiconductor boundary. Good ohmic contacts, with low specific resistance could be obtained.

For laser diodes is essential to define a thin lateral injection region on the surface of the GaAs cap layer. This stripe region is realised in various ways, the most common one being the deposition of an oxide on the GaAs surface. We used in this paper for electrical isolation an electrochemically grown anodic GaAs oxide .

The purpose of this article is to establish the formation mechanism of Au/Cr contact on GaAs and the behaviour of the anodic oxide/Cr/Au interface. Cr film is used in order to provide a good adherence on the isolation layers (anodic oxide in our case). Au layers are mainly used for the formation of the ohmic contact on GaAs, but only Au does not provide a good and reproducible ohmic contact.

Usually, before oxide deposition, a ridge etch is performed for lateral guiding of the laser emission. This etch is more effective when it is done very close to the active region. Eventual interdiffusions as a result of the heat treatment for ohmic contacts could act as nonradiative recombination centers and could reduce the laser efficiency, if the active region is reached.

Rutherford backscattering spectroscopy (RBS) is a valuable tool for analysis of thin film compositions. We used RBS technique to investigate the mass transport in Au/Cr/GaAs samples.

0-7803-3223-7/96/$5,00

In some previous RBS investigations devoted to the interdiffusion characterization of GaAs rectifying contacts containing W, Au or Pt, α particles of about 2 MeV were considered as the most convenient projectile. However, the analysis is hampered by the lack of mass resolution between Ga and As. There is an increasing trend to use ions heavier than ^{4}He. In the last two years we used a ^{7}Li beam at 4.5 MeV for RBS measurements and a noticeable improvement in depth and mass resolution was obtained.

2. EXPERIMENTAL PROCEDURE

Four different samples of (100) GaAs with Cr/Au contacts were used for determinations. The wafers were very well cleaned in triclorethylene and acetone and etched in HCl before processing.

The surface of two samples were electrochemical anodised in a tartaric acid:ethylen glycol/2:1 solution. The oxide was grown up to a voltage of 150 V at a constant current intensity of 1.5 mA/cm^2.

On all samples were deposited Cr and Au films by thermal evaporation. The Cr thickness was monitorised during the evaporation to be in order 400-600 A and the Au films were calculated to have 2000-2200 A.

Two of the samples -one with anodic oxide and one without- were annealed in vacuum at 400°C according to the temperature profile given in fig. 5.

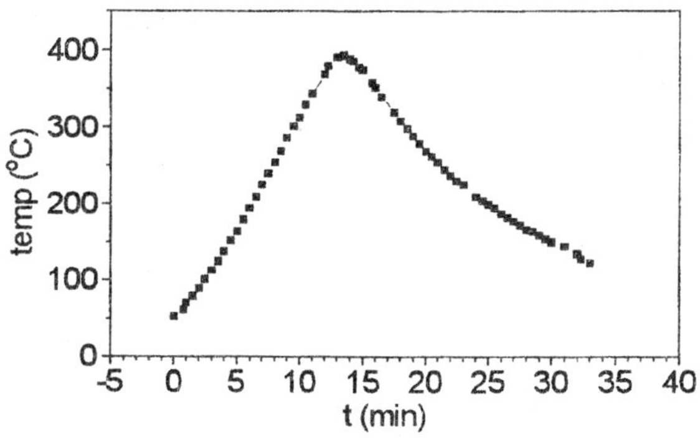

Fig.5

For the present measurement the $^{7}Li^{++}$ beam was provided by the Van der Graaff tandem accelerator of I.P.N.E. The backscattered particles were detected with a silicon barrier detector placed at 150° with respect to the beam. The energy resolution for ^{7}Li at 4 MeV was about 35 keV. This gives for Au a depth resolution of about 154A. The energy difference between Ga and As signals is 82 keV, so we are able to discriminate between them.

3. RESULTS

Four samples were measured. Experimental results of our study are summarised in figs. 1-4.

Fig. 1 shows the spectrum of an as deposited Au/Cr/GaAs sample. We determined the thickness of the film using RUMP simulation code /4,5/. The Au and Cr layer thickness were 1800A and 400A, respectively.

Fig. 2 illustrates the observed mass transport in measured RBS spectrum of an annealed Au/Cr/GaAs sample. The Au layer remains intact during the annealing process. No Au transport in the annealed sample was observed. The broad peak near channel 595 corresponds to the migration of Ga and As in the vicinity of the top surface Au layer, in a layer of 823 A thickness, with the maximum concentration at the depth of 137A. The small peak near channel 525 corresponds to a migration of Cr on the top surface of the Au layer. The diffused Cr signal is still visible as a peak on the top of the Ga and As signals. The Ga and As depletion in a layer of 1130A at Cr /GaAs is visible, too.

Fig. 3 shows the RBS spectrum of an unannealed sample prepared by anodically growth of an oxide film on GaAs prior to Cr and Au deposition. From the analysis of the spectrum using RUMP simulation program, we extracted information about Cr and Au layer thicknesses and about composition and thicknesses of the oxide film. The composition of the oxide film was found to be $GaAsO_3$ and the thickness is approximately 3500A. The Au and Cr layers were 2200A and 400A in thickness, respectively.

Fig. 4 shows the spectrum of annealed sample. No Au or Ga and As mass transport were observed. The small peak near channel 525 corresponds to Cr on the surface of Au layer. There is no noticeable change in the thickness of the oxide film.

4. CONCLUSIONS

The investigation of Au/Cr films deposited on GaAs surface and on GaAs covered with anodic oxide was realised by RBS technique. Using ^{7}Li ions as projectile in RBS scattering, we could separate the contribution of Ga and As in the RBS spectra. We could identify, in the samples with thermal treatment and without oxide deposition, the diffusion of both species Ga and As in the gold layer . Also the diffusion of Cr was observed in these samples through the Au layer. The oxide realized on the GaAs surface behaves like a barrier for Ga and As out-diffusion.

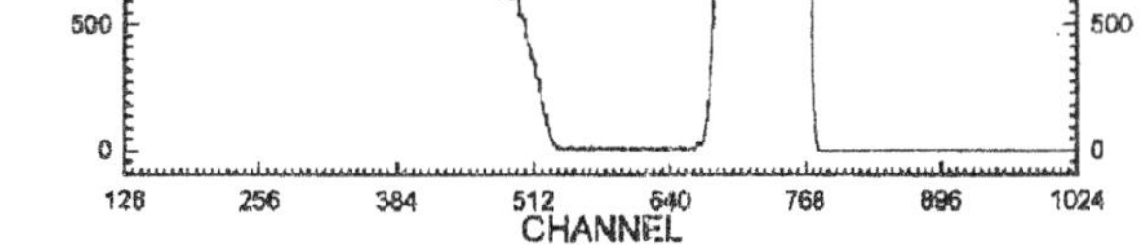

Fig.1

Spectra of ^{7}Li ions backscattered from a GaAs sample covered with 400A Cr and 1800A Au layers.

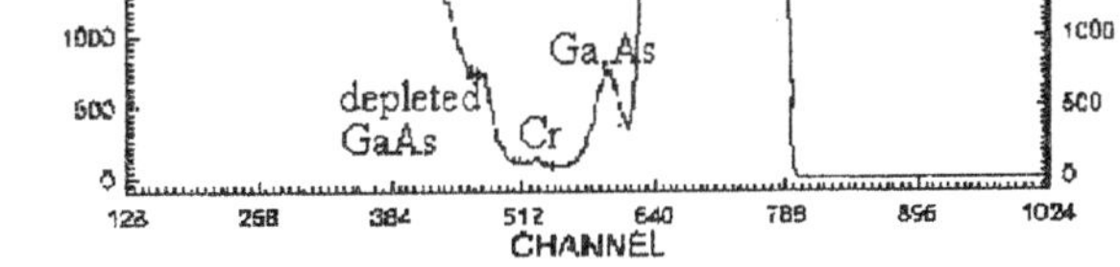

Fig.2

Spectra of ^{7}Li ions backscattered from a GaAs sample covered with 400A Cr and 1800A Au layers and annealed at 400°C for 1min.

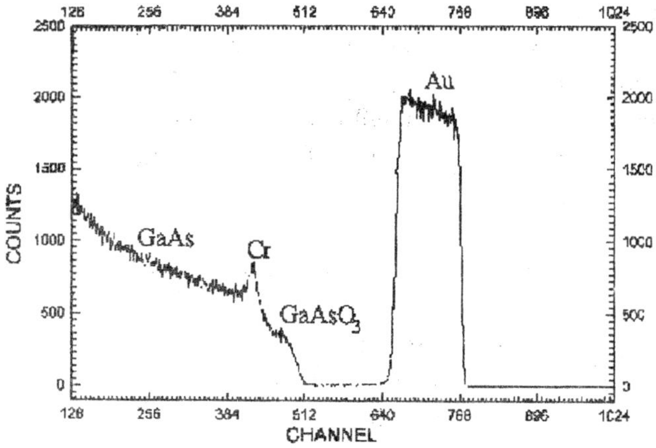

Fig.3

Spectra of ^{7}Li ions backscattered from an anodic oxidated GaAs sample covered with 400A Cr and 2200A Au layers.

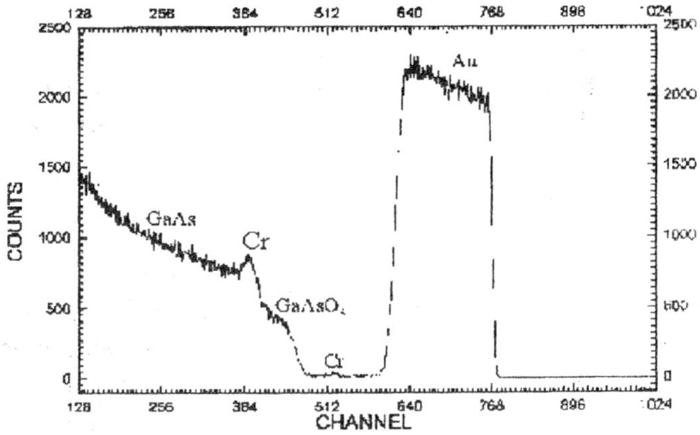

Fig.4

RBS spectra of an anodic oxidated GaAs sample covere with 400A Cr and 2200A Au layers and subsequentl annealed .

REFERENCES

/1/J. Gyulai, J. W. Mayer, J. Appl. Phys., Vol.42, no. 9, 3578(1971);
/2/C. S. Patuwathavithane et al., Nucl. Instr. Methods B56/57, 754(1991)
/3/G. Stareev et al., J. Appl. Phys., vol 74, no.12, 7344(1993);
/4/ L. R. Doolitle, Nucl. Instr. Methods B9, 344(1985);
/5/ L. R. Doolitle, Nucl. Instr. Methods B15, 227(1986);

MEASUREMENT OF CVD THIN FILMS THICKNESS BY SAMPLE WEIGHING METHOD

M. MODREANU, P.COSMIN, S.COSMIN, C.COBIANU, C.DUNARE
Institute of Microtechnology, Erou Iancu Nicolae 34 B, R-72996 Bucharest, Romania

Summary: **Several type of films obtained by Chemical Vapour Deposition technique-like silicon dioxide, phosphosilicate glass, silicon oxynitride, silicon nitride and polysilicon were investigated. A calibration technique for sample weighing method consisting in film density evaluation was developed. Sample weighing method was found to be a fast, precise and non-destructible method for the thin films thickness evaluation.**

1.Introduction

Chemical vapour deposition(CVD) is one of the most important methods of film formation. The importance of the CVD technique can be explained by its ability to deposit films of a wide range of elements and compounds at relatively low temperatures. Furthermore, the films are of good integrity and purity and their stoichiometry can very accurately be controlled.

In this paper we present a method for measurements of CVD thin film's thickness using a sample weighing method. The thin films were obtained in Atmospheric Pressure Chemical Vapour Deposition (APCVD) and Low Pressure Chemical Vapour Deposition (LPCVD) techniques.

In this method, the support is weighed before and after deposition and the film thickness can be determined if density is know.

In the first step knowing the geometrical characteristics of each layer(support area and film thickness) we determine the film density.

In the second step knowing the density and mass(by direct weighing) of the film we can determine the thickness of the as-deposited layer.

If A, ρ, h, Δm, are respectively the area, density, thickness and mass of the deposition layer, respectively, then $h*A*\rho=\Delta m$, hence $h=\Delta m/ A*\rho$.

The support area in the case of silicon wafers is calculated by the formula:

$$A = \pi R^2-A_{bf}-A_{sf} = R2\{\pi-\arcsin(l_{bf}/2R) - \arcsin(l_{sf}/2R) + \sin[2 \arcsin(l_{bf}/2R)] + \sin[2 \arcsin(l_{bf}/2R)]\}$$

where: A_{bf}=big flat area, A_{sf}=small flat area, l_{bf}=big flat length, l_{sf}=small flat length, R=radius of wafer.

The studied CVD thin layers were: silicon dioxide(SiO_2) and phosphosilicate glass(PSG)-in range of 0-9 wt.% P-obtained by APCVD technique and polysilicon(amorphous and polycrystalline), silicon nitride(Si_3N_4) and silicon oxynitride(SixOyNz-at different refractive index) obtained by LPCVD technique.

Silicon dioxide and PSG APCVD films can be grown1[1,2] by the simultaneous pyrolysis of silane, phosphine or both, in oxygen around 400°C.

Polycrystalline and amorphous(a-Si) silicon LPCVD films were grown[3,4] by pyrolysis of silane at different temperature(615°C in first case, 570°C and 590°C in the second).

Silicon nitride LPCVD films were growth[5] at 750°C from dichlorosilane(DCS) and ammonia(NH_3).

0-7803-3223-7/96/$5.00

Silicon oxynitride LPCVD films were growth[6] at 860°C from dichlorosilane,nitruos oxide(N_2O) and ammonia(NH_3).

2.Experimental details and results

CVD thin films presented above (0-1μm in thickness) were deposited on silicon substrate (p-type <100>, 16-24 Ω*cm sheet resistance) and on silicon dioxide(2500Å in thickness) thermally grown on the same substrate.

APCVD thin films were deposited in a cold wall rotary reactor at atmospheric pressure in a SiH_4-PH_3-O_2-N_2 system, at the temperature of 400°C. Silane, phosphine and diborane were diluted in nitrogen (5% SiH_4 in nitrogen, 2% PH_3 and B_2H_6 in nitrogen). The relative flow rate for SiH_4 to O_2 was 7,5, the flow rate of PH_3 was varied among 0-0,16 l/min.

Polysilicon films were chemically formed by pyrolysis of silane at three different temperatures 550°C, 570°C and 615°C. Flow rate of silane was 40 sccm and the total pressure during deposition was 400mTorr.

Silicon nitride films were obtained in an SiH_2Cl_2-NH_3 system at a temperature around 790°C. The relative gas flow rate ratio SiH_2Cl_2 to NH_3 was 1,35.The total pressure during the deposition was 430 mTorr.

SiO_xN_y samples studied in this work have different nominal compositions and refractive index.SiO_xN_y films were chemically formed by reacting dichlorosilane($SiCl_2H_2$) with nitrous oxide(N_2O) and ammonia(NH_3) at temperature around 860 °C. The total pressure during the deposition was 400 mTorr. The relative gas flow rate ratio N_2O to NH_3 in the deposition system, at a particular pressure and deposition temperature, determined the film composition. The relative gas flow rate ratios N_2O to NH_3 was varied among 2-7.

The uniformity of the film thickness was better than ±8% for APCVD layers and better 3% for LPCVD layers.

The phosphorus concentration in PSG films was determinate by using a standard plot of etch rate[7] in P-etch (A/s) vs. phosphorus concentration (weight per cent phosphorus). The error in the determination of the phosphorus concentration in the PSG films was less than ±5%.The weightings for as-deposited film were performed on a Sartorius assay balance-10^{-5} g precision.

The film thickness and refractive index were measured with an Elipsometer II (by using spectroelipsometry and reflectometry technique) for APCVD films,SixOyNz and Si_3N_4 films and with an α-Step profilometer for LPCVD films in 10 points on each wafer.

The LPCVD films were deposited on a 2500Å thermal SiO_2 in order to cover the holes of the back side of the wafer.

The used wafers have an area A=45,280± 0,01 cm^2.The error in the determination of the films density was less than ±4% for APCVD films and less than ±3% for CVD films. The error in the determination of the film thickness was found to be less ±7% for APCVD films and less ±5% for LPCVD films.

The experimental results are presented in Table 1.

2.1 APCVD thin films

As is presented in Table1 the silicon dioxide thin films has a density ρ=2,205 g/cm^3.

The PSG thin films have a bigger density than silicon dioxide, as is presented in Fig.1, in all range of phosphorus content.

The dependence of film density versus phosphorus content presents a minimum at 6 wt. % P. This minimum is also presented in literature[8].

Deposited film	Density(g/cm^3)
SiO_2	2,205
PSG 3 wt. % P	2,250
PSG 6 wt. % P	2,210
PSG 7wt. % P	2,220
PSG 9 wt. % P	2,230
SiO_2(LPCVD, T=860°C) n=1,45	2,420
SixOyNz n=1,60	2,760
SixOyNz n=1,72	2,810
SixOyNz n=1,80	2,880
Si_3N_4 n=2,00	2,940
a-Si (Tdep=550°C)	2, 310
a-Si (Tdep=570°C)	2,270
a-Si (Tdep=590°C)	2,265
Polycrystalline silicon (Tdep=615°C)	2,245

Table 1. Density of SiO_2, PSG, SixOyNz, Si_3N_4, a-Si and polycrystalline silicon films obtained in conditions mentioned above.

Fig.1 Density of the PSG thin layers vs. phosphorus content

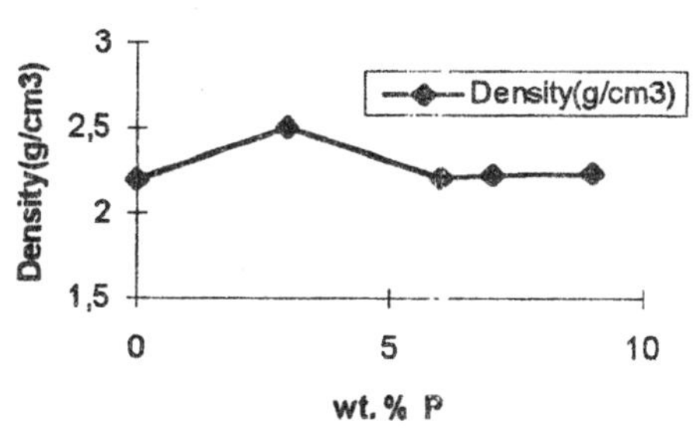

Fig.2 Density of the SixOyNz films vs. refractive index

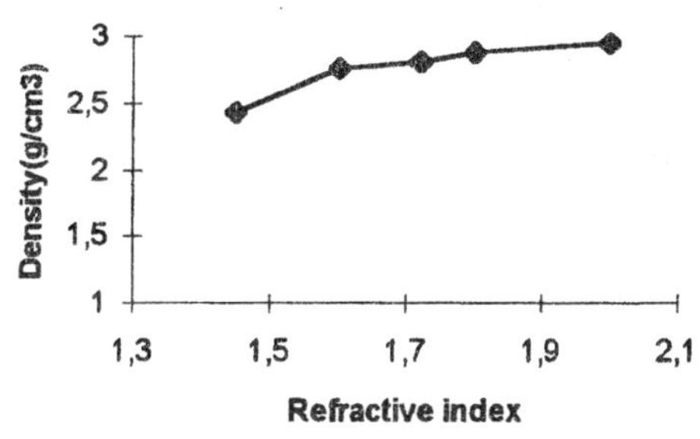

2.2 LPCVD thin films

2.2.1 Silicon oxynitride

In Fig.2 are presented the density of the SixOyNz films versus the refractive index, from pure silicon dioxide, n_{SiO_2}=1,45(high temperature LPCVD oxide) to pure Si_3N_4, n_{Si3N4}=2. The density of silicon oxynitride films increase with the refractive index from the pure silicon dioxide (ρ=2,420 g/cm^3) to silicon nitride (ρ=2,940 g/cm^3).

Usually, SixOyNz films are accepted to be a physical mixture consisting of two distinct phases, silicon dioxide and silicon nitride[9]. The increase of the refractive index of the silicon oxynitride is determined by the increase of the silicon nitride percentage in the film. As Si_3N_4 content increase, the silicon oxynitride film densities increase too.

2.2.2 Polysilicon

The refractive indexes for polysilicon films versus deposition temperature are presented in Fig. 3. The density of polycrystalline film was found to lower then of amorphous silicon. These are in accord with the fact that the amorphous states are more dense that polycrystalline state. The dependence the refractive indexes of polysilicon films versus deposition temperature are presented in Fig.4.

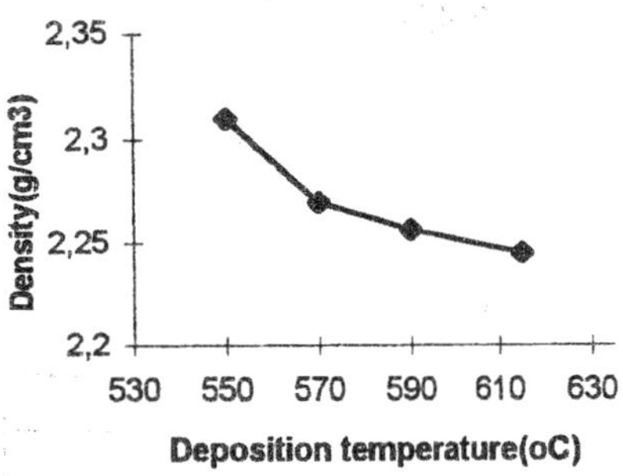

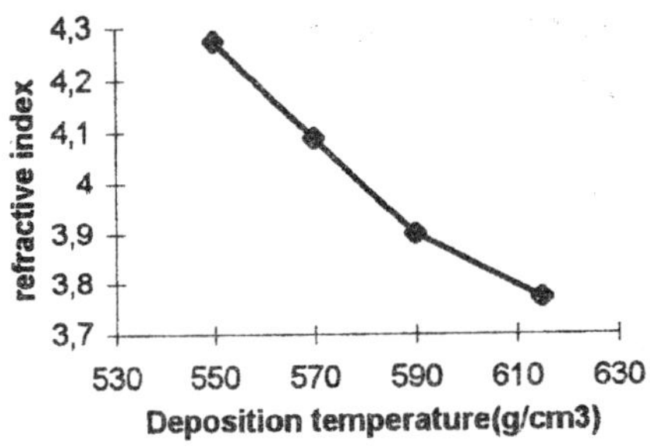

Fig.3 The density of polysilicon films vs. deposition temperature

Fig.4 The refractive index of the polysilicon films vs.deposition temperature

3. Conclusions

A calibration technique for sample weighing method consisting in film density evaluation was developed. This is a fast, precisely and non-destructible method for the thin film's thickness evaluation. The best results for LPCVD films were obtained, when the error in the evaluation of film thickness was less than $\pm5\%$. Also, this technique resolve successfully the problem of the amorphous silicon thin films thickness, who usually necessity a destructible method(alpha-step profilometer) and problem of thin films thickness greater then 0,5μm(specially for polysilicon, silicon nitride and silicon oxynitride).

REFERENCES:

/1./ W.Kern, *RCA Rev* 37 (1976) 55
/2./ W.Kern, *RCA Rev* 37 (1976) 78
/3./ R.Bisaro ,J. Magarino and N.Proust, *J.Appl.Phys.* 59, 1967(1986)
/4./R.S.Rosler, *Solid State Techn.*, 20 (4),63 (1977)
/5./K.F.Roenigk and K.F.Jensen, *J.E.S.*, 134 (7) ,(1987) 1777
/6./A.E.T Kniper et al., *J.Vac.Sci.Techn.*, 124, 599 (1977)
/7./ *Pyrox Instructional Manual*, 1980, pp.3-27
/8./ C.Pavelescu, C.Cobianu, *Thin Solid Films*, 196 (1991) 351
/9./ M.Modreanu, P.Cosmin, *CAS 95 Proceedings* (313)

Doping Properties of Pr_2O_3 Associate InGaAs Liquid Phase Epitaxial Growths

C.C. Liu, L.B. Chang and Y.C. Cheng
Department of Electrical Engineering, Chung-Cheng Institute of Technology,
Tashi, Taoyuan, Taiwan 33509, R.O.C

Abstract InGaAs epilayers were grown on the semi-insulating(SI) InP substrates by liquid phase epitaxy(LPE) with a rare-earth(RE) compound Pr_2O_3 doped into growth melt during each epitaxial process. Most grown layers yield mirror-like surfaces and good crystal quality. Without the prebaking process, the corresponding Hall measurements indicate that n-type background concentration of those InGaAs grown layers will decrease from a value of 1.6×10^{16} to 2.0×10^{15} cm^{-3}. Their correspondent 77 K mobility also significantly increases from a value of 15321 to 32171 cm^2/V-s. The photoluminescence(PL) spectra of Pr_2O_3 doped InGaAs epilayers display strong intensity ratios for band peak to the impurity peak, which also demonstrates the grown layers exhibit a pure crystal property.

1. Introduction

In recent years, doping III-V compound semiconductors with RE elements has received increasing interest and been grown by different epitaxial techniques, e.g., Molecular-Beam Epitaxy(MBE)[1], Metal-Organic Chemical Vapor Deposition(MOCVD)[2,3] and Liquid Phase Epitaxy (LPE)[4-9]. Because that RE elements have partially filled inner 4f shells, which are screened by their fully filled $5s^2$ and $5p^6$ shells; therefore, electrons in these shells are relatively insensitive to the host environment. Thus the characteristic inter-shell-transitions of these electrons may produce a sharp and stable line luminescence that is less sensitive to temperature variation and offers a suitable candidate for a light source [1-9]. Another reason for this interest is owing to RE elements' high reactivity with such residual impurities as sulfur, oxygen..., etc.. This property illustrates the possibility of utilizing RE elements in a compound semiconductors[4,9] purification process. Several previous investigations indicated that RE elements can reduce the n-type concentration of accidental impurities and increase the electron mobility[4,8]. The other investigations reported that RE elements will increase the p-type carrier concentration and function as a dopant itself[6,9]. In some other works[1,7], the sign of conduction type can be inverted by increasing the dopant amount of RE elements in the growth melt. Although many efforts have been devoted, the actual nature of RE in the III-V compound semiconductors still remains controversial.

In the meantime, the $In_{0.53}Ga_{0.47}As$/InP hetero-epilayer grown on the top of an InP substrate is an essential structure for optoelectronic devices and high speed transistors. However only one RE-doped InGaAs related work reported by Novak et. al.[4,6]. Therefore in this work we attempts to present a comprehensive and detailed account of the doping properties of InGaAs and InP LPE layers with Pr_2O_3 dopant during the growth process. To determine the relationships between those epilayers and the doping amount, we performed a number of measurements, such as X-ray diffraction measurements, secondary ion mass measurements(SIMS), and Hall measurements. Details are provided in the following sections.

2. Experimental

The InGaAs epilayers were grown on (100) oriented Fe doped semi-insulating InP substrates in a conventional horizontal sliding LPE system. Fe-doped InP substrate, high-purity polycrystalline solvent and Indium melt were loaded into the graphite boat after the standard cleaning processes. The furnace was heated to a temperature 20 ℃ higher than the correspondent saturate-temperature for 1 hour to allow the

0-7803-3223-7/96/$5,00

melts reach saturation. After that, the temperature was lowered with a cooling rate of 0.3 ℃/min.. Next, the melt was brought to contact the substrate for a period of 5 min. at 633.5 ℃.

After epitaxial growth the lattice-matching condition, between the grown InGaAs or InP layers and the substrate InP, was measured by double crystal X-ray diffraction along their (400) symmetrical plane. The thickness and surface morphology were measured according to the scanning electronic microscopic(SEM) inspection. Their depth profiles were examined by secondary ion mass spectroscope(SIMS). Carrier concentrations and mobility were characterized by Van der Pauw technique. The magnetic field used in this study was 5 kG while the current applied to the specimens was 1 mA. Photoluminescence measurements were taken at 20 K. The excitation source is the chopped 488 nm line of an Ar-laser focused to a spot of diameter about 3 mm, and the excitation intensity of laser is 50 mW.

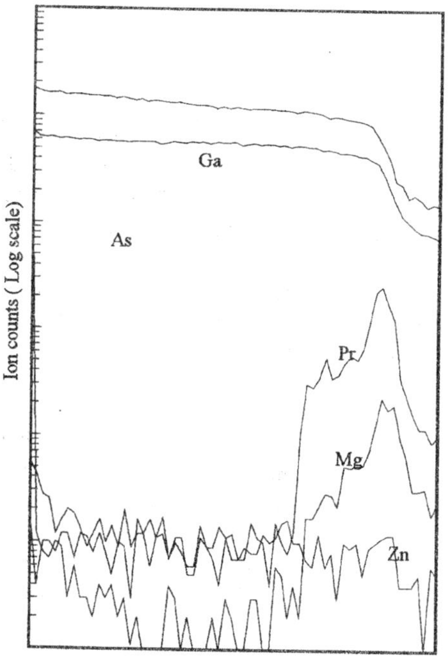

Figure 1. The SIMS profile of doped InGaAs epilayer.

3. Results and Discussion

3.1 Surface Morphology and Crystal Quality

Careful examinations were made to the surface morphology and the cross-sectional view between the InGaAs grown layer and the InP substrate, with Pr_2O_3 charge amounts W_{Pr2O3} of 0.3 and 2.245 (with milligram of Pr_2O_3mg per one gram of Indium; mg- Pr_2O_3 /1g In). The grown surface is general mirror-like and the interface is quite clear and free from inclusions. However, an irregular surface of the InGaAs-grown layer that had high W_{Pr2O3} value(2.245 mg- Pr_2O_3/1g-In) is observed. This irregular surface is different from the droplet surface reported previously[5], and is owing to the arsenic evaporation enhancement. The average growth rate per minute was about 1.12 um that seem not varied with doping of Pr_2O_3 . In Fig. 1, X-ray results show that the lattice constants of InGaAs-grown layers remained almost unchanged while W_{Pr2O3} ranging from 0 to 0.035. When the Pr_2O_3 charge amount increased to 0.9 (mg- Pr_2O_3/1g In), the grown surfaces became irregular and the lattice constant slightly increased.

3.2 SIMS analysis

Fig. 2 illustrates the SIMS depth profiles of these Pr_2O_3-doped InGaAs epitaxial layers. As revealed in this figure, under the smooth grown surface, the concentration of Pr atoms maintains a certain low level over most of the epitaxy region but piles up at the grown interface. Three results can be found from this measurement. First, whenever the Pr concentration increased, the corresponding Mg concentration also obviously increased. This is because that commercial rare-earth materials normally possess many impurities and Mg is frequently the dominator. Second, both Pr and Mg display a pile-up at the interface. Such a pile-up may be owing to the fact that defects easily occur around the interface area and make Pr atoms accumulate around the interface. The third, such a low Mg concentration observed by SIMS measurement would not turn the conductivity of InGaAs p-type and that will be discussed later.

3.3 Hall measurement

Hall measurement was performed to determine the relationships between the electrical properties of InGaAs and their corresponding Pr_2O_3 charge amount. Fig. 3 indicates that increasing the charged amount of Pr_2O_3 in InGaAs-growth melts would result in a significant enhancement of the electron mobility; from 15321 to 32171 $cm^2/V\text{-}s$; and a gradually decrease of their concentration; from 1.6×10^{16} to 2.0×10^{15} cm^{-3}; with respect to W_{Pr2O3} values varying from 0 to 0.9.

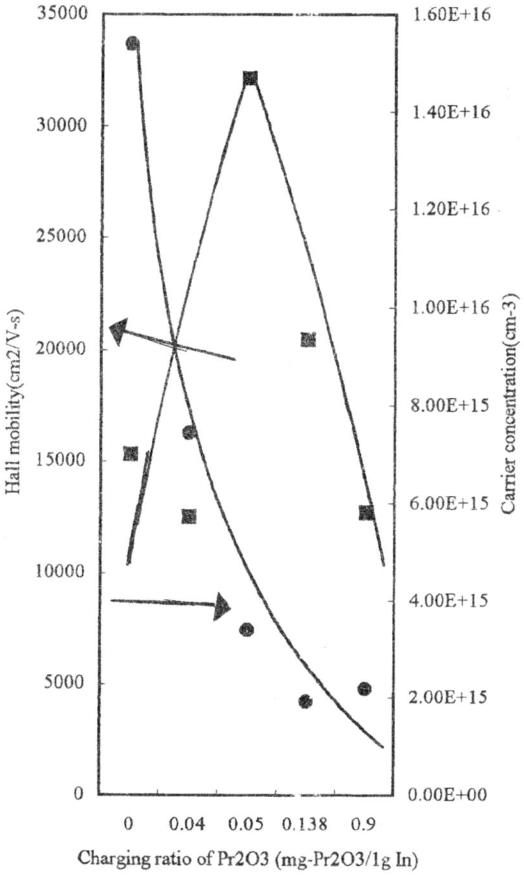

Figure2. The Hall results of doped InGaAs epilayers.

In the beginning of this doping process, some background impurities are gettered and the mobility is enhanced. When a specified doping amount is reached (with a W_{Pr2O3} value of 0.05), the highest mobility value of 32172 cm^2/V-s is observed. While W_{Pr2O3} further increased, the electron concentration of the epilayer is still decreasing but their correspondent mobility begins to decrease. This is because that heavily Pr_2O_3 doping produces defects, e.g., unwittingly doped Mg atoms and some other deep levels; which compensate the tendency of mobility increasing. Thus, properly doping Pr_2O_3 into the growth melt during high quality InGaAs LPE growth is very essential.

Notably, in all cases these grown layers exhibit n-type conductivity. Although, Novak[4] has reported for one p-type InGaAs-grown layer, this p-type conduction was not found in our experiment. From our results, we conclude that due to the different segregation coefficient, Pr_2O_3 doped InGaAs, InGaP and InP grown layers are all n-type which are grown from In-rich melts. While AlGaAs and GaAs are p-type which are grown from Ga-rich solutions.

3.4 Photoluminescence measurement

Photoluminescence was adopted to measure the optical properties of those Pr_2O_3-doped InGaAs grown layers. Fig.4 & Table 1 shows the correspondent 20 K PL spectra. These spectra basically consist of two peaks: B1 and B2. The major peak B1 is due to the band edge emission, and the side peak B2 is due to an impurity related donor-acceptor(DA) transitions or electron-acceptor(eA) recombination. The full width of half maximum(FWHM) value of the PL spectrum of Pr_2O_3 -doped grown layer has a good value around 7 meV and is much narrower than that of undoped counterparts (as illustrated in Table 1). And reveals that the grown layers doped with Pr_2O_3 have better optical quality than the undoped one. Besides, the intensity ratio of major peak over side peaks, B2/B1, can also be treated as an index of the grown layer's optical quality. The B2/B1 values of Pr_2O_3-doped grown layers are all smaller than that undoped one. Thus, PL results support the fact that doping with Pr_2O_3 has the capability of reducing the background impurities and, thereby, enhancing their optical properties.

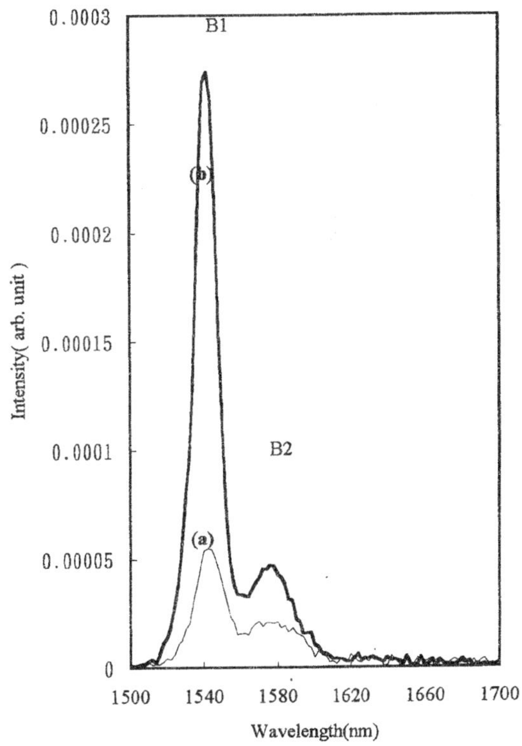

Figure3 The PL spectra of Pr_2O_3 doped InGaAs epilayers.

4. Conclusion

The Pr_2O_3 doped InGaAs layers were grown on semi-insulating InP substrates by LPE method. By doping these grown melts with Pr_2O_3 properly, the morphology of those grown layers possessed a mirror-like surface and flat interface. Hall and photoluminescence measurements revealed that Pr_2O_3 doped InGaAs grown layers may both reduce the background impurities(as low as 2.0×10^{15} cm^{-3}), without the conventional prebacking process, and enhance their carrier mobility(as high as 32171 $cm^2/V\text{-}s$). In all circumstances, the grown layers exhibit only n-type conductivity, which is thought to be the major gettering effect of Pr element accompanied by a slight Mg ion compensation effect than to be introduced a new p-type Pr dopant itself.

Acknowledgments

The authors would like to express their thankfulness to the National Science Council of the Republic of China for their financial support. (under Contract No. NSC-84-2215-E014-003).

Reference

[1]. R. S. Smith, H. Ennen, P. Wennekers and M. Maier, Appl. Phys. Letters 50 (1987) 49.

[2]. K. Uwai, H. Nakagome and K. Takahei, Appl. Phys. Letters, 50 (1987) 977.

[3] . K. Uwai, H. Nakagome and K. Takahei, Appl. Phys. Letters, 51 (1987) 1010.

[4] J. Novak, S. Hasenorl and M. Kulittayora, J. Crystal Growth, 110 (1991) 862.

[5] N. Nakagome, K. Takahei and Y. Homma, J . Crystal Growth, 85 (1987) 345.

[6] J. Novak, M. Kuliffayova, M. Morvic and P. Kordos, J. Crystal Growth, 96 (1989) 645.

[7] J. Raczynska, K. Frone, J. M. Langer, A. Le manska and A. Stapor, Appl. Pyys. Letters, 53 (1988) 761.

[8] M. Z. Lai and L. B. Chang, J. Appl. Phys., 72(4) (1992) 1312.

[9] G. C. Jiang, L. B. Chang, G. H. Chang and L. S. Lu, J. Crystal Growth, 152 (1995) 127.

Table 1. Summary of the PL properties of Pr_2O_3 doped InGaAs epilayers

No.	The doping amount (Pr_2O_3/1g In)	PL spectrum		Carrier concentration	The ratio of B1/B2
		Position	FWHM		
41	0mg	1.544um	11meV	$1.5*10^{16}$	3.1
44	0.04mg	1.543um	8meV	$7.4*10^{15}$	5.9
45	0.05mg	1.541um	7meV	$3.4*10^{15}$	7.3

DISTRIBUTION IMPROVEMENT FOR MAIN IMPURITY, OXYGEN AND CARBON CONCENTRATION IN CZOCHRALSKI SILICON SINGLE CRYSTALS

V. Vanca, *Z.Schlett, M.Dinu, V.Enache
MATPUR S.A. Bucuresti
*UNIVERSITATEA DE VEST Timisoara

SUMMARY

In order to improve the axial distribution for main impurity, oxygen and carbon concentration in Cz silicon single crystals the authors have established a computer numeric method to reduce the scattering area for its values along the ingot by using different pulling rates.

INTRODUCTION

The most important impurities in Cz silicon crystals are:P, B, Sb, As, O and C. Their alloying with silicon is closely related to segregation laws and hence we should meet three groups of elements (GRABMAYER):

- elements with segregation coefficient greater then 1 (O K_0=1,4);
- elements with segregation coefficient smaller then 1 (B K_B=0,8; P K_P=0,35);
- elements with very small segregation coefficient (C K_C=0,05; Ca,Fe,Cu,Au).

Consequently for a Cz crystal the main impurity is to be unhomogeneously distributed along the ingot axis as follows:

- the doping element concentration is greater at the final part of crystal;
- oxygen content is larger at the initial part of ingot;
- carbon and metals content will grow at the end of it.

Main impurity concentration (P or B) was determined by rezistivity measurements with a 4 point VSM4 rezistivimeter at MATPUR S.A. The effective segregation coefficient (K_e) differs from the theoretical one (K_0) as it is determined by:
a) pulling speed(V); b) limit layer thickness (thin liquid layer neighbouring the growing interface that represents area with some special characteristics as no convection currents and a great amount of rejected impurities from crystallisation front)(KURODA):

$$K_e = \frac{1}{1+(\frac{1}{K_0}-1)*e^{-\frac{V*\delta}{D}}} \quad (1)$$

where: D - diffusion coefficient for impurity in melted silicon and δ - limit layer thickness.

If the crystal is grown at constant pulling speed, by analytic calculus we reach the following equation for axial distribution of impurity concentration :

0-7803-3223-7/96/$5,00 © 1996 IEEE

$$C(x) = (1-x)^{(K_e-1)} = (1-x)^{\frac{1}{1+(\frac{1}{K_0}-1)*e^{-\frac{v*\delta}{D}}}-1} \quad (2)$$

where x - solidification rate
with:

$$x = \frac{m}{m_0}$$

m - solidified fraction;
m_0 - initial amount of melt.

This paper is to carry out an optimum concentration (a narrow range of values) for P,O and C in the Cz Si single crystal by establishing a speed program related to the above mentioned elements.

RESULTS AND DISCUSSIONS

As we can see in eq.2, C(x) depends on $v*\delta/D$ term. Our attention was focused on adjusting only the v pulling speed. δ/D term used in calculus were the one required from experimental data (a set of Cz ingots pulled at constant speed at MATPUR-Bucharest site at constant speed with computed value $\delta/D=0.46$).

If we shall use a range or a set of pulling speeds, during the same growth process, eq.2 wouldn't match. To avoid this fact we have perfected an iteration computing program for establishing impurities' concentration. The iteration follows solidified fraction x (relative length).

$$L(x_i) = \frac{C_0-\Delta x_i*S_t^{i-1}}{(1-x_i)} \quad (3)$$

with:

$$S_t^i = \sum_{j=0}^{i} S(x_i)$$

$$\Delta x_i = (x_i-x_{i-1})$$

and the solid impurity concentration is:

$$S(x_i) = K(x_i)*L(x_i) \quad (4)$$

where: $K(x_i)$ - effective segregation coefficient corresponding to relative length x_i.

Our pulling speed program is following three steps schedule:

$$V(x) = V_0+a*p*(x-c)+b*q*(x-d) \quad (5)$$

where: V_0 - basic speed;
c,d - inflexion points;
p,q - decreasing slant for pulling speed

$$a = \begin{cases} 1 , & x \le c \\ 0 , & x > c \end{cases}$$

$$b = \begin{cases} 0 , & x < d \\ 1 , & x \ge d \end{cases}$$

Afterwards we have defined a working function m_{pq} as follows:

$$m_{pq}=\sum_{i=0}^{1} \{ [C_D(x_i)-C_D(0.5)]U+[C_O(x_i)-C_O(0.5)]W+[C_C(x_i)-C_C(0.5)]Z\} \quad (6)$$

and it calculates a summa for concentration deviations corresponding the doping element (D), oxygen (O) and carbon (C) away from mid length of ingot. U,W and Z ponder the interest for a specific impurity.

Table 1 exhibits values for m(p,q) function by using: $K_0=0.35$ phosphorus segregation coefficient; $R_0=1.4$ oxygen segregation coefficient; $P_0=0.05$ carbon segregation coefficient; $\delta/D = 0.46$; $C_0=1$; $c = 0.25$; $d = 0.75$; $V_0= 1$ mm/min; U=1,W=1,Z=1.

For computing "m" function values the authors have accomplished an iteration program with computer aid.

Table 1

m(0,0)= 57.91	m(0,1)= 57.87	m(0,2)= 57.86	m(0,3)= 57.89
m(1,0)= 57.77	m(1,1)= 57.73	m(1,2)= 57.72	m(1,3)= 57.75
m(2,0)= 57.54	m(2,1)= 57.50	m(2,2)= 57.49	m(2,3)= 57.52
m(3,0)= 57.27	m(3,1)= 57.23	m(3,2)= 57.22	m(3,3)= 57.25
m(4,0)= 57.00	m(4,1)= 56.95	m(4,2)= 56.95	m(4,3)= 56.98

In eq. (6) we have considered fraction $x<=0.9$ as is well known that about 10% of initial weight of the melt is left in the crucible due to pulling technology.

As we can see in Table 1, the lowest value for function "m"(minimal variation for axial concentration) has been reached for p=4 and q=1. Figure 2 presents the axial variation concentration of main impurities for two different ingots: one pulled with constant speed and the other pulled by following the speed programm hinted by "m" function minimal values.

CONCLUSIONS

For finding a pulling speed program able to minimise axial rezistivity variations and oxygen and carbon concentration variations for Cz Si single crystals, the authors have accomplished an iteration computer program that is about establishing a desired impurity concentration area along the pulled ingot. Absolute scattering concentration value for the grown crystal was computed by using m(p,q) function.

The experimental results related are ready to be used to improve the Cz Si single crystal fabrication process as our first ingots grown using the above mentioned results led us in confirming our calculus.

Even if aspects as crucible dissolving in melted silicon and evaporation phenomena during growth were neglected, the results of our research do not indicate large errors or false results.

REFERENCES

1. Grabmayer, J., in Crystals: Growth, Properties, and Applications, vol.8, Springer, Berlin 1982, p.68;
2. E.Kuroda , H.Kozuka, J.Cryst.Growth, 63(1983), 276.

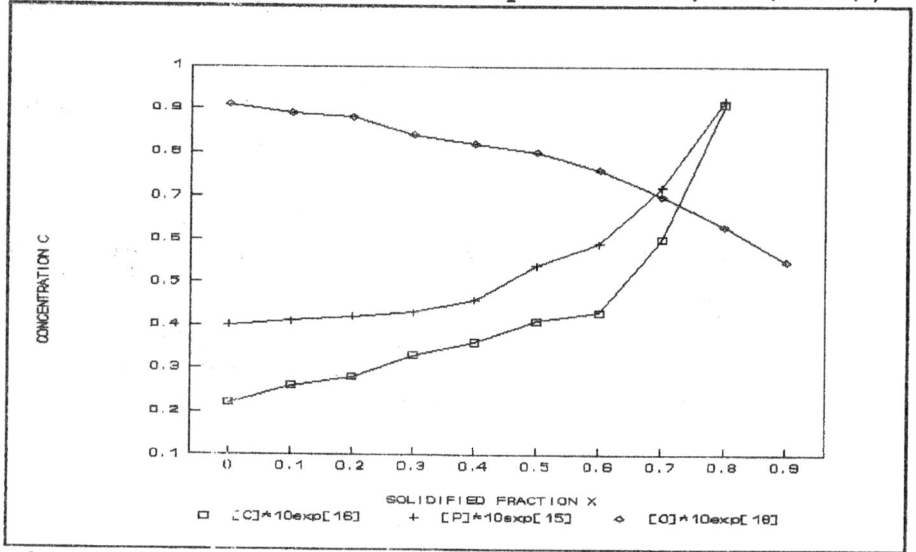

Figure 1. Typical axial distribution of P,O and C content for a 12.5 Kg Cz Si single crystal ingot grown from 14 Kg melt at 80 mm diam. and 950 mm length

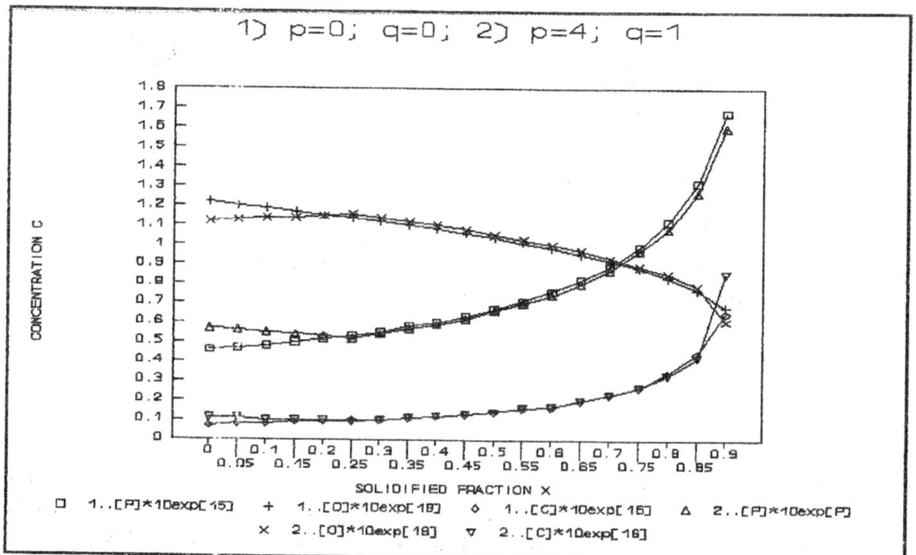

Figure 2. Impurity concentration for two sets of (p,q) values: 1) p=0, q=0; 2) p=4, q=1

DOPING-RESTRUCTURATION PROCESS DURING PHOSPHORUS DIFFUSION IN POLYSILICON LAYERS ON SILICON

F. Găiseanu[1*], D. Krüger[2], J. Stoemenos[3], C. A. Dimitriadis[3**], C. Postolache [4,**]
H. Richter[2*], W. Schröter[5], M. Seibt[5*]

[1.] Research Institute for Electronic Components (ICCE) Erou Iancu Nicolae 32B, R-72996 Bucharest, Romania
[2.] Institute of Semiconductor Physics, W.Korsing-Str., PO Box 409, 15204 Frankfurt (Oder), Germany
[3.] Department of Physics, University of Thessaloniki 54006 Thessaloniki, Greece
[4.] ROMES S. A., Erou Iancu Nicolae 32B, R-72996 Bucharest, Romania
[5.] IV Physikalisches Institut der Universität Göttingen, Bunsenstrasse 11-15, 37073 Göttingen, Germany

SUMMARY

TEM, SR and SIMS investigations of LP-CVD polySi layer deposited at 620^0 C on Si substrate, doped by P prediffusion at 900^0 C and subsequently annealed at temperatures in the range 900^0 C-1000^0 C allow to propose a doping -restructuration mechanism based on the self-interstitial injection during the P diffusion.

1. INTRODUCTION

It was previously reported the behaviour at low temperature treatments (750^0 C $\leq$ T $\leq$ 950^0 C) of the polySi layers "in-situ" doped with P from a spike-type P atom distribution for heterobipolar transistor (HBT) applications [1]. We extend in this paper the investigation on the basis of analysis by cross section transmission microscopy (TEM), secondary ion mass spectroscopy (SIMS) and spreading resistence (SR) to the polySi layers doped with P from liquid $POCl_3$ source and used as local diffusion sources in the range T $\geq$ 900^0 C, useful for power high-frequency bipolar transistors (H FBT) [2]. It is proposed a doping-restructuration mechanism based on the Si self-interstitial injection which contributes to the doping and structural modifications of the polySi layer during the P diffusion.

---

* Electrochemical Society Active Member

** IEEE Active Member

0-7803-3223-7/96/$5,00

2. EXPERIMENTAL CONDITIONS

LP-CVD polySi layers were deposited at 620^0 C on (111)-oriented, n-type Si with a resistivity of (6-12) Ωcm, RCA-cleaned [1] prior the deposition. The polySi layer was P doped during a $POCl_3$ prediffusion at 900^0 C for 10 min. The phosphosilicate glass (PSG) grown during the P prediffusion was maintained at the top of the polySi external surface during the subsequent drive-in annealings at 900^0C - 1000^0 C for periods of time no longer than 60min. The samples ere TEM analysed to investigate the structure properties during multisequential annealing steps. The P doping profiles were determined by SIMS and SR to compare the chemical and electrical profiles and to correlate with the TEM observations.

3. RESULTS AND DISCUSSION

Cross-section TEM micrographs of samples after LP-CVD and P diffusion are shown in Fig.1a and 1b respectively. A typical SIMS and SR P profile is shown in Fig.2 a and b after a diffusion step of 30 min. at 950^0 C. The main results derived from above mentioned determinations are : (i) the as-deposited polySi layer consists in columnar [110] preferentially oriented grains (Fig.1a)) ; a thin layer near the polySi/Si interface is composed by small grains with randomly distributed orientations (coalescence grain layer) ; (ii) the restructuration process of the polySi in large size grains starts from the polySi/Si interface and continues toward the external polySi surface even during short-time 900^0 C prediffusion (Fig.1b)); (iii) the polySi bulk layer of 400 nm is fully and uniformily doped during P prediffusion process (a junction depth of 0.15 μm was obtained in Si after the prediffusion) ; (iv) a P concentration peack is observed after the drive-in annealing (Fig.2a) in the range 900^0C -1000^0C ; (v) the electrically active P concentration increases toward the polySi/Si interface excepting long-time durations (60 min.).

We propose a mechanism based on the Si self-interstitial injection to explain the enhancement of the doping - restructuration process during P diffusion in polySi, which can be described as follows. The P diffusion itself as well as the PSG growth at the polySi interface injects Si self-interstitials [3,4] in the polySi layer, which determines the enhancement of the grain growth [2,5]. During the growth process the surface of the graind boundaries decreases and thereby the inactive P amount, segregated at the grain boundaries, decreases correspondinly. However, accordingly with the proposed mechanism, the enhancement of the grain growth determines an enhancement of the P atom incorporation in the grains, and their diffusion in the Si bulk furthermore generates Si self-interstitial atoms which additionally contributes to the growth enhancement by their rordonation at the grain boundary, in a continuously self-promoted doping-restructuration process. The elementary operating cell of such a chain multiplicative reaction can be schematically represented as follows :

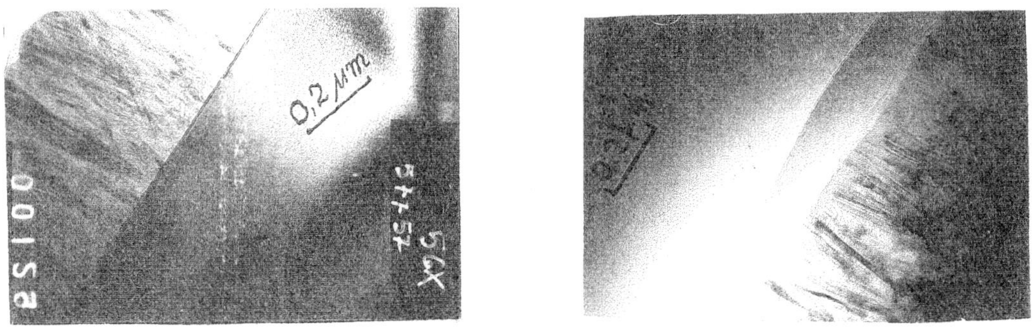

a b

Fig.1 Cross-section TEM micrographs of samples after : a) LP-CVD deposition; b) P prediffusion at 900°C for 10 min.

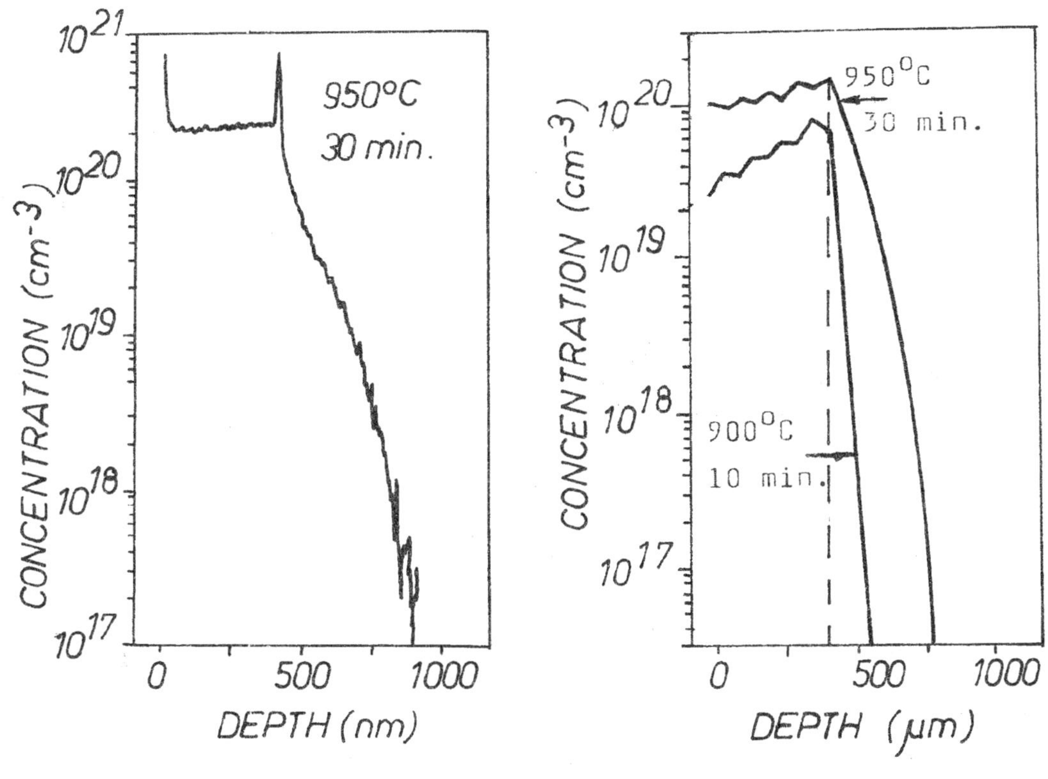

a b

Fig.2 Phosphorus diffusion profiles after 30 min. diffusion at 950°C determined by : a) SIMS; b) SR technique (the P profile after the prediffusion of 10 min. at 900°C is shown for comparison).

$$\begin{array}{l} \quad (I)_{PSGg} \\ \quad\downarrow \\ \longrightarrow (E)_{Gg} \longrightarrow (E)_{Pinc} \longrightarrow (I)_{Pdiff} \longrightarrow \end{array} \qquad (1)$$

where (I) are the Si self-interstitials generated by PSG growth ($(PSG)_g$) and P diffusion (P_{diff}), and (E) is the enhancement of grain growth (Gg) and of the P incorporation (Pinc) respectively. In oppozition with the previously described reaction, the flux of $(I)_{PSGg}$ is countinuously consumed for the grain growth and the further continuation of the doping-restructuration process is assured by means of the chain-type reaction as-proposed in rel (1).

Specifically, the above described mechanisme can explain the expreimental results as follows. The doping-restructuration process is initiated in the as-deposited columnar grain structure by a fast P diffusion along the short-path grain boundaries, allowing the uniform distribution of the P concentration in a short time (10 min.) at 900^0C ; the inactive P atoms segregate at the grain boundaries and at the polySi/Si interface, including the coalescence grain layer. During a subsequent annealing of 30 min. at 950^0C the electrically active P amount is substantially increased from about 1.2×10^{15} cm^{-2} to about 4×10^{15} cm^{-2} due to the P incorporation and diffusion in the Si large size grain bulk. The gradual increasing of the electrically active P concentration near the polySi/Si interface is a result of the rstructuration process which starts from the polySi/Si interface and continues toward the polySi external surface.

4. CONCLUSIONS

The proposed doping-restructuration mechanism, based on the self-interstitial injection, well explains the experimental results on the P diffusion from a $POCl_3$ source in the 400 nm polySi layer, deposited by LP-CVD at 620^0 C on Si, and on the subsequent drive-in P diffusions in the range 900^0C -1000^0C.

REFERENCES

[1] D.Krüger, P.Gawcrzewski, R.Kurps and J.Schlote, Semicon. Sci. Technol., 10, 326 (1995).

[2] F.Găiseanu, C.A.Dimitriadis, J.Stoemenos, C.Postolache, International Semiconductor Conference (CAS), Sinaia Oct. 9-12 (1996).

[3] U.Gösele and T.Y.Tan, in Material Sciences and Technology, vol.4, W.Schröter, Editor, p.198, VCH Weinheim (1991).

[4]. F.Găiseanu and W.Schröter, J. Electrochem. Soc., 143, 361 (1996).

[5] P.E.Proano, D.G.Ast, J. Appl. Phys., 66, 2189 (1989).

Effect of Alloy Scattering on the Magnetotransport in Narrowgap Semiconductors at Low Temperature

Pallab Banerji*, Chaitali Chakraborty and C. K. Sarkar**, *Senior Member, IEEE*, K. K. Mallik , *Member IEEE*, M. N. Roy, *Senior Member, IEEE*

Department of Electronics and Telecommunication Engineering, Jadavpur University, Calcutta-700 032, India

**Department of Physics, University of Calcutta, 92 Acharya Prafulla Chandra Road, Calcutta-700 009,India*

*** Department of Physics, B. E College (D. U.), Shibpur, Howrah-711 103, India*

SUMMARY– Magneto transport in narrow band-gap semiconductors at low temperature in the presence of a quantising magnetic field has been theoretically investigated assuming the electron scattering by the acoustic phonon, the ionized impurities and the alloy disorder. The model includes band nonparabolicity, free carrier screening and nonequipartition of acoustic phonon. It is seen that the low field longitudinal mobility in n-$Hg_{0.8}Cd_{0.2}Te$ in the extreme quantum limit (EQL) at 4.2K is totally governed by the scattering of carriers due to alloy disorder potential. The other scattering mechanisms such as the acoustic phonon scattering via deformation potential or the scattering due to the ionized impurities do not have any bearing on the mobility. The total effect of all such scattering mechanisms is very much close to that when the alloy disorder scattering is considered alone. However, at microwave frequencies both the real and imaginary parts of the mobility decrease with increase in frequency above 1 GHz. Longitudinal drift velocity in the high field regime has also been investigated in the EQL to study the effect of alloy scattering reduces the hot electron longitudinal drift velocity. The theoretical values fit well with the experimental results.

I. Introduction

In the present days, the transport coefficients are studied under strong magnetic and electric fields. These fields, when applied, change the energy spectrum of electrons from continuum to discrete levels, known as band quantization. Under such conditions, the transport is termed as quantum transport [1]. One reason for studying the quantum magneto transport is the widespread availability of inexpensive superconducting magnets capable of generating fields ~ 15T. Another reason is the progress in semiconductor physics and technology and the need to obtain more precise information about the band structures and scattering processes in technically relevant or promising materials. In the present paper transport in narrow-gap compound semiconductor such as $Hg_{0.8}Cd_{0.2}Te$ under low and high steady electric fields have been investigated. Further more, the effect of time varying electric field on the electron mobility has also been investigated in the low electric field regime.

II. Theoretical Formulation

(a) Low Field Regime

The theoretical calculations include the inelastic acoustic phonon scattering with nonequipartition of phonons, the ionized impurity and the alloy disorder scattering. The model also includes the band non-parabolicity since n–$Hg_{0.8}Cd_{0.2}Te$ is a narrow band gap semiconductor. The effect of free carrier screening on the scattering mechanisms is also incorporated since magnetic quantization causes the screening function to be magnetic field dependent and anisotropic [2]. Carriers are assumed to occupy the lowest Landau level. The assumption may be justified for materials $Hg_{0.8}Cd_{0.2}Te$ having very low effective mass given rise to large Landau subband separation. The optical phonon scattering is not considered because it is not important at low temperatures. In order to obtain longitudinal mobility, the relaxation times for acoustic phonon scattering via deformation potential, ionized impurity scattering and alloy scattering mechanisms are required and the expressions for them can respectively be written as [3]

$$1/\tau_{ac} = \frac{(m^* a_0)^{1/2} F_1{}^2}{8\sqrt{\pi}\,\hbar\, l^3 \rho\, u_{ac}} (1+\beta)^{5/2} \times (1+\beta+\alpha_s)^{-2} \left(\frac{N_R}{\sqrt{E+\hbar\omega_0}} + \frac{N_R+1}{\sqrt{E-\hbar\omega_0}} \right) \quad (1)$$

$$1/\tau_{imp} = \frac{N_I e^4 l^2}{8\pi(\epsilon_s \epsilon_0)^2 \hbar^2} (2 m^* a_0)^{1/2} \int_0^\infty (u+\beta v+\alpha_s)^{-2} e^{-u}\, du \quad (2)$$

0-7803-3223-7/96/$5,00

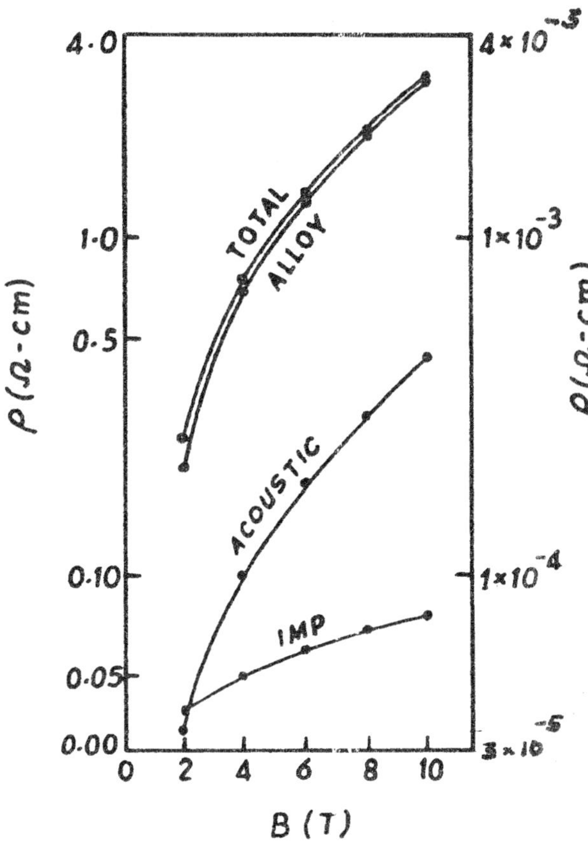

Fig. 1. Variation of the theoretical longitudinal resistivity in $Hg_{0.8}Cd_{0.2}Te$ in the extreme quantum limit as a function of magnetic field at a lattice temperature of 4.2K for $n=5\times 10^{20}$ m^{-3}. It also shows the magnetic field variation of the longitudinal resistivity for the acoustic phonon, the ionized impurity, and the alloy disorder scatterings

$$1/\tau_{\text{alloy}} = \frac{3\pi^2}{16\,N_A}\,x\,(1-x)\frac{E_0^2}{2\pi(\hbar 1)^2}\left(\frac{1+\beta}{1+\beta+\alpha_s}\right)^2(2m^*a_0)^{1/2}\frac{1}{\sqrt{E}} \quad (3)$$

For calculating low field microwave mobility, the distribution function of the electrons in the lowest sub-band can be written as $f = g(k_z) + \phi\exp(j\omega t)$ where $g(k_z)$ is the equilibrium distribution function and is the perturbation distribution function, odd in k_z. Taking such a distribution function, the real and imaginary parts of the mobility have been calculated [4] and those can be written as

$$\mu_R = \frac{2e}{\sqrt{\pi}\,m^*a_0}\int_0^\infty \frac{\sqrt{x}\,e^{-x}\,\tau(x)}{1+\omega^2\tau^2}\,dx \quad (4)$$

$$\mu_I = \frac{2e}{\sqrt{\pi}\,m^*a_0}\int_0^\infty \frac{\sqrt{x}\,e^{-x}\,\omega\,\tau^2(x)}{1+\omega^2\tau^2}\,dx \quad (5)$$

where $x = \frac{E_Z}{k_BT}$ and $E_Z = \frac{\hbar^2 k_z^2}{2m^*a_0}$

(b) High Field Regime

Longitudinal conductivity of electron gas in alloy semiconductors subjected to high electric field in the presence of a quantising magnetic field where both the fields being parallel to each other has also been studied. The inelastic acoustic phonon scattering is used to calculate the energy loss rate while the elastic acoustic phonon, ionized impurity and alloy disorder scatterings contribute to the mobility via momentum relaxation. Assuming that the electric field is applied in the same direction as **B** and taking a Maxwell-Boltzmann distribution function for carriers characterized by an electron temperature T_e, the average energy loss rate per electron through the emission and the absorption of acoustic phonons is expressed as [5]

$$\dot{P}_{ac} = W_{ac}\,f_{ac}(v)\,\log[\,exp(-C)\,\frac{4k_B\,T_e}{E_c}\,] \quad (6)$$

where C is the Euler's constant and f_{ac} is $f_{ac}(v)$ at v=1.

The electric field is given by the relation $E = \sqrt{\frac{P_{ac}}{e\mu}}$ where μ is the mobility and its value is calculated from the momentum relaxation times of carriers.

III. Results

The low field longitudinal resistivity has been calculated using eqn.4 and eqn.5 for n–$Hg_{0.8}Cd_{0.2}Te$ at 4.2K in the **EQL** using the parameters of HgCdTe [3]. The theoretical variation of the resistivity as a function of magnetic field for $n= 5\times10^{20}$ m^{-3} is shown in *Fig.1.* It also shows the magnetic field variation of the longitudinal resistivity and the alloy disorder scattering. The theoretical values of the resistivity at 4T and 6T have been compared with the experimental results which are 0.7 and 1.5 Ω-cm respectively [6]. The alloy scattering strength required to fit the above experimental values is found to be equal to the difference resistivity at 4.2K in the **EQL**, is found to increase with the magnetic field which is in agreement with the experimental observation. The magnetic field dependence of the longitudinal resistivity is obtained as B^n where n=0.24.

The variations of drift velocity with electron temperature are shown in *Fig.2.* The upper two curves, at B=4T & 6T respectively, exhibit the contribution without alloy scattering, while the lower curves for B=4T & 6T exhibit the same with alloy scattering. It is seen that the inclusion of alloy scattering reduces the magnitude of drift velocity quite significantly. The drift velocity without alloy scattering contribution appears to increase steadily with electron temperature. However, the lower curves show a tendency for saturation. It is also observed from the figure that the drift velocity of hot electrons in the **EQL** increases with the increasing magnetic field.

IV. Discussion

The present analysis of the longitudinal resistivity of $Hg_{0.8}Cd_{0.2}Te$ in the **EQL** is based on the relaxation time formalism of the Boltzmann transport equation. The comparison of the theoretical results with the experimental results at T=4.2K are made on the consideration that the system behaves as a free electron gas. The experimental results for n-$Hg_{0.8}Cd_{0.2}Te$ used for the present analysis is taken from the work of Nimtz and Gebhardt [6]. The authors clearly indcate that the Wigner crystallization is only observed in n-$Hg_{0.8}Cd_{0.2}Te$ at temperatures lower than T=4.2K and it behaves as a free electron gas at 4.2K. The occurrence of the Wigner crystallization is observed at very low temperatures, in millikelvin region. The present model for the longitudinal resistvity at low temperature in the **EQL** has been utilized to calculate the alloy scattering strength in n-$Hg_{0.8}Cd_{0.2}Te$, which is found to be the difference in the band-gap energies of the constituent elements of HgCdTe alloy.

At low temperature the power loss as well as the drift velocity increase with electron temperature. The effect of ionized impurity scattering decreases with increasing electron temperature because it is of electrostatic nature. As the effect of Coulomb scatttering decreases with increasing electron temperature, the drift velocity increases steadily. But when the alloy scattering is considered, the drift velocity decreases because of the enhancement in scattering rate. However, the power loss due to hot electron in the extreme quantum limit is found to show a weak magnetic field dependence and it increases slowly with the magnetic field. This gives rise to electron drift velocity as an increasing function of magnetic field.

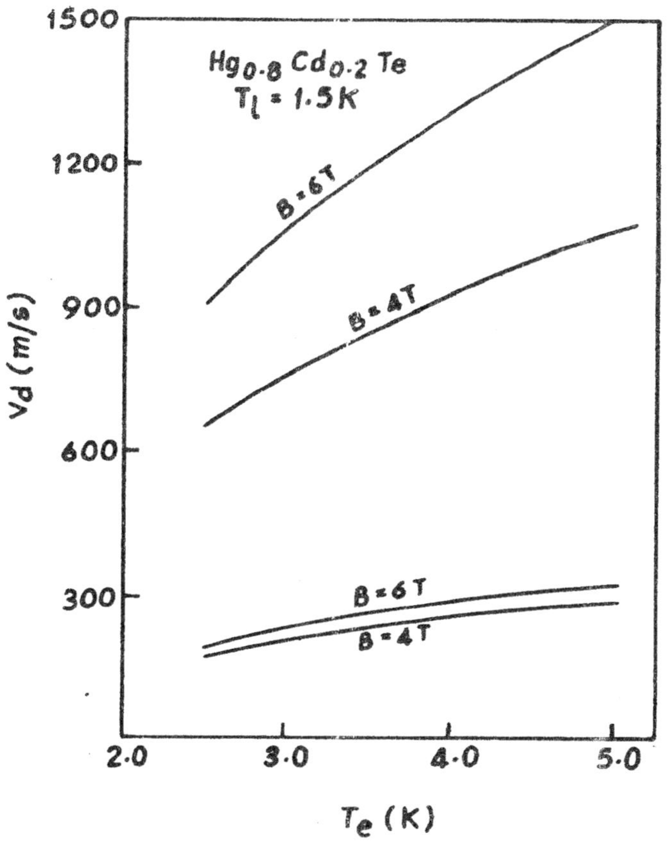

Fig. 2. Variation of the drift velocity of hot electrons in n-$Hg_{0.8}Cd_{0.2}Te$ in the extreme quantum limit as a function of electron temperature of 1.5K for magnetic fields of 4 and 6T. The lower two curves and upper two curves, respectively, represent the variations with and without alloy scattering contributions.

The effect of alloy scattering is found to dominate for both low and high field mobility in the **EQL** at low temperatures. The low temperature high field transport is primarily determined by the momentum relaxation time of nonphonon type of scatterings such as ionized impurity and alloy disorder, since the effect of electron temperature (energy relaxation time) on the mobility is not so significant due to small energy loss rate when acoustic phonon scattering is considered. In the present analysis, the effect of ionized impurity scattering on the high field mobility is diminished due to low impurity concentration (10^{20} m^{-3}) as well as due to high electric field and leaving the effect of alloy scattering to be most dominant.

Acknowledgement

The work is financially supported by CSIR (India) and UGC (India).

References

[1] L. M. Roth and R. N. Argyres, *In Semiconductors and Semimetals ed. R. K. Willardson and A. C. Beer (Academic, New York)*, p. 159, 1966.

[2] A. Fortini, *phys. stat. sol. (b)*, vol. 125, p. 259, 1984.

[3] P.Banerji and C.K.sarkar, *J. Appl. Phys.*, vol. 70, p. 1467, 1991.

[4] P. Banerji, *"Electronic Transport in Semiconductors under Energy Band Quantisation" [Ph.D. Thesis, Jadavpur University, Calcutta, India], 1995.*

[5] P. Banerji and C. K. Sarkar, *J. Appl. Phys.*, vol. 75, p. 1231, 1994.

[6] G.Nimtz and J.Gebhardt, in *Proceeding of the 18th International Conference on the Physics of Semiconductors*, Stockholm, 1986, edited by O.Engstrom (World Scientific, Singapore, 1987) p. 1197.

Session 16

RELIABILITY (Poster)

DEEP TRAPS AND BRIGHTNESS DEGRADATION ON THE OXYGEN-RICH CONCENTRATION IN ZnS:TbOF THIN FILM ELECTROLUMINESCENT DEVICES

C. W. Wang, T. J. Sheu
Electronic Engineering Department
Kaohsiung Polytechnic Institute
Ta-Hsu Hsiang, Kaohsiung County, Taiwan.
AND
Y. K. Su, M.Yokoyama
Electrical Engineering Department
National Cheng Kung University, Tainan, Taiwan.

SUMMARY

The purpose of this work is to study the relationship between the oxygen-rich concentration and brightness degradation in ZnS:TbOF green A.C. thin-film electroluminescent devices. The results showed that higher oxygen-content (O/Tb>1) in the phosphor layer not only created the deep hole traps (E_{t1} and/or E_{t2}) but also yielded the EL devices to get moisture. As a result, lower brightness were obtained.

1. INTRODUCTION

In the recent decade, the efficiency and brightness of Tb-doped ZnS green electroluminescent (EL) devices fabricated by sputtering have reached a commercial level. Nevertheless, the mechanism responsible for this efficiency improvement is not yet clearly understood . In 1968, Kahng proposed that the TbF_3 molecular center could improve the EL efficiency if its large cross section was utilized for direct impact excitation by hot electrons.[1] On the other hand, it has been reported that panels having a TbF complex center exhibit bright green EL.[2,3] Recently, some experimental results have indicated that oxygen doping is essential to improve the EL brightness of the ZnS:TbF thin films.[2,4] Okamoto and his coworkers have demonstrated that the TbOF complex center with the structure of the O occuping the nearest substitutional site and the F the nearest interstital site to the Tb is more efficient than the TbF complex center in ZnS thin-film EL devices. As a result, how oxygen doping affects the brightness and efficiency of ZnS:TbOF EL devices becomes an important issue.

The purpose of this work is to study the relationship between the oxygen-rich concentration and deep defect levels in ZnS:TbOF materials and to provide a new insight into how the brightness of ZnS:TbOF could be affected by the oxygen doping.

2. EXPERIMENTAL ARRANGEMENTS

ZnS:TbOF phosphor layers were prepared by RF magnetron sputtering method. A typical concentration of Tb atoms in the resultant thin film were fixed at 4.0 wt%. During the growth process, a mixed gas with different ratios of argon and oxygen was used as the sputtering gas to obtain different oxygen contents in the phosphor layer. All

0-7803-3223-7/96/$5,00

sputtered film conditions and the measured properties are illustrated in Table 1. The Tb, F, and O concentrations were measured by electron probe microanalysis (EPMA). The EL devices were prepared in doubled-stacked structure of ITO/ SiO_2 (100 nm)/ Ta_2O_5 (350 nm)/ ZnS:TbOF (600 nm)/ SiO_2 (100 nm)/ Ta_2O_5 (350 nm)/ Al. [5] The B-V characteristic of EL device were investigated with a 1 KHz alternating pulse voltage excitation.

Growth condition		sample (a)	sample (b)	sample (c)	sample (d)
Ar flow rate	(sccm)	50	49	48	45
O_2 flow rate	(sccm)	0	1	2	5
Substrate temperature	(°C)	175	175	175	175
O/Tb ratio		1.0	1.4	1.9	4.2
Maximum brightness	(cd/m^2)	320	98	60	25

Table 1. Growth conditions of ZnS:TbOF EL devices for RF magnetron sputtering.

3. RESULTS AND DISCUSSION

The relationship between brightness and different O/Tb ratio in ZnS phosphor layer was shown in Fig.1. It is obvious that for samples with heavier O/Tb ratio (O/Tb ratio >1), a lower brightness was obtained. Furthermore, from Deep Level Transient Spectroscopy (DLTS) measurements [6] we found out that as long as O/Tb ratio greater than 1, there exists the deep hole traps E_{t1} and/or E_{t2} in the ZnS:TbOF phosphor film, which was shown in Fig.2. And no any deep defect signal could be found in sample (a) which O/Tb ratio equals to 1. We suspect that this E_{t1} deep hole trap is responsible for the poor brightness in sample (b), (c), and (d). And we also have confidence in believing that E_{t1} is an oxygen-related hole trap since the DLTS signal intensity is in proportion to defect concentration. The E_{t2} peak in sample (d) is also believed to be an oxygen-related deep hole trap with a different lattice relaxation compared with E_{t1} trap.

Another significant result is that the oxygen concentration in the ZnS:TbOF phosphor layer has some dependence for the EL devices getting moisture, which was proved by using Fourier Tansform Infrared Spectroscopy (FTIR). In Fig.3, the FTIR results showed that the more O/Tb ratio, the stronger H-O-H stretching vibration can be obviously detected in the range 3600-3200 cm^{-1}. It implies higher oxygen content in the phosphor layer, easier for the EL devices to get moisture and lose their luminescent characteristics.

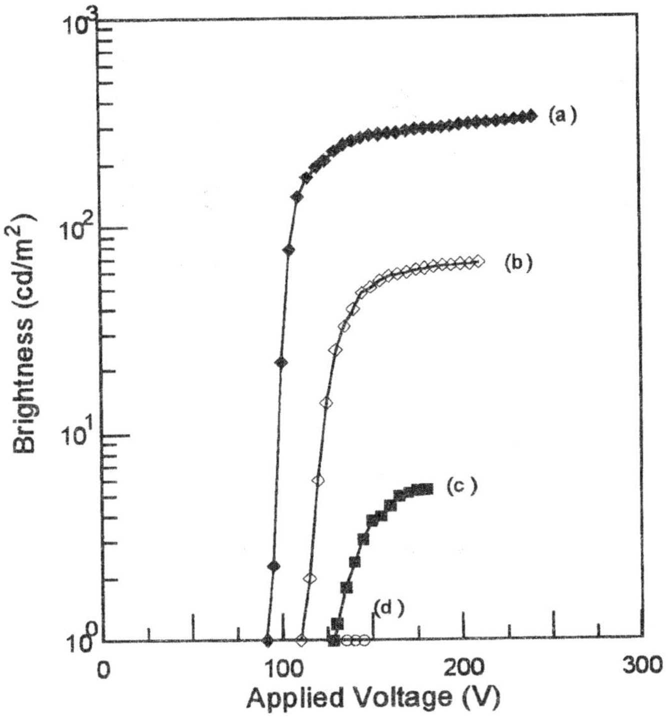

Fig. 1. Brightness-Voltage characteristics of ZnS:TbOF ACTFEL devices with different oxygen-doping in the phosphor layer. The O/Tb ratio for samples (a), (b), (c), and (d) are 1.0, 1.2, 1.5, and 2.8, respectively. It is noted that the brightness for sample (d) is hardly to be detected and always below 1 cd/m^2.

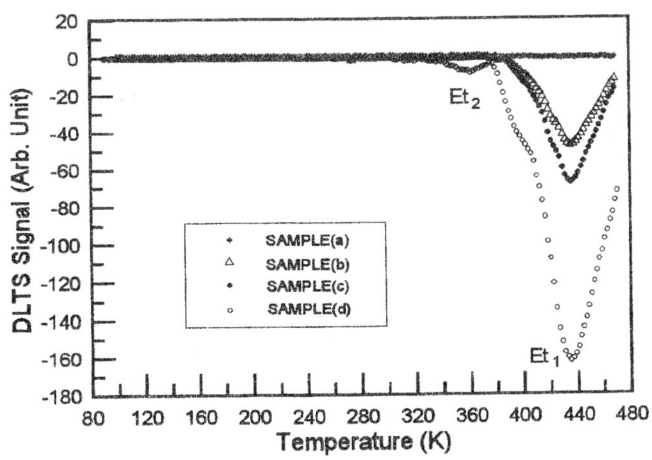

Fig. 2. DLTS signals of different oxygen-doping in ZnS:TbOF phosphor layer. The O/Tb ratio for samples (a), (b), (c), and (d) are 1.0, 1.2, 1.5, and 2.8, respectively.

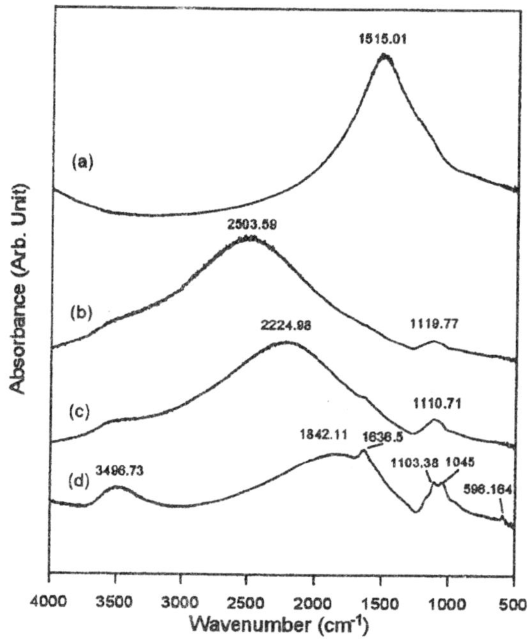

Fig. 3. FTIR spectra of different oxygen-doping in ZnS:TbOF phosphor layer. The O/Tb ratio for samples (a), (b), (c), and (d) are 1.0, 1.2, 1.5, and 2.8, respectively.

4. CONCLUSIONS

A novel evidence is reported here that too much oxygen content (O/Tb ratio greater than 1) in the ZnS:TbOF phosphor layer will reduce the brightness of the EL device. The mechanisms are suggested that the excess oxygen content could create the deep hole traps E_{t1} and E_{t2}, which were demonstrated by DLTS measurement. The origins of E_{t1} and E_{t2} were also believed to have some dependence for the EL devices getting moisture.

5. ACKNOWLEDGEMENT

The authors wish to thank National Science Council (NSC) in Taiwan for receipt of financial assistance (project number: NSC85-2215-E-214-003) for one of us (C. W. Wang) to undertake this work.

6. REFERENCES

[1]. D. Kahng, Appl. Phys. Lett. 13, 210 (1968).
[2]. T. Ogura, A. Mikami, K. Tanaka, K. Taniguchi, M. Yoshida, and S. Nakajima, Appl. Phys. Lett. 48,1570 (1968).
[3]. K. Okamoto, and K. Watanabe, Appl. Phys. Lett. 49, 578 (1986).
[4]. K. Okamato, T. Yoshimi, and S. Miura, Appl. Phys. Lett. 53,678 (1988).
[5]. S. K. Tiku, and S. H. Rustonji, IEEE Transactions on Electron Devices, Vol. ED-36 (9), 1947 (1989).
[6]. D. V. Lang, Journal of Appl. Phys. 45, 3014 (1974).

AN INVESTIGATION OF DEEP ELECTRON TRAP AT THE DIFFERENT INSULATOR-PHOSPHOR INTERFACES IN ZnS:TbOF ACTFEL DEVICES

C.W. Wang, T.J. Sheu
Electronic Engineering Department
Kaohsiung Polytechnic Institute
Ta-Hsu Hsiang, Kaohsiung County, Taiwan.
AND
Y.K. Su, M. Yokoyama
Electrical Engineering Department
National Cheng Kung University, Tainan, Taiwan.

SUMMARY

Engergy-resolved DLTS method was utilized to measure the interface electron energy distribution at the insulator-semiconductor interface. The results showed that Ta_2O_5/ZnS:TbOF has shallower interface state engergy distribution and higher averaged interface state density than SiO_2/ZnS:TbOF.

1. INTRODUCTION

A.C. thin film electroluminescent (ACTFEL) devices are of commercial and scientific importance because of their application in flat panel displays (FPD). These devices also serve as a vehicle to study the physics of insulator-semiconductor interfaces. Several authors have analyzed the ACTFEL device behavior. Light generation in the phosphor layer is attributed to either direct impact excitation [1-3] or modified energy-transfer of activator centers [4-6] by high energy electrons. The source of these "hot" electrons at the semiconductor-insulator interfaces are responsible for the luminescent output of the device. In spite of this critical role of interface electrons in ACTFEL displays, few measurements on their energy distribution have been reported.[7,8] In this paper we present a method where the interface electron energy distribution is determined by using energy-resolved Deep Level Transient Spectroscopy (DLTS) [9-11] method.

2. METHOD

The energy-resolved DLTS technique is specially designed for analyzing the interface states in MOS devices. When the injection pulse hight is small enough, the DLTS signal coming from the interface states within the very narrow interval (E_r, E_p) is a rather sharp spectrum. This behavior is one of the characteristic features of a quasi-discrete level localized at the energy $E_t \approx \left(E_r + E_p\right)/2$. In this situation, the ΔC that comes from the interface states can be reduced to [12,13]

$$\Delta C = \frac{-C(t_1)^3}{\varepsilon_s N_D(w)} \cdot \frac{KT}{C_{OX}} N_S(E_t) \cdot \ln\left(\frac{t_2}{t_1}\right) \qquad (1)$$

0-7803-3223-7/96/$5,00 © 1996 IEEE

Which is the same expression as for a discrete level. Furthermore, the activation energy of the interface state ΔE_t is given by[9]

$$\Delta E_t \equiv E_c - E_t = qV_S + qV_F - 1/2 \cdot \Delta E_F \qquad (2)$$

Where V_s is the surface potential, qV_F is the Fermi level measured from the conduction band edge in the bulk semiconductor. ΔE_F, the variation of the surface Fermi level, which relates to the change of the applied pulse voltage can be approximately writen as [14]

$$\Delta E_F \cong q \cdot [1 - C_T/C_{OX}]\big|_{V_r} \cdot (V_p - V_r) \qquad (3)$$

Where C_T is the total capacitance. Moreover, the interface state density $N_S(E_t)$ can also be calculated from Eq.(1).

3. EXPERIMENTAL RESULTS AND DISCUSSION

For a comparative evaluation of the electron ejection mechanisms from different insulator-phosphor junctions, the variations in interface-state energy distribution for SiO_2/ZnS:TbOF and Ta_2O_5/ZnS:TbOF structures were investigated. Fig.1 shows the traditional DLTS measurements for these two structures. From Fig.1, we observed that for both SiO_2/ZnS:TbOF and Ta_2O_5/ZnS:TbOF devices, there exist continuously distributed interface states at the interface-phosphor junction. And the electron-energy distribution calculated by energy-resolved DLTS method were situated at 0.76～1.04ev and 0.55～0.92ev, respectively. It is in accordance with the results that using Ta_2O_5 as the insulator layer for ACTFEL device has a smaller threshold voltage than that by SiO_2 since Ta_2O_5/ZnS:TbOF has shallower interface state energies than SiO_2/ZnS:TbOF.

Furthermore, the interface state density $N_S(E_t)$ for both SiO_2/ZnS:TbOF and Ta_2O_5/ZnS:TbOF were calculated by using Eq.(1) and the results were shown in Fig.2. It is obtained that Ta_2O_5/ZnS:TbOF has higher averaged interface state density than SiO_2/ZnS:TbOF.

4. CONCLUSIONS

A new evidence is reported here that by using energy-resolved DLTS technique, the interface state energy distribution at the insulator-semiconductor interfaces can be measured. The experiments showed that Ta_2O_5/ZnS:TbOF has shallower interface state energy distribution and higher averaged interface state density than SiO_2/ZnS:TbOF. This is in accordance with the results that using Ta_2O_5 as the insulator layer for ACTFEL device has a smaller threshold voltage than that by SiO_2.

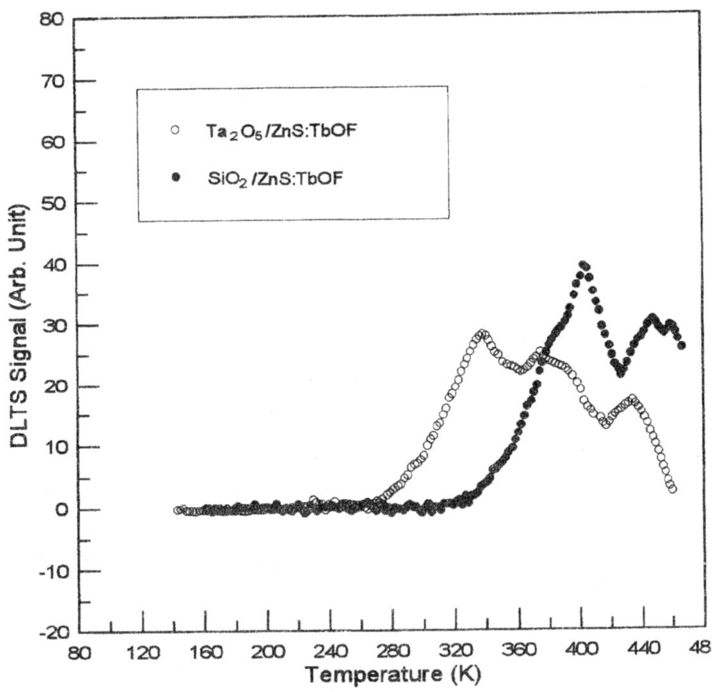

Fig. 1. DLTS signals for Ta_2O_5/ZnS:TbOF and SiO_2/ZnS:TbOF structures.

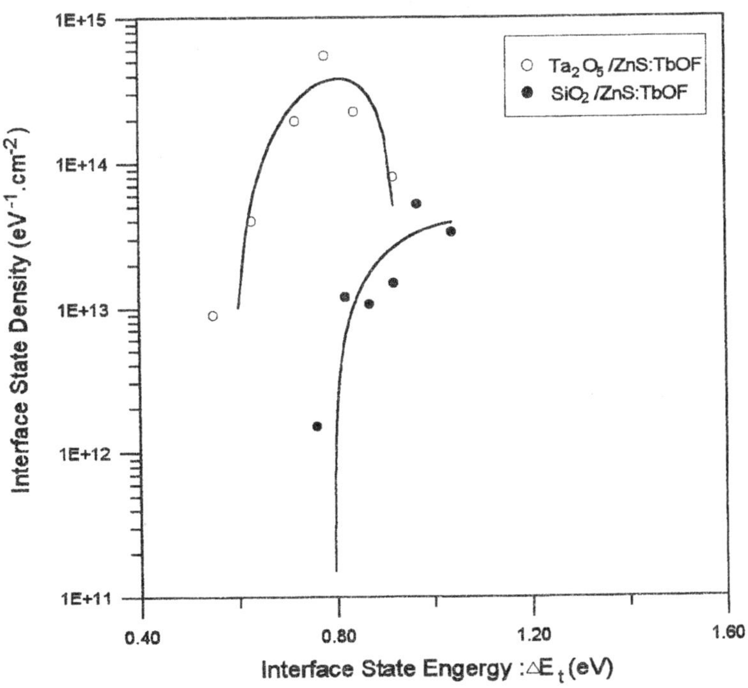

Fig.2. The interface state densities determined by energy-resolved DLTS for both SiO_2/ZnS:TbOF and Ta_2O_5/ZnS:TbOF interfaces.

5. ACKNOWLEDGEMENT

The authors wish to thank National Science Council (NSC) in Taiwan for receipt of financial assistance (contract number: NSC85-2215-E-214-003) for one of us (C. W. Wang) to undertake this work.

6. REFERENCES

[1] D. Kahng, Appl. Phys. Lett. 13, 210 (1968).
[2] D.C. Krupka, and D.M. Mahoney, J. Appl. Phys. 43, 2314 (1972).
[3] D.C. Krupka, J. Appl. Phys. 43, 476 (1972).
[4] K. Okamoto, and S. Miura, Appl. Phys. Lett. 49, 1596 (1986).
[5] A. Mikami, T. Ogvra, K. Taniguchi, M. Yoshida, and S. Nakajima, J. Appl. Phys. 64 (7), 3650 (1989).
[6] V. Marrello, L. Samvelson, and A. Onton, J. Appl. Phys. 52, 3590 (1981).
[7] T. Inoguchi and S. Mito, Electroluminescence, Edited by J. I. Pankove, Topics in Applied Phys. 17, 197 (1977).
[8] S.Kabayashi et al, Proceedings of the Sixth International Workshop in Electroluminescence, Cinco Puntos Press, E1 Paso, TX. July 1992, p.234.
[9] T. Katsube, K. Kakimoto, and T. Ikoma, J. Appl. Phys. 52, 3504 (1981).
[10] W.D. Eades and R.M. Swanson, J. Appl. Phys. 56, 1744 (1984).
[11] N.M. Johnson, J. Vac. Sci. Technol. 21, 303 (1982).
[12] K. Yamasaki, M. Yoshida and T.Suganon, Jpn. J. Appl. Phys. 18(1), 113 (1979).
[13] F. Murray, R.Carin and P.Bogdanski, J. Appl. Phys. 60(10), 3592 (1986).
[14] I. Thurzo and E.Pincik, Phys. stat. Sol. (a)86, 795 (1984).

STRESS AND STRAIN IN AUTOMOTIVE DIODES - A RVT, IR AND XR STUDY

L.Galateanu[1], Mihaela Stoica[1], C.Bozdog[2], E.Popa[2], A.D.Stoica[3]

1 - ICCE Bucharest, 2 - SC Baneasa SA Bucharest, 3 - IFTM Bucharest

SUMMARY

The methods of achieving the needed reliability for the automotive rectifier CAN diodes are searched. The RVT tool needed for investigation was found, a new stress relief shape was designed and IR and XR investigations were performed toward the complete elimination of the residual strain left behind the grinding and diffusion processes. An IR method of the rough surface layer characterization is for the first time employed and the optical configuration for XR experiments was improved in angular resolution.

1. INTRODUCTION

The automotive semiconductor devices must survive in an adverse climatic, dynamic and electromagnetic environment /1/. Different types of failure may occur: changes in electrical parameters, microscopic cracks, fractures, intermittent functioning, seal failure and others.

In this paper three steps toward the required reliability of CAN diodes, I_F = 25 A and U_R = 200 V, manufactured by SC Baneasa SA, F2300 are presented. The devices were subjected to over 20% failures in operation.

The first step was to isolate and to reproduce the failure type in experimental conditions. The second one, to lower the stress applied on the chip by packaging design modification and the third one, to search for process induced strain sources that might affect the reliability. The results: a powerful tool to simulate the harsh operational conditions that exactly reproduces the failure; a reduction of failure below 4% after the assembling induced stress relief and the identification of two processes influencing the reliability - the grinding and the diffusion.

A novel IR method for the characterization of roughly damaged layer on the silicon surface is used here. The method based of Mie's theory of single scattering /3/ is presented elsewhere /4/. The XR diffraction double crystal method used for the study of residual strain of ground and diffused samples has an improved angular resolution of 4.5×10^{-3} degrees.

2. RELIABILITY INVESTIGATIONS

Failure percentages above 20 % during operation were recorded on CAN automotive rectifier diodes. The failure consisted in strong degradation of the reverse characteristic. SEM analysis pointed out microcracks at the edge of the semiconductor chip. In order to reproduce the operational environment various experiments were performed /2/ on the automotive devices, but no failure appeared after vibration and mechanical shock tests, high (+175 ^{0} C) and low (-55 ^{0} C) temperature storage and high temperature reverse biasing (T_j = +175 ^{0}C, U_R = 150V). Only after performing the rapid variations of temperature RVT test (15 cycles of 30 min. at extreme temperatures of +150 ^{0} C and -55^0 C) the same failure type in the same percentage as in operational conditions were obtained.

As a result the RVT test was accepted as a tool to evaluate the reliability. It was also concluded that the extreme temperature cycles causes the failure because some internal stress sources already exists inside of the packaged device.

3. ASSEMBLING IMPROVEMENT

The assembling system of the diodes was analysed. The soldering of the stress relief shape of the internal connection wire on the whole area of Al metallization of the chip (figure 1a)

0-7803-3223-7/96/$5,00 © 1996 IEEE

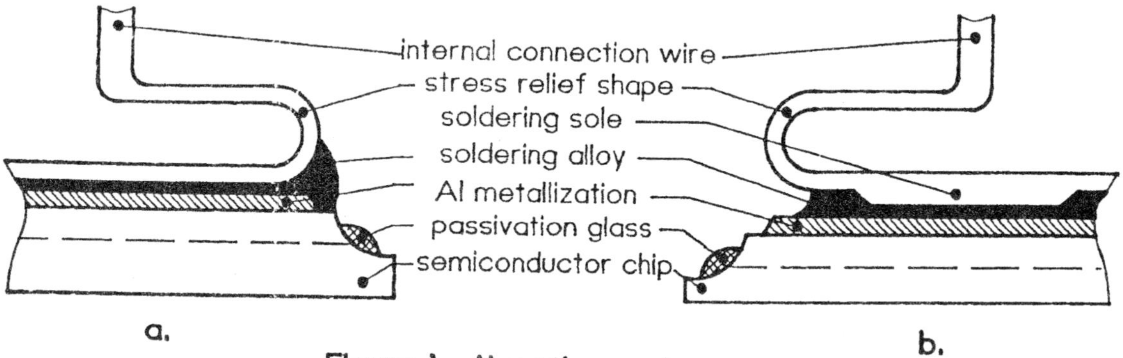

Figure 1. Mounting systems

leads to a strong internal stress at the edge of the chip, where microcracks were observed after the solidification of the soldering alloy (Pb-Ag-In). A new design for the soldering system of the internal connection wire was performed The new stress relief shape contains a sole which performs two functions: (i) confines the solder area to the middle zone of the chip and (ii) allows the release of the stress relief shape curved zone.

In this way, the failure percentage on RVT test diminished below 4%. The failure was of the same type as before the improvement. Since the planar-epitaxial chips, with higher mechanical strength, tested in the same packaging system, suffered no failures on the RVT test, it was concluded that the reason of the 4% failures must be searched inside the chip.

4. SURFACE STATUS AFTER MECHANICAL-CHEMICAL PROCESSING

For reasons pertaining to surface wettability the CAN diodes are made using ground wafers. The grinding process leaves behind a damaged layer which can be viewed as consisting of two overlapped areas. The upper one is represented by the so-called roughness and microcracks advancing toward the bulk. It is followed by the second area of internal strain which slowly diminishes in depth. For a complete characterization of these two areas, two complementary methods were used: IR transmission spectroscopy and XR diffraction.

All investigated silicon wafers were N, (111), 30 - 60 Ωcm, 210-280 μm.

IR spectral traces of silicon rough surface were obtained in the 2.5 - 40.0 μm wavelength range on a Specord M 80 spectrometer. The bulk absorbtion of silicon lattice was subtracted from each sample's spectrum. Surface profile obtained on a Perthen perthometer are presented as reference (figure 2a). The XR-diffraction curves were obtained on the same wafers using the double crystal technique, CuK_α radiation, reflex plane (333). The breaking force, F, measured as in /5/ is also presented in Table 1.

Table 1 - Parameters obtained for ground (GW) and etched (EW) wafers

Sample	R_H (μm)	$A_{2.5}$	F(kgf)	σ_1 (10^{-3}deg)	σ_2 (10^{-3}deg)	$R_{2/1}$
1. GW (8μm)	<2.5	1.8	1.4-1.7	10.45(2)	39.1(1)	0.40
2. GW (18μm)	5.5	2.0	0.7-0.9	10.47(2)	33.6(1)	0.7
3. EW (alkali)	12.5	1.2	4.0 - 4.5			

Two parameters were extracted from the IR spectra (figure 2a): the position of maximum absorbtion which stands for the depth of roughness R_H and the absorbtion amplitude at the

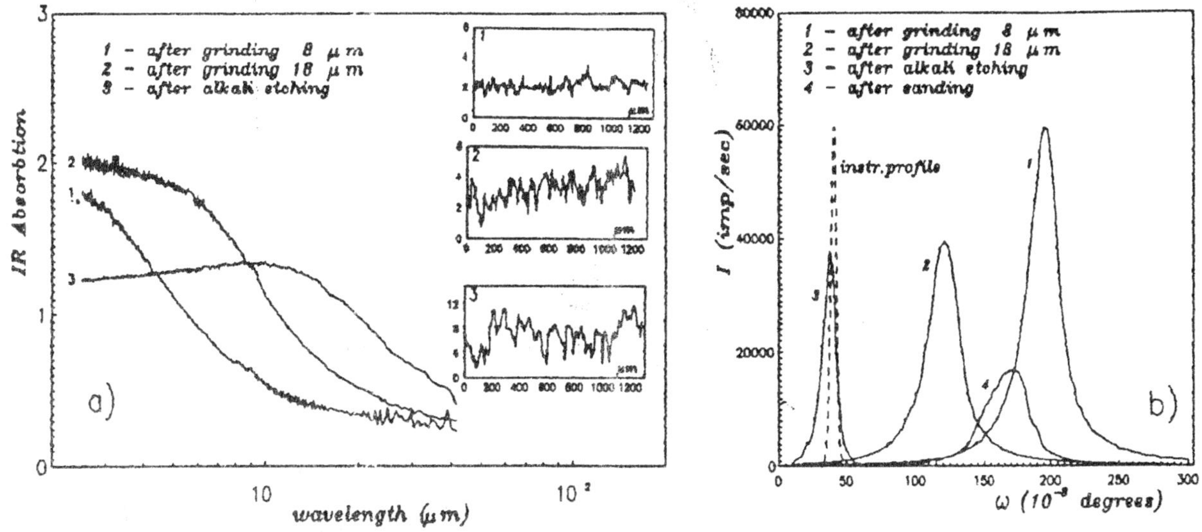

Figure 2. IR spectra (a) and XR rocking curves (b) of rough surface silicon wafers.

lowest wavelength $A_{2.5}$ (Table 1) depending upon the presence of the surface fine perturbations as shown inserted in figure 2a .

The XR rocking curves (figure 2b) obtained on these wafers were approximate by the sum of two gaussian profiles .The narrow profile component may be assigned to the diffraction on the bulk, while the broader one seems to be produced by the scattering on the surface roughness. The dispersion σ_1 and σ_2 and integral intensities ratio $R_{2/1}$ are given in Table 1.

From Table 1 it can be seen that the residual strain below the surface depends upon the density of the fine perturbations, being almost independent of the roughness depth. The R_H value does not completely explain the variation of F, which seems to be dependent upon the fine perturbations density, also measurable by $R_{2/1}$ from XR data.

5. INTERNAL STRAIN AFTER BORON AND PHOSPHORUS DIFFUSION

The high doping concentration N_c induces a corresponding lattice misfit m /6/. If m is above the elastic host lattice limit, misfit dislocations can be generated. In topographic XR experiments /7/ was shown that boron diffusion generated dislocations are oriented along <110> directions and glides on a {111} plane toward the strate-substrate interface.

For the automotive diodes a double painting technique is used, comprising B-Al and P liquid sources. The diffusion conditions are: 1210^0 C or $1255\ ^0$ C, 40 or 24 hours , N_2/O_2.

The rocking curves obtained allow us to isolate two XR profile types for the boron diffused side (figure 3a and 3b), while for phosphorus diffusion all XR profiles were similar to the one in figure 3c. The two peaks in boron profiles stands for the reflexion on substrate (the weaker one) and on diffused layer (the larger one). The boron diffusion exhibit obvious dissimilarities in the two presented case. The distance between the two peaks, greater in the case (a), is the result of a deeper strained layer. By using the values extracted from the gaussian spectral fits, the medium elastic deformation, ε, normal to surface was calculated as $8.40\text{x}10^{-4}$ in the case (a), and $3.81\text{x}10^{-4}$ in the case (b). This medium deformation ε is related to the misfit factor m and a proportionality factor which varies between 1 and (1+ν)/(1-ν), where ν is the Poisson coefficient . Thus, the multiplication of ε value for (b) profile by a factor of two might be

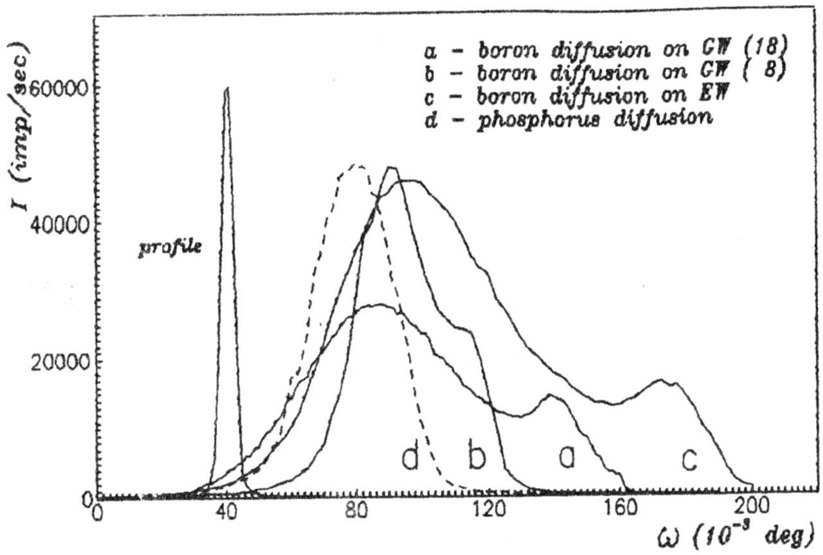

Figure 3. Rocking curves obtained by XR diffraction on diffused wafers

explained by an increase in N_c from 4 to $8x10^{19}$ cm^{-3}. A comparison of the absolute values of ε and m leads us to conclude that in both cases the stress did not relaxed by dislocation generation. Moreover, (a)-type samples present the concavity at boron diffused side, while in the (b)-type samples, a reverse curvature appears. So, a coupling between B and P concentration must be taken into account. It is likely to suppose that the subsequent stress will enhance the diffusion induced strain until the elasticity limit or even to crystal fracture, with larger probability in (a)-type samples.

6. CONCLUSIONS

The causes leading to 20% failures in operation of CAN automotive diodes were investigated. The tool for reliability investigation was established, the soldering system of internal connection wire was redesigned and a diminution to 4% failures was achieved. For diminishing the failures below 4% the MESA processed chip strength is to be increased. The surface preparation was found to have great influence over the internal strain below the surface of wafers. The technological conditions of B diffusion, or moreover, the coupling conditions between B and P diffusions seem to play an important role in the strong strain induced in the silicon crystalline substrate. Further investigations are needed for quantitative estimation.

IR and XR were found as being valuable methods of perturbations characterization in the crystalline silicon case for the depth and strength of the coarse damaged layer and elastical-plastical disturbed layer, respectively.

Acknowledgements:

The authors are very much indebted to M.Udrea for many helpful discussions and to A.Miut for technological advice.

REFERENCES

1. O. T. McCarter, General Motors Corp. Doc2316, AOD01NO1
2. M. Udrea et al., Reliability Report, SC Baneasa SA, 1995
3. G. Kortum, "Reflectance Spectroscopy", N.Y., 1969
4. A. D. Stoica, M. Stoica, C. Bozdog., to be publ.
5. M. Udrea et al., Internal Report, SC Baneasa SA, 1995
6. A. Fukuhara, Y. Takano, Acta Cryst. A38,137, 1977
7. D.Sachelarie,M.Stoica,M.Sachelarie,M.Tazlauanu,A.Diaconu, MIEL'85, Liubljana,1985

CAN YOU DELIVER THE SEMICONDUCTOR DEVICES AFTER PERFORMING PERIODIC TESTS ?

Maria Dragan and Marius Bazu
Research Institute for Electronic Components (ICCE)
Str. Erou Iancu Nicolae 32B, 72996, Bucharest, Romania

SUMMARY

A method to detect the character of a periodic test, destructive or nondestructive, was developed. This method allows the delivery of those semiconductor devices which undergo nondestructive periodic tests. Beneficial effects for the device manufacturers are obvious.

1.INTRODUCTION

A manufacturer of semiconductor devices must perform each year and for each type of device periodic tests on 400-500 items. Afterwards, these devices are not delivered because one cannot say if the tests are destructive or not. For small serie production or for expensive types, the loose of these devices can be economically significant. In this paper, a method to detect the character of a periodic test (destructive/ nondestructive) is presented. With this method, a manufacturer of semiconductor devices can diminish the cost of the periodic tests.

2.METHOD

A test is considered to be destructive if it modifies the specified properties or appearance of the tested device or if it stresses the device in such a way that the intrinsic quality and/or the opperational reliability of the device are likely to be reduced after the test/1/. In practice, if it is performed at a value not exceeding an absolute limiting value and if the aspect of the device is not affected , such a test is destructive 'if the same test cannot be performed again on the same device with the same probability of success as before the first test was performed' /1/.

Consequently , a method to detect the character of a periodic test (and of a lot by lot test, too) must compare the probability of success at the first and at the second performance of the test.

The probability of success was defined in the hypothesis that the electrical parameter values are normally distributted , with mean (m) and standard deviation (s). The probability for a parameter value to be within the interval (LSL,USL), where LSL = Lower Specification Limit and USL = Upper Specification Limit, is given by:

$$P=P(x\leq(USL-m)/s)-P(x\leq(LSL-m)/s), \text{ with } x\sim N(0,1). \quad (1)$$

0-7803-3223-7/96/$5,00

The probability of succes is calculated at the beginning of the test (P_i), after the first (P_I) and after the second performance of the test (P_{II}). A test is considered destructive if:

$$(P_I - P_{II})/(P_i - P_I) > 2 \qquad (2)$$

3.EXPERIMENT

Four types of semiconductor devices (rectifier diode 1N 4001 in F 126 case, switching diode 1N 4446 in DO 35 case, NPN transistor BC 107 in TO 18 case and NPN transistor BC 171 in TO 92 case) undergo mechanical and climatic tests (robustness of terminations, resistance to soldering heat, vibration followed by acceleration-steady state,salt mist, damp heat and storage at high temperature). Each test was performed twice, on samples of 18 or 25 items. After each performance of a test, the devices were electrically characterized.

The measured electrical parameters were: i) forward voltage (V_F) and residual current (I_R) for the diodes, ii) current gain (h_{21E}) and residual currents (I_{CEO} and I_{CES}) for the transistors.

4.RESULTS AND DISCUSSION

At the beginning of the tests, the residual currents have very low values. Consequently, the probability of success P_i is very closed to 1 (between 0.99972 and 1). After testing, P was not bellow 0.9983 (value obtained only for the rectified diode 1N 4001, at resistance to soldering heat). Therefore, the residual currents cannot be used as degradation indicative.

The forward voltage of the diodes has P=1 initially and also after each test, because the standard deviation is very small.

The current gain of the transistors is much more sensitive to the degradation of the devices at periodic tests. For the transistors BC 107 and BC 171, h_{21E} was measured for two conditions: h_{21E}^1 (I_C= 10μA, V_{CE}= 5V) and h_{21E}^2(I_C=2mA, V_{CE}=5V). For BC 171 the periodic tests where the drift of the current gain was significant were the storage at high temperature (T=150°C) and the damp heat (T=85°C, RH=85%).

The drift of the current gain Δh_{21E} for the transistor BC171, at storage at high temperature and at damp heat, is presented in figures 1 and 2, respectively. At I_C=10μA, the current gain h_{21E} is more sensitive to the quality of the structure surface. The shape of the drift curves can be explained as follows : i) at the storage at high temperature, the surface of the structure is stabilized by the heat and, consequently, after an initial increase, the drift is saturated at 25% (see figure 1), ii) at the damp heat, the humidity requires a certain time to attain the structure surface (500 hours, in figure 2), afterwards the increase is accelerated.

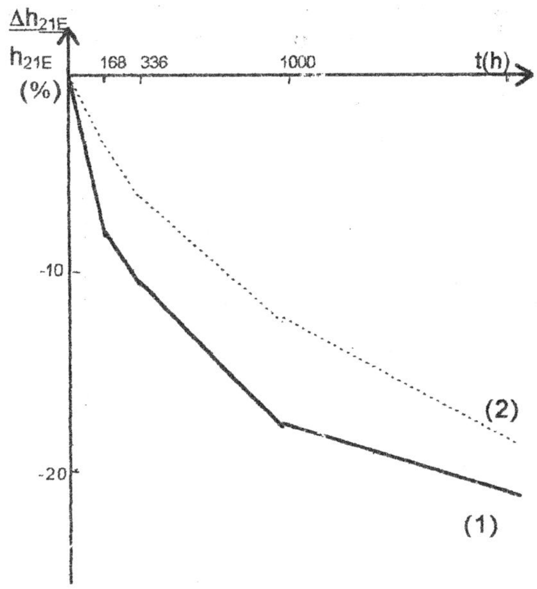

Figure 1

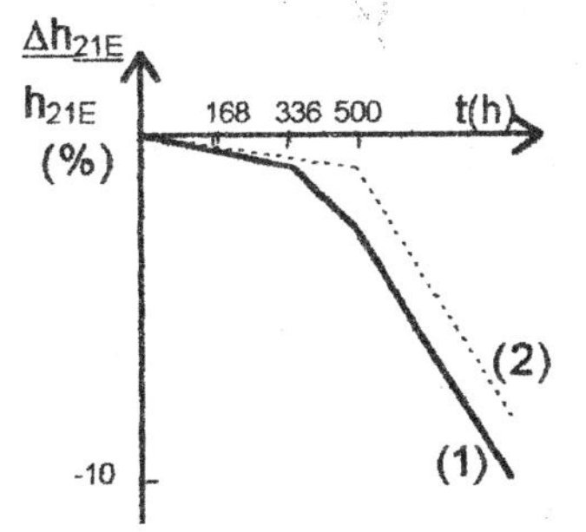

Figure 2

The probabilities of success were calculated and the results obtained for BC 171 at storage at high temperature and at damp heat are presented in tables 1 and 2, respectively.

Table 1. The probability of success for BC 171, at storage at high temperature

t(h)	0	168	336	1000	2000
P	P_i			P_I	P_{II}
$P(h_{21E}^1)$	0.986	0.988	0.989	0.992	0.985
$P(h_{21E}^2)$	0.977	0.978	0.978	0.973	0.963

Table 2. The probability of success for BC 171, at damp heat

t(h)	0	168	336	500	1000
P	P_i			P_I	P_{II}
$P(h_{21E}^1)$	0.963	0.969	0.966	0.967	0.963
$P(h_{21E}^2)$	0.977	0.977	0.976	0.976	0.965

The durations of the tests were 1000 hours for the storage at high temperature and 500 hours for the damp heat. For the first performance of the test, intermediate results at 168h (duration of the lot test) and 336h were also given.

With relation (2), the condition for a test to be considered destructive was verified for the data from the tables 1 and 2. The results are presented in table 3.

Table 3. Determination of the destructive character of a test

Periodic test	Electrical parameter	$(P_I-P_{II})/(P_i-P_I)$	Character
Storage at high temperature	h_{21E}^1 h_{21E}^2	-1.17 2.5	N D
Damp heat	h_{21E}^1 h_{21E}^2	-1 11	N D

N=nondestructive test, D=destructive test

The parameter h_{21E}^2 was more affected by the tests because the initial values were closed to LSL.

In the situation presented in table 3, both tests are destructive, because the rule is that a test is considered to be destructive even if only one electrical parameter (h_{21E}^2 in this case) sustains this character.

All the other periodic tests for the rest of devices were found to be nondestructive, therefore the tested devices can be delivered.

5.CONCLUSIONS

A method to detect the character of periodic tests, destructive or not, for semiconductor devices was presented. With this method, the devices which passed through nondestructive tests can be delivered, diminishing the cost of the periodic tests. Consequently, the method has beneficial effects for the manufacturers of semiconductor devices.

6.REFERENCES

1.47/1389/FDIS, IEC, October 1995.

THE VARIATIONS OF InP SINGLE CRYSTAL MECHANICAL PROPERTIES INDUCED BY ELECTRON IRRADIATION

D.Grabko, M.Medinskaya, N.Pyshnaya, I.Tiginyanu

Institute of Applied Physics, 2028 Kishinev, Moldova Republic

SUMMARY

The influence of electron irradiation (E_{ir}= 3.5- 4 MeV, D = $5\cdot10^{15}$ - $3\cdot10^{17}cm^{-2}$) on InP:Fe and InP:Zn single crystal microhardness (H) was investigated in the present work. It was established that microhardness dependence on irradiation dose has a non-monotonous character. The presence of three regions at H(D) curves is explained by internal defect structure modifications.

1. INTRODUCTION

Structural defects appearing in crystals after irradiation affect essentially upon their main physical properties including mechanical ones [1,2]. The modification of these properties as a function of irradiation type and dose gives an important information about the type and the concentration of generated point defects. In this connection present work describes an electron irradiation influence on microhardness (H) of semi-insulating InP:Fe and p-type InP:Zn single crystals.

2. EXPERIMENTAL PROCEDURE

Liquid encapsulated Czochralsky grown InP single crystals were used in experiments. Their room temperature parameters were: $n = 2\cdot10^{8}$ cm^{-3}, μ = 1700 cm^{-2}/V s for InP:Fe , while $p = 2.8\cdot10^{17}$ cm^{-3}, μ = 130 cm^{-2}/V s for InP:Zn. An electron irradiation was performed at target temperature maintained near 70^{0}C and beam energy of 3.5 - 4 MeV. Irradiation doses were varied from $5\cdot10^{15}cm^{-2}$ to $3\cdot10^{17}cm^{-2}$.

The (001) crystal face was deformed at room temperature by

0-7803-3223-7/96/$5,00

using the concentrated load (indenter - Vickers diamond pyramid). The microhardness tester PMT-3 and usual formula

$$H = 1.854\ P/d^2$$

were used for microhardness determination by penetration method [3]. Here P is the indenter load, d is the size of indentation diagonal. Indenter loads were egual to 20, 40 and 100 g.

3. RESULTS AND DISCUSSION

It was established that the electron irradiation leads to the variation of InP:Fe and InP:Zn single crystal microhardness. An example of H(D) dependence for d || <100> indenter orientation is presented in the Fig.1.

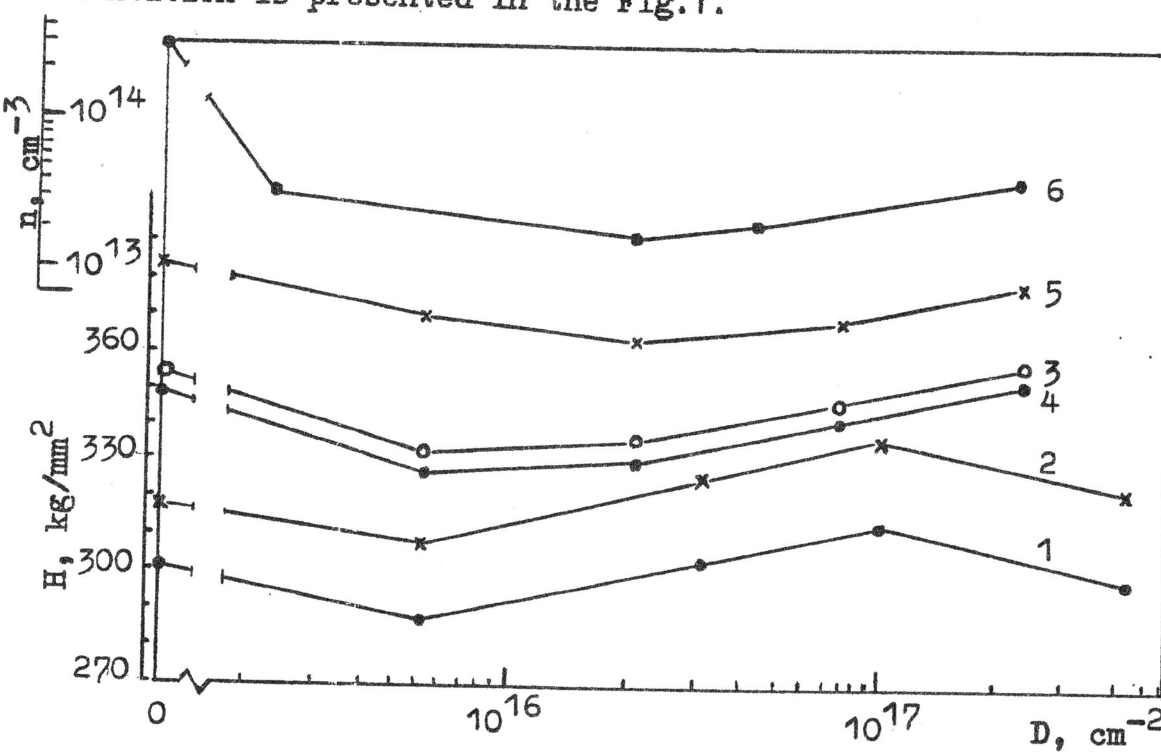

Fig.1. Microhardness dependence on irradiation dose for InP:Fe (1,2), InP:Zn (3-5) single crystals. P, g: 1,4 -40; 2,5 - 100; 3 - 20; 6 - free electron concentration at T = 300 K for electron-irradiated n-InP.

It is evident that all the curves are not monotonous. The curve for InP:Fe crystal may be divided into three monotonous parts (curv.1,2). At small irradiation doses microhardness values diminish comparing with the same for unirradiated samples (this means the softening). But microhardness increases in the dose interval $(5\cdot10^{15} - 8\cdot10^{16})$ cm^{-2}, exeeding H value for the as grown sample. At last a microhardness decreases again at irradiation doses from $8\cdot10^{16}$ to $3\cdot10^{17}$ cm^{-2}. As it is seen from Fig.1, H versus D functions behave analogously for loads of 40 g (curve 1) and 100 g (curve2).

As far as InP:Zn single crystals are concerned, only two monotonous parts appear at H(D) curve in the investigated dose interval. It is especially typical for 20 g and 40 g loads. Perhaps the softening stage will take a place at $D > 1\cdot10^{17} cm^{-2}$.

It is interesting to note that according to [4] (see Fig.1) free electron concentration curve versus irradiation dose for n - InP epilayers also has a minimum. Let's note that this minimum is observed at small doses too. The electron concentration decrease in the interval to $(1\cdot10^{15} - 2\cdot10^{16}) cm^{-2}$ was explained in [4] by the residual impurity connection with radiation defect centers. When dose increases from $2\cdot10^{16}$ cm^{-2} to $1\cdot10^{17} cm^{-2}$, donor and acceptor type defect accumulation occurs, donor type defects being predominant. This process leads to free electron concentration increase.

Such explanation correlates with the results obtained in the present work. It may be supposed that an interaction between impurity atoms with the radiation defects leads to the neutralization of some part of formers. In this case the resistance to the dislocation motion by residual defect centres will become lower. This will result in hardness diminution at small irradiation doses.

In contrast, due to the enhancement of charged defect concentration at doses to 10^{17} cm^{-2} moving dislocations will be braked more effectively, this will cause a microhardness increase.

Microhardness diminution in InP:Fe single crystals at the highest irradiation doses may be explained by creation of different point defect associates. Then the effective concentration of charged defect formations decreases.

Thus the microhardness value variation is a reflection of the changes in InP structural defect ensemble induced by fast electron bombardment.

CONCLUSION

1. The electron irradiation of InP:Fe and InP:Zn single crystals leads to the variation of their mechanical properties (microhardness and dislocation mobility).

2. H versus D curves have non-monotonous character.

3. The microhardness diminuition at small irradiation doses ($1\cdot10^{15}$ - $5\cdot10^{15}$ cm^{-2}) may be explained by the connection of residual impurity centers with radiation defect ones.

The H increase in a dose interval from $5\cdot10^{15}cm^{-2}$ to $1\cdot10^{17}cm^{-2}$ may be connected with an intensive accumulation of donors and acceptors generated by irradiation.

REFERENCES

1. F.P.Korshunov, Yu.Bogatyrev and V.A.Vavilov. Effect of Radiation on Integrated Circuits, Minsk, 1986, 254 p.(in Russian)

2. M.G.Milvidskii, V.B.Osvenskii. Structure defects in semiconductor single crystals, Moskow: Metallurgia, 1984, 255 p. (in Russian)

3. Yu.S.Boyarskaya, D.Z.Grabko and M.S.Kats, Physics of microindentation process, Kishinev: Shtiintsa, 1986, 294 p. (in Russian)

4. N.B.Pyshnaya, I.M.Tiginyanu and V.V.Ursaki. Reduction of the degree of conductivity compensation in epitaxial n-InP films by fast-electron bombardment //Semiconductors, 28, 1, 1994, p. 1-3.

DEVELOPMENT OF THE AMORPHOUS, HIGH RESISTIVITY Ge:(O,N) FILMS FOR RADIATION-HARDENED MIS DEVICE APPLICATIONS

D. Jishiashvili, Z. Shiolashvili, R. Janelidze, V. Gobronidze,
E. Kutelia, L. Mosidze, I. Nakhutsrishvili, M. Katsiashvili

Institute of the Cybernetics Academy of Sciences of the Georgia.
5, Euli Str., 380086, Tbilisi, Georgia.

SUMMARY

Application of amorphous, high resistivity O- and N- doped Ge films is proposed to solve some radiation hardness problems. The new type of insulating a-Ge:(O,N) films makes it possible to increase the radiation hardness of metal-insulator-semiconductor (MIS) transistors at least by 4 orders of magnitude. Fabrication of MIS integrated circuits on the basis of a-Ge:(O,N) films can enlarge the utilization of these advantageous devices in radioactive environment (space, nuclear power-plants etc.)

INTRODUCTION

A significant limitation of current MIS devices is their susceptibilities to ionizing radiation. While the mechanism of damage is reasonably well understood, a rational approach to producing radiation-hardened MIS devices has yet to emerge.This is an important issue for a wide range of scientific and industrial applications, including instrumentation in nuclear power reactors, medicine and space systems.

The main advantages of MIS ICs as compared with bipolar ones are: low energy consumption, high system complexity, low cost and relative simplicity of the MIS technology. However, under ionizing irradiation the MIS device characteristics strongly degrade due to charge generation and trapping in the traditionally used insulating films (SiO_2, Si_3N_4, Al_2O_3). Low radiation hardness is the main limiting factor for utilization of advantageous MIS ICs in radioelectronic systems which are operating in radioactive environment.

A variety of charge instabilities in MIS structures are known to occur under ionizing irradiation (ionic, dipolar, electronic). As a result, the fixed charge increases in the bulk of the dielectric films and new surface states appear at the insulator-semiconductor and metal-insulator interfaces. In fact, the insulator turns into conducting the so called electretic state which is characterized by long relaxation times. In such conditions the MIS transistor becomes fully insensitive to applied gate voltage. Therefore, it is impossible to create radiation-hardened MIS IC on the basis of traditional insulating films. Various improvements, such as addition of auxiliary devices, decreasing SiO_2 films thickness etc. make possible to increase the radiation hardness only up to 10^5-10^6 rad.

The purpose of this work is to develop an essentially new type of insulating material for radiation hardened MIS ICs.

0-7803-3223-7/96/$5,00

EXPERIMENTAL

The films of amorphous O- and N-doped Ge [a-Ge:(O,N)] were deposited on n- and p-type Si (N_D=1.3 x 10^{15}cm^{-3}, N_A=3.0 x 10^{15}cm^{-3}) by triode reactive sputtering in mixture of Ar, O_2 and N_2 at the total pressure of 3 x 10^{-3} Torr. During the deposition process the substrate temperature was held at 150°C. The chemical composition of the films was determined by Auger electron spectroscopy to an accuracy of $\pm$5 at.%. Capacitance-voltage (c-v) characteristics were taken at 1 MHz and the voltage ramp of 0.5 V/s.

Radiation testing was performed under the condition which were established by the military standards of the FSU.

RESULTS AND DISCUSSION

As it was mentioned above, radiation induced instabilities in traditional dielectrics are caused by long relaxation time processes. Bradley [1] proposed to eliminate these kind of instabilities by the substitution of an insulating film for a crystalline semiconductor film with a wide energy gap and a high resistivity. He proceeds from the assumption that long relaxation time processes will be compensated by movement of free carriers whose relaxation times are significantly shorter. Subsequently this kind of MIS structures was realized on the basis of such crystalline semiconductor films as ZnP_2, Ga_2Se_3, GaAlAs, CdP_2, Ga_2S_3 grown on Si and GaAs [2-5].

For normal operation of these MIS structures the following criteria should be satisfied:

- the thickness of the I-layer should be substantially less than the effective screening distance. In this case the capacitance of the I-films will remain constant at all measuring conditions thus insuring surface conductance modulation of the semiconductor;
- the density of states at the I-S interface should be minimal (10^{10}-10^{11}cm^{-2}eV^{-1});
- to form the true inversion region the leakage current through the I-layer should not exceed the current generated at the surface of the semiconductor.

However, the application of crystalline semiconductor I-films is connected with difficulties in crystal growth processes and impossibility to vary the composition of the I-films and thus minimize the density of surface states at the I-S interface.

Our approach is a further extension of Bradley's theoretical assumption. We suppose that high resistivity amorphous doped semiconductor films can serve as an insulator films in MIS devices, provided that all the above mentioned criteria are satisfied.

Amorphous doped semiconductors seem to be a more flexible material as compared with the crystalline semiconductor films with their fixed composition and fixed interatomic distance. Indeed, the deposition process of an amorphous film is much easier. Besides, the concentration of a dopant in amorphous semiconductor can be varied to modify their physical properties and minimize the density of interface states and traps.

We select amorphous O- and N-doped Ge films as an insulator for Si based MIS structures and transistors. The composition of the film was not optimized and all the experimental results presented below are obtained for the films with the fixed composition $Ge_{0.3}N_{0.5}$ and the resistivity of (5-7) x 10^8 Ohm·cm.

Fig.1 shows typical 1 MHz c-v characteristics of Al/a-Ge:(O,N)/Si MIS structures and the corresponding theoretical curve. Some samples exhibit clockwise hysteresis with the maximal width of 0.2 V due to uncontrollable deviations in the film deposition technology.

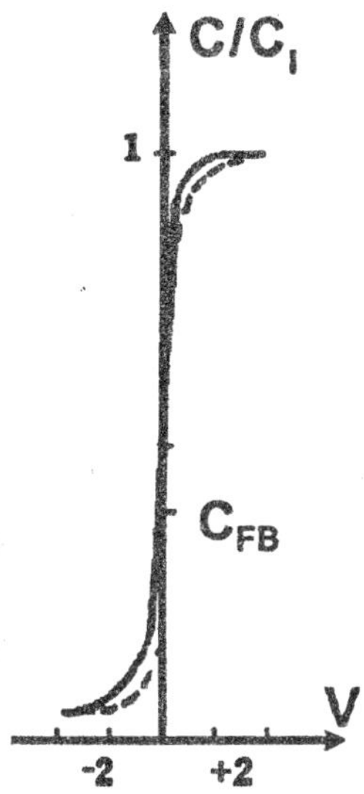

Fig.1.Theoretical (the dashed line) and experimental C-V curves of Al/a-Ge:(O,N)/Si structure

Comparison of the theoretical and experimental curves indicates that the inversion-to-accumulation transition is very sharp. No bulk traps are observed in the insulator film. The experimental characteristics are nearly parallel to the theoretical curve and slightly shifted along the voltage axis. It is clear that this shift is caused by a positive charge fixed in the insulator and that the interface itself is relatively perfect. Calculated flat-band charge is low and for different samples varies in the range of (8-11) x $10^{10}cm^{-2}$. The experimental and calculated minimum inversion capacitance values are in good agreement, as is illustrated in Fig.1. This fact along with the low surface state density and MIS field effect transistor (MISFET) characteristics strongly suggests that true inversion and accumulation have been reached at the Si surface covered with the a-Ge:(O,N).

No changes of the flat-band voltage values were observed after bias-temperature stress (300°C, $\pm 5 \times 10^5$V/cm, 10 min) thus confirming high stability of our films.

All these prove that a-Ge:(O,N) forms a high quality interface with Si and can serve as a good insulator layer in MIS devices. On the basis of the a-Ge:(O,N) film, n- and p-channel enhancement mode MISFETs were fabricated with the nominal gate dimensions (length x width) of 5 x 50 μm^2 and the threshold voltages (V_T) of -0.7 and +1.0 V, respectively.

No V_T shift was observed during the radiation testing of MISFETs under the condition established by the military standard "Klamat 6" which includes:

1. n-γ irradiation of the nuclear power-plant - 5 x 10^{14} neutron/cm², 5 x 10^6 rad (γ);

2. n-γ irradiation of the nuclear explosion (pulse irradiation) - 5 x 10^{14} neutron/cm², 5 x 10^7 rad (γ);

3. Imitation of space radiation fields:

a) alpha-particle irradiation with E = 8-10 MeV - 10^8 rad;

b) electron irradiation with E = 4.5 MeV - 10^9 rad;

Application of higher doses and fluencies was limited by unreasonably long exposure times and technical complexity. During these experiments the gate electrodes of MISFETs were opened.

As is generally known, the radiation damage is most severe with positive gate bias applied during the exposure. In our experiments the V_T shifting was evaluated only for ± 5 V gate bias applied during the electron bombardment at a fluency level of 3 x 10^{16}

(4.5 MeV electrons)/cm^2 - (10^9 rad). Under this conditions the V_T shift does not exceed (5-8) % of its initial value. During a similar irradiation the SiO_2 based MOSFET fails (i.e. enhancement mode device becomes of depletion type) with +5 V applied bias at a fluency level of 10^{12} e/cm^2 (3×10^4 rad). As is generally known, the maximum absorbed doses for SiO_2 based devices fabricated by the radiation-resistant technology does not exceed 10^6 rad.

CONCLUSIONS

In conclusion, the a-Ge:(O,N) films show promise of considerable radiation hardness. Utilization of these films in MIS devices offers an increase in hardness by 4 orders of magnitude over that of conventional devices with SiO_2.

The results reported in this paper clearly show two major achievements:

1. The experimental verification of the theoretical assumption that amorphous, high resistivity semiconductor film can serve as an insulator layer in MIS devices.
2. The first demonstration that with the utilization of a-Ge:(O,N) films radiation hardness of MIS IC can be significantly increased (by 4 orders of magnitude as compared with the SiO_2 based devices).

The purpose of this work was to show the capabilities of a new type insulator material. Considerable work remains to be done, especially in deposition technology and radiation testing.

The deposition technology should be adjusted to meet the needs of the conversional MIS IC fabrication processes. Further systematic investigation are to be carried out to improve the properties of the a-Ge:(O,N) films and to optimize their composition. This will allow to fabricate radiation-hardened MIS IC and enlarge their application in space electronic, safety and control system of the nuclear power stations, etc.

REFERENCES

1. J. Bradley. Electron Lett., 3, 526 (1967)
2. Б.И. Сысоев, В.Ф.Сыноров. Физ. и тех. полуп. 6, № 10, 1856, 1972.
3. Б.И. Сысоев, В.Ф.Антюшин, В.Д.Стрыгин. Физ. и тех. полуп. 18, № 10, 1739, 1984.
4. Э.А. Ильичев, Э.А.Полторацкий. Физ. и тех. полуп. 20, № 10, 1782, 1986.
5. Б.И. Сысоев, Н.И.Безрядин. Микроэлектроника. 9, № 2, 121, 1980.

ON THE NATURE OF LEAKAGE CURRENT OF FAST RECOVERY SILICON pn JUNCTIONS

V.V.N. Obreja, Gh. Dinoiu

ICCE

Abstract Some experimental facts related to electron irradiated or gold doped silicon PN junctions are presented . After junction irradiation, the reverse recovery time decreases more than order of magnitude but the reverse current increase is not more than twice near room temperature. The behavior of such junctions from the room temperature up to 320 ^{o}C reveal that present-day manufactured fast recovery junctions, may be dominated by a higher peripheral surface leakage current than for similar standard junctions, both at reverse bias and low forward bias.

1. INTRODUCTION

It was shown, [1] , that a fast recovery silicon diode has a significantly higher reverse leakage current than a similar standard diode. The leakage currents were measured at a junction temperature value, T_j = 180 ^{o}C since at the room temperarture the surface leakage was dominant. According to [1] it is understood that a significant higher reverse leakage current for fast diodes is due to a large number of generation - recombination centres which intentionally are introduced in the junction bulk depletion layer. An analysis of present - day catalogue data of leading companies in the field, reveals that a fast recovery diode has at higher T_j values a reverse current well in excess of that corresponding to a similar standard diode. The maximum operating junction temperature, $T_{j\ max}$ for fast diodes do not exceed 150°C whereas the coresponding values for standard diodes may reach 175-190°C. Otherwise for the same total power dissipation a fast recovery diode may suffer a catastrophic failure whereas a similar standard diode needs an additional increase of its dissipated power in order to be degraded. It was also shown a long time ago, [2], that a fast recovery diode may have at low forward bias (0÷0,5 V) significantly higher values of forward current than a normal diode. The ideality factor may reach values as high as 2. Such a behavior is also attributed to a high density of recombination centres in the junction bulk space charge region.

The purpose of this work, is to show that some experimental facts cannot be understood if the leakage current of fast recovery diodes is primary governed according to the known theory, by junction bulk generation-recombination phenomena.

2. EXPERIMENTAL RESULTS. DISCUSSION

Silicon N^+PP^+ or P^+NN^+ mesa passivated rectifier structures with a breakdown voltage value in a range of 140-2000 V were 5 - 7 MeV electron irradiated at a dose in excess of 1 Mrad. Before irradiation a reverse recovery time, t_{rr} , more than 30 µs has been measured for structures having a breakdown voltage greater than 1000 V. After electron irradiation, values less than 2 - 2,5 µs have been found for t_{rr} and a slight increase of reverse leakage current (no more than two times) has been observed at the room temperature and low bies voltage, V_R = 1-10 V. No change in the breakdown voltage value has occured excepting a softer knee of the current-voltage characteristic. Nevertheless, at higher T_j values a significant excessive reverse leakage current has been observed after electron irradiation. Typical experimental results are shown in Fig.1. The temperature dependence No.1 has been

0-7803-3223-7/96/$5,00

obtained for an irradiated silicon rectifier diode structure which before irradiation had trr = 35 μs and after irradiation a trr = 2 μs has been measured at the room temperature. One can see from Fig.1 that the dependence No.1 exhibits below Tj = 150°C activation energy (Ea) values of approximate 0,7 eV and 1.20 eV. At higher T_j values towards 200 - 250°C, an Ea = 0.97 eV is constantly obtained but over 250°C again a lower Ea value is possible. After the dependence No.1 has been completed by reaching T_j = 300 - 320°C, the N^+PP^+ structure has been stored about 8 hours at this temperature. After this storage time, a temperature dependence No.2 have been obtained for the same structure. Now Ea = 0.62 eV below 100°C whereas above this temperature, constantly, Ea = 0.93 eV. It has been found that trr has increased from 2 μs to 16 μs after high temperature storage while the reverse current, I_R , may have a lower value than the preirradiation one near room temperature but not at high T_j. The dependence No.3 is typical for nonirradiated similar structures. One can see that this dependence has E_a= 0,76 eV above 200 °C whereas below this temperature, 2 portions of ≈ 0,6 eV are wisible. Between these portions again there exists one with a higher activation energy, Ea≈ 0,8 eV. In Fig.1 it is nòt shown the temperature dependence corresponding to the No.3 dependence after high temperature annealing of the noniradiated structure. We have found that such a dependence has Ea ≈ 0.8 eV at high T_j values and Ea ≈ 0,6 eV towards the room temperature. At lower T_j values, a significant reduction of reverse current has taken place ,[3], but no change in the trr value (≈ 40μs) has been observed after 10 hours storage time.

Such experimental facts give rise to some doubt regarding the bulk nature of leakage current in fast recovery silicon PN junctions. Apparently, the significant reduction of reverse leakage current after the high T_j storage time, could be attributed to a decrease of the density of generatin recombination-centres in the bulk depletion region of irradiated structure. This seems to be supported by the increase of trr from 2μ s to 16 μs. Nevertheless if we compare the typical dependences No.1 and 2 for the irradiated structure with the dependence No.3 which is typical to a non- irradiated structure, then one can observe that towards the room temperature, a nonirradiated and a irradiated structure may have comparable reverse currents at low V_R although the values of trr differ at least with an order of magnitude. The activation energy value remains almost unchanged near the roon temperature. At high Tj values one can see that the three temperature dependences have Ea less than 1.0 eV though, the reverse current tend to reach saturation, [3]. Especially for the No.3 dependence, Ea would have to reach 1.2eV because of the expected bulk diffusion component dominance while such a value appears for the No.1 curve below 100°C, where a dominance of the bulk generation current component is expected. Such a behaviour is in agreement with recent results for advanced silicon N^+P planar junctions, [4] , where an E_a around 1.0 eV has been found.at high T_j and 0.5 eV below 80 °C when the junction surface component is dominant. Even a slope E_a = 1.2 eV exhibited at high T_j , [5] , cannot be used as undoubt experimental evidence for the bulk diffusion current because such a behavior may be caused by insufficient controlled junction surface phenomena.

A reduction of reverse current after high temperature storage of an electron irradiated PN junction seems to be an effect of surface phenomena from the junction peripheral interface as in the case of nonirradiated junctions, [3]. Such a conclusion is supported by recent work, [6] , regarding the behavior of the drain junction of γ irradiated power MOSFETs after thermal annealing, when a reduction of interface states has been observed. Gold doped diodes having trr around 1μs have also been investigated. The curve No.4 and 5 are typical examples. Again one can see at lower T_j values that Ea = 0.55 - 0.68 eV whereas at elevated Tj , Ea reaches 0.9 - 1.0 eV. This time, no change of t_{rr} have occured after 10 hours storage time at T_j = 300°C whereas a significant increase of I_R has occured at higher V_R values. The dependence No.6 (the dotted curve) is for a standard recovery low power rectifier diode, having a similar semiconductor structure as for the No.4 diode but with trr = 32μs instead of 1 /us. One can see an intermediate portion with a higher Ea value (1.0 eV) than at T_j values

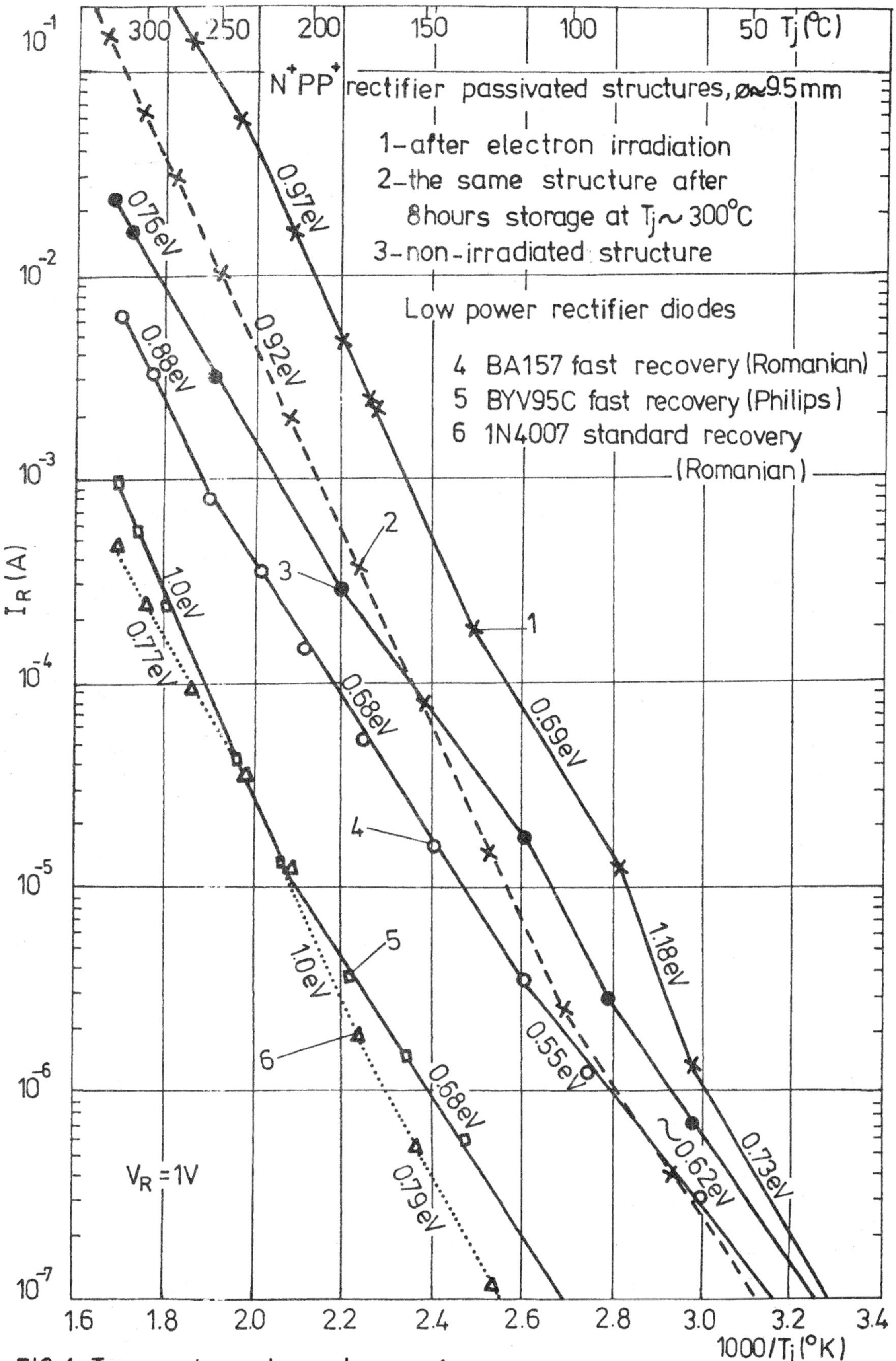

FIG. 1. Temperature dependence of reverse current.

above 250°C. It has to be mentioned that a Philips diode used for the curve No.5, is glass passivated whereas the junctions used for the other dependences were silicone rubber passivated.
It has been observed that to a reduction or an increase of I_R for fast recovery junctions, corresponds also a reduction or an increase, respectively, of forward current at low bios voltage (0 - 0.5 V) as in the case of standard junctions. This is valid not only at the room temperature but aven at high Tj values.

As a consequence of the above experimental results, we are tempted to think that the leakage current of fast recovery silicon PN junctions is dominated as for the standard diodes, [3-5] , by the surface peripheral component both at reverse bias and low forward bias (under 0.5 V). Thus the surface nature of excessive reverse current of fast junctions would explain their poor stability at high T_j values. It remains a decisive experiment to be carried out to validate such a conclusion. This is related to a comparison of leakage currents of junctions with the same perimeter but significat different areas.

3. CONCLUSION

The bulk nature of leakage current of a fast recovery silicon PN junction both at reverse and low forward bias due to a large density of generation-recombination centres in its depletion region, is not in agreement with some experimental facts. Thus after the electron radiation of a silicon rectifier passivated structure with a dose of 1Mrad, a decrease of t_{rr} with more than an order of magnitude has occured. Nonetheless, the reverse current at V_R = 1 - 10 V has increased no more than twice near room temperature where a dominance of junction bulk generation component is expected. I_R has increased with more than an order of magnitude at higher T_j values towards 200 - 300 ° C where a tendency to saturation is visible but the activation energy E_a is not greater than 1.0 eV. After at least 5 - 8 hours annealing time at T_j = 300 - 320 °C of an irradiated structure, an increase of t_{rr} up to a half of the preirradiation value occurs. This is accompanied by a significant reduction of I_R which near room temperature may reach values less than the preirradiation ones . At higher T_j , I_R reduction does not reach the preirradiation values and E_a may have a slight decrease up to the preirradiation value of 0.8 - o.9 eV. For gold - doped or standard junctions no change in the t_{rr} value occurs, although a significant change of I_R is possible after high T_j annealing. The surface nature of I_R for fast recovery junctions would explain lower permissible values of T_{jmax} than for similar usual junctions.

ACKNOWLEDGMENT

Part of of this work has been financially supported by the grant No. 3793 / 1995 awarded by Romanian Academy which is gratefully acknowledged. Thanks are especially due to Mrs. E. Iliescu from IFTAR - Bucharest for her support with the electron radiation facility.

REFERENCES

[1] B.J. Baliga et al. , Comparison of Gold, Platinum and Electron Irradiation for Controlling Lifetime in Power Rectifiers, *IEEE Trans. ED-24, 1977, 685*

[2] C.T. Sah, Effect of Surface Recombination and Channel on PN Junction and Transistor Characteristics, *IRE Trans. on Electron Devices, January, 1962, 94*

[3] V.V. Obreja, Experiments on Surface Peripheral Leakage Reverse Current of a High Voltage Silicon PN Junction, *Proceedings CAS' 93 , 265*

[4] M. Horiuchi et al. , Reduction of PN Junction Leakage Current by Using Poly-Si Interlayered SOI Wafers, *IEEE Trans. ED-42, May , 1995, 876*

[5] V.V. Obreja, Experimental Disproofs Related to the Current-Voltage Characteristic Theory of a Semiconductor PN Junction, *Proc.20th Intern. Conf. on Microelectronics(MIEL),1992,Slovenia, 169*

[6] A. Jaksic et al., Effects of γ Irradiation and Postirradiation Thermal Annealing in Power VDMOSFETs , *Proc. MIEL' 95, Nis, Serbia, 1995, 249*

THE NATURE OF M-InP CONTACTS AGING

G.S. KOROTCHENKOV, V.A. MICHAILOV, V.A. BLAJE

Technical University of Moldova, Laboratory of Microelectronics.
Bld. Stefan cel Mare, 168, Chisinau, 2012, Moldova.
Tel.:(373)-2-247143. Fax: (373)-2-237509.

Summary

In this report the causes of low temperature aging of M-InP contacts (Au, Ag, Cu, Al, Ni) were examined. On the basis of analysis of M-InP contacts Auger-profiles a conclusion on dominated processes, taking place at the M-InP interface was drawn. We had found that M-InP Schottky barriers aging was defined by proceeding of diffusion processes at the interface. The main of these diffusion processes is indium diffusion from InP into metal and intermediate layer.

1.Introduction

The main processes that can be met at the interface and can influence the temporal stability of Schottky barriers parameters are processes of mutual diffusion at the interface, ingress of oxygen through the metal coating with creation of oxide pellicle at the interface, and chemical reactions changing physical-chemical properties of contacting materials [1,2]. As it was shown in [1], for M-Si Schottky diodes which were keeping at the room temperature, the most essential process was the oxygen diffusion through the metal. At the same time for GaAs Schottky barriers the dominant processes of aging was a mutual diffusion. At that, the main process was a diffusion of gallium from GaAs into metal [2]. In this report we solved the similar problem for M-InP contacts.

2.Methods of investigation

Regarding the processes which occur in M-InP contacts during the time of prolonged keeping in atmosphere we aredetermined by changing in distribution profiles of the main elements. They were determined with the help of Auger-spectroscopy with layer by layer etching by the Ar ion beam. The comparison was conducted for contacts formed by deposition on the InP surface of the metals (Au, Ag, Cu, Ni and Al) that, according to their chemical activity [2,3], relate to different groups. The Schottky barriers were created by thermal vacuum evaporation of metals onto substrates of n-InP (100) (N_d-N_a ~1.10^{16} cm^3) after its chemical treatment. Auger-spectroscopy analysis was made by the method which was described in [6].

To accelerate the processes of interaction at the interface , the M-InP Schottky barriers were sustained at the temperature 90-100°C. As it was shown in [7], these temperatures, accelerating the transformation of Schottky barriers parameters, do not influence yet on the regularities of change of M-InP Schottky diodes parameters (φ_b, n), that are typical for M-InP Schottky diodes, keeping at room temperature. The time of withstanding was 36 days.

3.Results and their discussion

The Fig.1 shows Auger-profiles that reflect change of interface condition of the

0-7803-3223-7/96/$5,00

examined M-InP contacts in the process of their aging. One can see that aging of Schottky diodes with intermediate layer of native oxides and degradation of their parameters may be connected with the following processes:

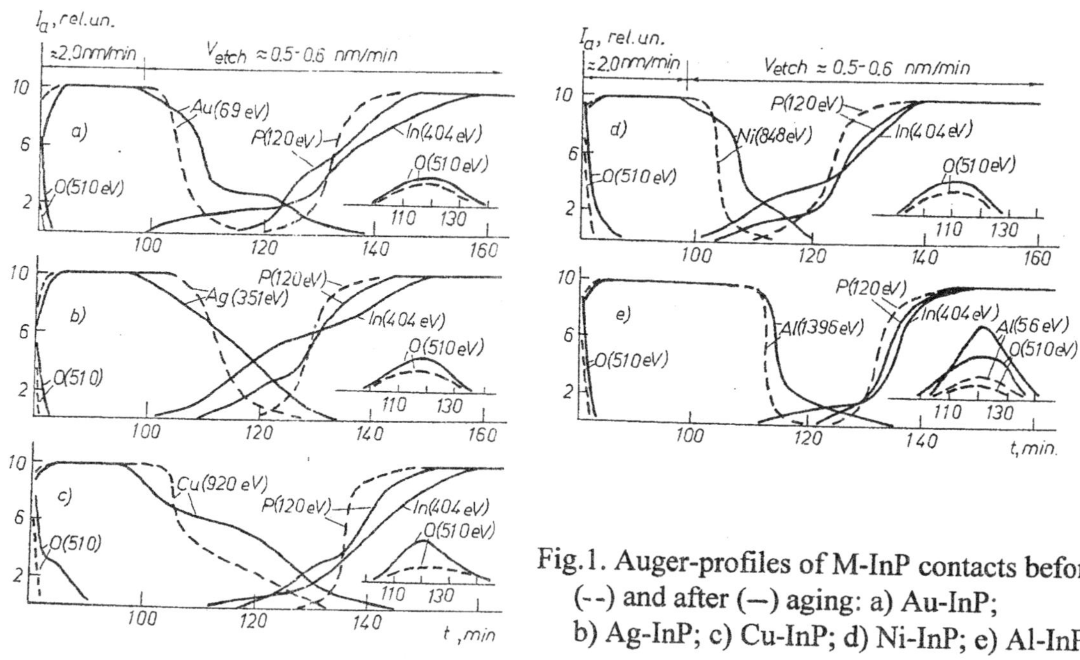

Fig.1. Auger-profiles of M-InP contacts before (--) and after (—) aging: a) Au-InP; b) Ag-InP; c) Cu-InP; d) Ni-InP; e) Al-InP.

Firstly, one can see the increasing in oxygen concentration in the intermediate layer. It depicts additional oxidation of InP and, possibly, creating of the oxides of contacting metals. According to the value of oxygen concentration transformation in intermediate layer which was determined by change of area under the curves of oxygen (510 eV) Auger-profiles, the examined Schottky barriers can be located in the following line: Au, Ni, Ag, Cu, Al. Using this criteria we can expect that the contacts on the base of Au, Ni and Ag will have the highest temporal stability of the parameters. However, the experimental research of regularities of changes in electric characteristics of M-p-InP Schottky barriers in the process of their aging [4, 5] gives the other succession of contacting metals, which is as follows: Al, Ni, Cu, Au. Ag. In this line Al-InP contacts posses heightened stability and Ag- and Au-InP contacts are characterized by significant transformation of initial electric characteristics. Hence, it appears that in case of Schottky barriers created on the real InP surface with native oxides, the penetration of oxygen through metal covering does not determine the change of φ_b . At the same time, it is impossible to exclude that this very effect can be caused by the observed increasing in the non-ideality coefficient of I-V characteristics in the process of aging for some diodes with chemically active metals [6].

Secondly, it is observed the penetration of contacting metal atoms into the intermediate layer. Earlier we thought [7] that this effect is one of the major in degradation mechanism of electric characteristics of M-InP Schottky barriers. However, the received results have shown that within the limits of mistake in experiment the penetration of contacting metals (excluding Au) is limited by the intermediate layer. This allows us to conclude that though diffusion of contacting metals atoms can contribute to change of intermediate layer' parameters [7], it cannot not influence on the φ_b value of M-InP contacts, especially, on the first stage of its aging.

Thirdly, the strong interface diffusion has been taken place at the M-InP interface. At that, the main display of this process is penetration of In from InP into the intermediate layer.

For different metals this process goes with different intensity. If for Al-InP contacts the depth of In diffusion is limited by thickness <10 nm, then for Ag-InP contacts indium is observed even in metal and on its surface. Penetration of phosphorus into the intermediate layer unlike indium is weaker and in the process of aging, the spatial position of phosphorus for all contacting metals is practically at the same level. As a result because of primary leaving of In from InP the interface diffusion is accompanied by violation of stoichiometry of InP subsurface layer. The last one is displayed in forming of under-contact area enriched by phosphorus.

It is necessary to pay attention to the fact that the degree of violation of InP subsurface layer stoichiometry and the thickness of the layer with surplus phosphorus strongly depends on chemical activity of contacting metals. In particular, these effects are more obvious for metals (Au, Ag, Cu) that create the high potential barriers at the interface with n-InP. For Ni and Al metals that are to another group of metals and create low potential barriers at the M-InP interface the violation of InP stoichiometry in the intermediate layer is practically absent. This kind of correlation allows to confirm the suggestion done in [3] that surplus of phosphorus in the under-contact area of InP namely provides receiving in M-n-InP contacts higher height of potential barriers. The explanation like this corresponds to small φ_b for In-n-InP contacts and data of theoretical calculations [8] that depict acceptor character of levels created by vacancies of indium (V_{In}) and anti-structural defects of phosphorus (P_{In}).

If one examine metals to locate in a line, according to the extent of interface diffusion and to the degree of violation of InP stoichiometry in under-contact area, taking place in the process of M-InP contacts aging, it can be depicted as follows: Al, Ni, Cu, Au, Ag. As a generalized parameter that quantitatively characterize the changes of the M-InP interface condition the area between curves of In and P Auger-profiles was used. In this line the extent of violation of stoichiometry increases while passage from Al to Ag and it is in the exact conformity with the row of metals that reflects succession in decreasing of temporal stability of M-InP Schottky barriers electrical parameters [3,4]. It is necessary to point at that intensification of interface diffusion in M-GaAs contacts during the changing of contacting metal type, has another succession for GaAs [2,3] than for InP. It confirms specific features of the processes, taking place at the M-InP interface, caused by peculiarities of physical-chemical properties of contacting materials.

The find of a criteria reflecting a reason for the increasing of interface stability and temporal stability of Schottky barriers parameters, formed on InP by chemically active metals, allowed us to examine several thermodynamic parameters that characterize chemical activity of contacting metals (Table 1).

Table 1. Thermodynamic parameters of contacting metals

Metal	$\Delta H_f(M_xP_y)$, eV/mol' [9]	ΔH_r, eV/at [9]	ΔH_{sol} (M, In) eV/mol [9]	$\Delta H(M_xO_y).10^{-6}$ Dj/(cal.mol)[10]
Ag	-0,46 (AgP_2)	+(0,80-1,10)	-0,15	30 (Ag_2O)
Au	-1,00 (Au_2P_3)	+(0,70-0,30)	-0,70	80 (Au_2O_3)
Cu	-1,30 (Cu_3P)	-(0,17-0,22)	-0,15	166 (Cu_2O)
Ni	-4,12 (Ni_5P_2)	-(0,51-0,71)	-0,50	244 (NiO)
Al	-1,70 (AlP)	-0,92	+0,32	1675 (Al_2O_3)
In	-0,78 (InP)	-	-	926 (In_2O_3)
P	-0,78 (InP)	-	-	1531 (P_2O_3)

Note: $\Delta H_f(M_xP_y)$ - standard heat of creation of stable metal phosphide; ΔH_r - interaction heat on M-InP interface, normalized at one metal atom; $\Delta H_{sol}(M,In)$ - heat of indium dissolution in metal; $\Delta H(M_xO_y)$ - heat of oxide creation.

We can draw the conclusion that interface diffusion and appearance of subsurface area with surplus phosphorus exists for metals having negative heat for indium dissolution and inclining to create a uninterrupted row of solid solutions with the In. The standard warmth of stable phosphide creation that is usually connected with stepped conduct of $\varphi_b(X_m)$ dependence for M-InP contacts [2] correlates in general with the temporal stability of Schottky barriers parameters. But, probably, it is not the main factor determining temporal and thermodynamic stability of M-InP contacts containing intermediate layer of their native oxides. For example, as it shown in the Tab.1 for Ni that has the maximum ΔH_f value interface diffusion is more than at Al-InP contact where Al has lower ΔH_f value. It appears as more exact criteria, characterized thermodynamic stability of real interface can serve warmth of interaction at M-InP interface (ΔH_r) and warmth of oxides creation $\Delta H(M_xO_y)$. In accordance with this criteria more stable are contacting pairs are characterized by maximal values of the indicated parameters. Contacts where the warmth of oxides creation and warmth of interaction at the M-InP interface are minimal, have minimal both temperature and temporal stabilities of its parameters.

4.Conclusions

1. Aging of M-InP Schottky barriers is conditioned by proceeding of diffusion processes at the interface. The main process is indium diffusion from InP into metal. Primary leave of indium from InP leads to interface diffusion of contacts and violation of InP subsurface layer stoichiometry. The last one leads to forming an under-contact area, enriched with phosphorus.

2. Extent of interface diffusion and violation of stoichiometry decreases in the row of Ag, Au, Cu, Ni, Al and is in direct accordance with warmth of interface interaction and warmth of oxides creation by contacting metals. Contacts that have maximal values of indicated parameters have the maximal stability of electrical characteristics and phase composition of the interface.

References

[1]. J.P.Ponpon, P.Siffent, J. Appl. Phys. 50(7), p.5050 (1979);

[2]. Thin Films - Interdiffusion and Reaction, edited by J.M.Poate, K.N.Tu, J.W.Wayer, (A Wiley-Inter-science Publication, New York, Chichester, Brisbane, Toronto, 1978);

[3]. G.S.Korotchenkov, I.P.Molodian. Schottky Barriers, Native Oxides and MOS Structures on the Base of Indium Phosphide (Shtiintsa, Chisinau, 1984);

[4]. G.S.Korotchenkov, V.A.Blaje, V.I.Tsvitsinsky, J. Electronnaya Technika, Ser.2, 1986, No.4, p.67, (Russian);

[5]. G.S.Korotchenkov, V.A.Blaje, J. Electronnaya Technika. Ser.2, 1988, No.2, p.58, (Russian);

[6]. G.S.Korotchenkov, V.I.Tsvitsinsky, V.A.Michailov, J. Vac. Sci. and Technology. B. 1985, V. 3(4), p.981;

[7]. G.S.Korotchenkov, V.A.Blaje, Rus. J. Microelectronics, 1986, V. 15(5), p.415;

[8]. T.Krendelewicz, W.G.Petro, I.Linday, W.E.Spicer, J. Vac. Sci. and Technol. A. 1984, V.2(2), p.542;

[9]. J.F.McGilp, J. Phys. C: Sol. St. Phys. 1984, V. 17(12), p.2249;

[10].Reference book. Fiziko-Himicheskie Svoistva Okislov, edited by G.V.Samsonov (Metallurgeya, Moscow, 1969).

Transient Photocapacitance Spectroscopy of Al-As_2Se_3 junction

S.D.Shutov and I.A.Vasiliev

Institute of Applied Physics
5 Academiei Str,.Chisinau MD-2028, Moldova Republic

Summary. *From long-term (20 to 1000 s) photocapacitance relaxation measurements on an Al-As_2Se_3 junction the spectral dependence of density and photoionization cross section of the gap states was derived in the range from 1.35 to 1.92 eV. Two photoionization thresholds at 1.36 and 1.53 eV have been found and ascribed to the gap states of charged defects.*

1. Introduction

The model of charged defects with discrete energy levels located in the gap symmetrically respective the Fermi level appeared to be able to explain several experimental results for glassy a-As_2Se_3. On the other hand the study of transient photocurrents and induced absorbtion revealed wide and featureless distribution of localized states [1]. To clarify the character of the distribution it is expedient to apply sensitive and effective spectroscopy, which allows to probe the localized state distribution in a wide range of energies. Among these is the isothermal transient photocapacitance spectroscopy [2]. In the present work an attempt is made to apply that technique for a-As_2Se_3. An adaptation of the technique for capacitance measurements at infra-low frequences (10^{-3} to 10^{-1} Hz) was demanded due to high Maxwell's relaxation time.

2. Experimental

The photocapacitance was measured at the surface barrier of Al-As_2Se_3 diodes obtained by successive vacuum deposition onto glassy substrates of an ohmic electrode (Sb), a film of a-As_2Se_3 of thickness 1.2 to 1.5 μm and of a semitransparent aluminium cover, which formed a contact barrier of 1.14 eV in height with As_2Se_3. The capacitance relaxation was monitored both in the dark and under illumination with photon energy 0.83...1.96 eV and photon flux $10^{15}...10^{16}$ cm^{-2} s^{-1}. The capacitance in dependence of time was measured by quasistatic technique with electrometric registration of displacement current through a reverse

0-7803-3223-7/96/$5,00

biased junction under slowly varying (with a period 20 to 1000 s) triangular voltage pulse. The measurements were made at room temperature.

3. Experimental results

The difference in square capacitance of the junction under illumination (C_l) and in the dark (C_d) is shown in Fig.1 in dependence of time (t) for various photon energies (hν) of optical excitation. The character of the relaxation curves is changed with increasing of photon energy from nearly linear function of time at hν <1.35 eV (Fig.1a) to an exponent at higher energies (Fig.1b). This fact shows that the photocapacitance relaxation rate varies over wide limits determined by photogeneration rate of charge carriers from localized states and increases with photon energy. If at low photon energies only the initial portion of relaxation is seen in the experimental time window, at high energies the relaxation manages to complete and the photocapacitance comes to equilibrium.

The existence of exponential form in photocapacitance relaxation at hν >1.35 eV allows to use the parameters of the relaxation curves for determination of density (N) and photoionization cross-sections (χ) of ionized gap states, assuming that for each photon energy the photocapacitance relaxation is described by the equation [2]

$$\Delta C^2 = C_l^2 - C_d^2 = \Delta C_{st}^2 \,[1 - \exp(-t/\tau)] \qquad (1)$$

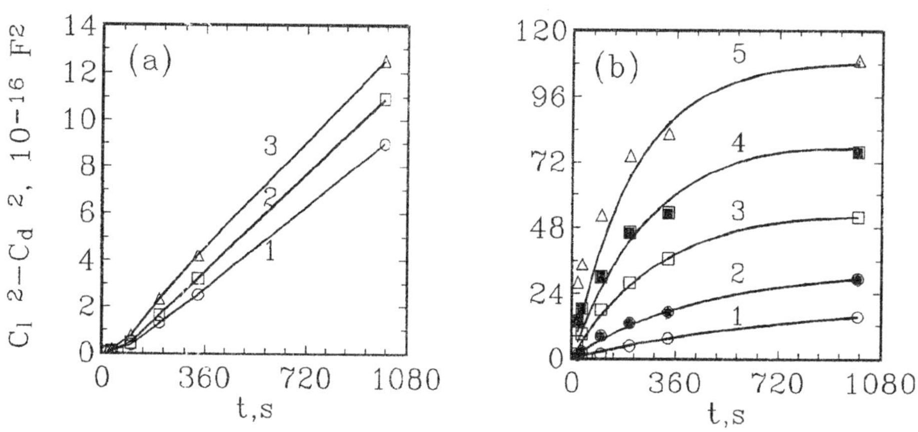

Fig.1. Isothermal photocapacitance relaxation curves for photon energies hν, eV: (a) 1-0.88, 2-0.95, 3-1.13; (b) 1-1.35, 2-1.55, 3-1.63, 4-1.72, 5-1.82

with steady-state value $\Delta C_{st}^2 \cong BN$ and the time constant $\tau^{-1}=\chi\Phi$ (here Φ is the incident photon flux and $B= \varepsilon\varepsilon_0 A^2 q/2(V+V_k)$ is the junction parameter, where $\varepsilon\varepsilon_0$ is the dielectric constant of As_2Se_3 film, A is the contact area, q is the electron charge, V is the applied bias and V_k is the build-in potential). The parameters of the exponent were found from graphical solving of equations (1) obtained for the two values of photocapacitance at t_1 and t_2 in the range from 100 to 1000 s for each hv. The spectral dependence of N and χ are shown in Fig.2a and 2b.

4. Discussion

The product of the values N and χ has the meaning of absorbtion coefficient and is determined by that part of it, which contribute to photocapacitance due to optical transitions between the gap states and extended states. That is why the spectrum N(hv) shown in Fig.2a reflects the density-of-states pattern in regions of the gap located near the mobility edges of conducting bands. As it is seen the density of states takes values characteristic for the band tail regions in disordered semiconductors, which decrease towards the middle of the gap.

Unlike to the curve in Fig.2a the spectral dependence of photoionization cross-section (Fig.2b) has features at about 1.35 and 1.60 eV. Those features may be analyzed using the well known Lucovsky's model for photoionization of an impurity centre. This model proved to be a good approximation in the case when a deep centre becomes neutral or positive due to optical excitation. Such situation is characteristic for optical transitions with participation of defects in chalcogenide glasses ($D^- \rightarrow D^o$, $D^+ \rightarrow D^0$ for example).

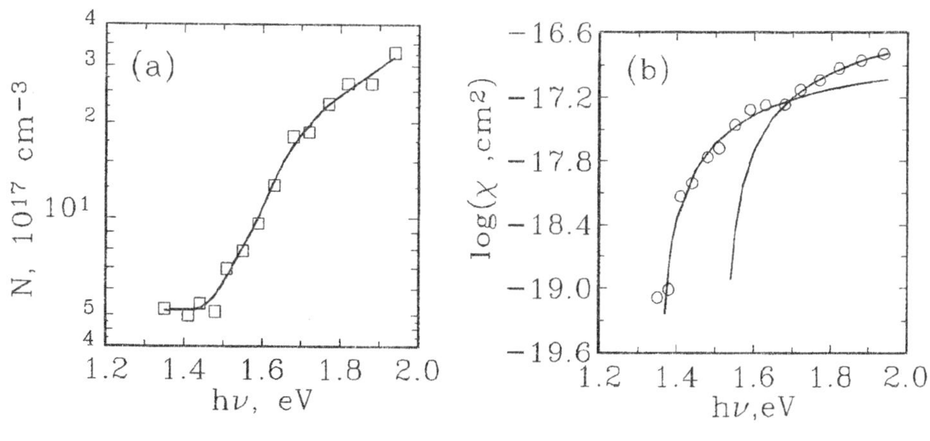

Fig.2. The spectra of density (a) and photoionization cross-section (b) of optically excited states in a-As_2Se_3 film

We recognized that the spectrum $\chi(h\nu)$ in Fig. 2b plotted in Lucovsky's co-ordinates $\chi^{2/3}(h\nu)^2$ vs. $h\nu$ is well approximated by two straight lines with photoionization thresholds at 1.36 eV and 1.53 eV. The spectra calculated by the Lucovsky's formula are shown by full lines in Fig.2b. We may assume that the observed photoionization cross section spectra correspond to the energy scheme of charged defects presented in Ref.[1]. In particular, the thresholds at 1.36 eV and 1.53 eV are in accordance with optical excitation of A^+ and A^- centres, respectively, with generation of a hole in the valence band (in the latter case with additional thermal excitation).

5.Conclusion

In conclusion, we have applied for the first time the photocapacitance spectroscopy to study gap states in chalcogenide glassy semiconductors. In the energy range from 1.35 to 1.92 eV the spectra of density of states from 5×10^{17} to 3×10^{18} cm^{-3} and photoionization cross-section between 2×10^{-19} and 3×10^{-17} cm^2 have been obtained. In the photoionization cross section spectrum two thresholds are defined at 1.36 and 1.53 eV; the magnitude of the cross-section is about 2.8 times larger in the latter case. These differences apparently suggest the dissimilar nature of the localized centres under consideration.

REFERENCES

[1] G.T.Adriaenssens, Phil.Mag.B, 1990, 62, 79-87

[2] L.S.Berman and A.A.Lebedev. Capacitance spectroscopy of deep centers in semiconductors (in Russian). Leningrad,1981, 179 p.

PLENARY SESSION 2
(Invited Papers)

INVITED PAPER

The New Generation of Power Semiconductor Devices

Gehan A.J. Amaratunga[a] and Florin Udrea[b]

a. Department of Electrical Engineering and Electronics, Liverpool University, Liverpool LG9 3BX, England

b. Department of Engineering, Cambridge University, Cambridge CB2 1PZ, England

Summary

The next generation of power devices are likely to extend MOS controlled bipolar (MCB) device concepts to cover very high voltage (up to 6kV) applications. Such devices will be based on utilising the advantages brought about by trench gate MOSFETS to control bipolar current flow. The semiconductor phenomena which can be accessed to optimise power device performance through use of trench gate MOS channels are discussed. Device structures which use PIN diode and thyristor type carrier distributions to reduce power loss within the device are presented. The next generation of power switching devices is proposed as being based on trench gate devices which have hybrid PIN diode and dynamic thyristor structures.

1. Introduction

Control of electric power switching using solid state electronic devices has a history extending over three decades, commencing from the development of the power PIN diode and the thyristor. At present there are diodes and thyristors which are capable of handling voltages and currents up to 6kV and 1kA. Examination of the history of power devices reveals two distinct routes which have been followed to advance power devices to their present state. First, the evolutionary optimisation of a given device type to reduce power losses within it, and second, the invention of new device structures and concepts which have improved efficiency and easier controllability. The former trend is also seen in the development of low power devices for integrated circuits, and has been successful in achieving revolutionary progress towards todays ULSI circuits. However in power devices, evolutionary developments in PIN diodes and thyristors have not led to dramatic increases in the current and voltage ratings at which they can be used. It is the invention of new device structures, unlike in low power devices, which has led to the significant advances in the controllability of power devices, and hence their wide spread use. The first major advance in power device structure was the development of the Gate Turn Off (GTO) thyristor, which allowed control over both the turn on and turn off process of the device [1]. The other major conceptual innovation which has had the most impact on power devices is the merging of MOS switching and bipolar current flow in the insulated gate bipolar transistor (IGBT) [2,3]. Over the last decade the IGBT has been through a number of technological optimisation stages and is now an efficient power switch up to ratings of 2.5 kV and 1kA [4,5]. It now seems inevitable that the MOS-controlled bipolar (MCB) concept will extend to include the highest voltage and current ratings at which GTO thyristors are currently used. In this paper we will examine a number device structures based

0-7803-3223-7/96/$5,00

on the MCB concept which are likely to have the lowest specific loss (watts cm-2) in high power switching. These structures for the next generation of power devices are all based on utilising a trench gate MOSFET for device control. We will not examine the more speculative area of extending the upper voltage and current ratings at which solid state devices are viable by using materials other than Si.

2. Trench gates and MCB devices

The generic trench gate IGBT (TIGBT) structure and its equivalent circuit are shown in Fig. 1.

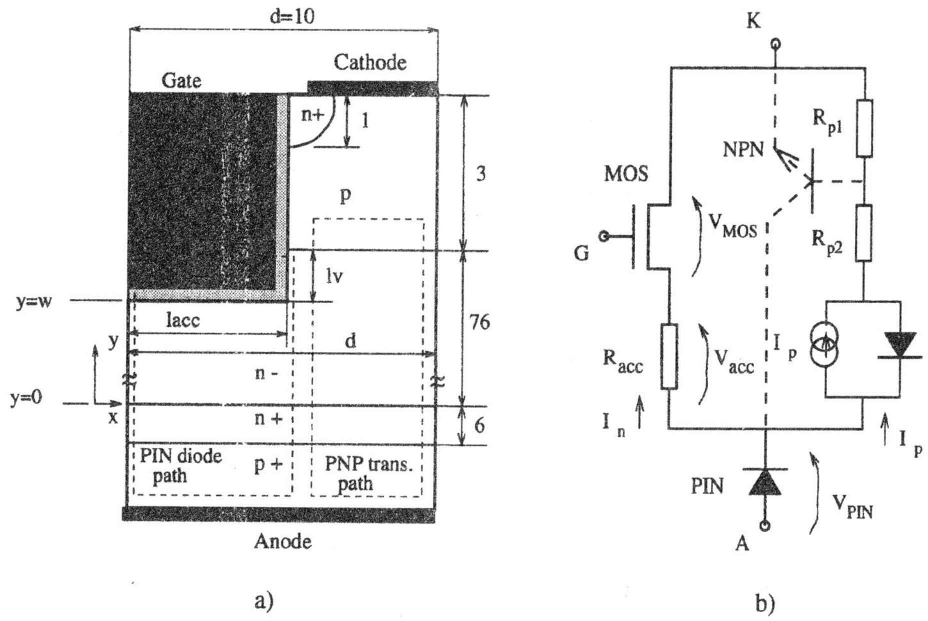

Figure 1: (a) A 900 V Trench IGBT structure and (b) the Trench IGBT equivalent circuit. The circuit model [9] is comprised of a PIN diode, a MOSFET transistor which in the linear region acts as a channel resistance, and an accumulation layer resistance. To take into account the contribution of the hole current flowing through the p base to the cathode contact a current source in parallel with a diode is included in the equivalent circuit. It is as if the PNP transistor were split in an Ebbers - Moll model using a PIN diode with modified boundary conditions to describe the high-level injection. Thus, the boundary condition at the cathode side is given by the coupled presence of the PIN diode and PNP transistor.

Originally it was thought [6,7] that the main advantage was a geometric one of being able to increase the gate perimeter to cell area by having the MOS channel placed vertically into the device. Additionally there was the advantage that, unlike in the conventional DMOS geometry, the main anode current and the gate current flow are vertical. This ensures that the hole current distribution in the p-base/collector is such that turn on of the parasitic upper npn transistor (and hence loss of gate control) is suppressed. Recently, with the increasing availability of technology for trench gate formation, operation of the TIGBT has been analysed in detail [8]. What these studies have revealed is that the TIGBT structure allows access to a number of device phenomena, normally less significant in a DMOS geometry, which can contribute to a substantial reduction in on-state resistance.

2.1 Effect of p-body hole current

The first was the effect of having hole current in parallel to the channel in the p-body/collector. It has been shown [9] that with such a hole current, the distributed rise of the p-body potential leads to a reduction in the depletion charge under the gate. But, since the surface potential at the gate is fixed after inversion, and the charge on the gate(for a given gate voltage) has to be balanced by the sum of the depletion and inversion charge, the reduction in the depletion charge is accompanied by an increase in the inversion charge. This in turn leads to a reduced MOS channel resistance over and above that due to geometry factors alone, and hence a reduced total voltage drop across the device at a specified current density . The analytical expression for the anode current in a TIGBT, including the effect of hole current in the base on the channel charge, can be expressed as [9]:

$$\begin{aligned} J \;=\; & \frac{Z\mu_{ch}C_{ox}}{AL_{ch}(1-\alpha_{pnp})}\{(V_G - V_{FB} - 2\cdot\phi_F)(V_{on} - V_{PIN}) - \frac{1}{2}(V_{on} - V_{PIN})^2 \\ & - \frac{(2/3)K(V_{on} - V_{PIN})}{V_{on} - V_{PIN} - R_{p2}\alpha_{pnp}JA}[(V_{on} - V_{PIN} - (R_{p1} + R_{p2})\alpha_{pnp}JA + 2\phi_F)^{3/2} \\ & - (2\phi_F - R_{p1}\alpha_{pnp}JA)^{3/2}]\} \qquad (1) \end{aligned}$$

where J is the on-state anode current density, V_{on} the on-state anode voltage, V_{PIN} the PIN diode voltage drop, V_G the gate voltage, V_{FB} the flat-band voltage, ϕ_F the potential difference between intrinsic level and Fermi level, R_{p1} the cathode short resistance, R_{p2} the MOS substrate resistance, α_{pnp} the static current gain of the PNP transistor, Z the channel width, A the device active area μ_{ch} the channel mobility, C_{ox} the specific oxide capacitance and $K = \sqrt{2\varepsilon_s q N_A}/C_{ox}$, N_A is the p base average doping.

As can be seen from (1) the enhancement in anode current due to the hole current in the base is related to the static gain of the main pnp transistor α_{pnp} in the TIGBT structure. In general this effect is significant provided α_{pnp} is in the range 0.4 - 0.6. In general this applies for n-base widths $\leq 100\mu$m or TIGBTs rated up to 1.2kV.

2.2 PIN diode-like carrier distribution

The key factor in reducing the on resistance in power devices is the effective modulation of the n-drift conductivity. Thyristors and PIN diodes have low on state resistance due to increased storage of carriers due to injection of carriers into the drift region at both the lower (anode) and upper (cathode) end. In MCB devices such as the IGBT injection only occurs from the anode end. Therefore in the conventional IGBT the carrier distribution in the drift region is similar to that in a wide base bipolar transistor, with the excess carrier distribution decaying to zero at the upper end. The two different types of distribution are shown in Fig. 2a. The carrier profile at the cathode side in a TIGBT is given by a combined PIN diode-PNP transistor carrier distribution (Fig. 2b).

In a TIGBT it is possible to improve the modulation of the drift region by optimum design of the device to achieve a carrier distribution similar to that of a PIN-diode. This is achieved by increasing the accumulation layer length at the bottom of the trench, see Fig. 1a and 2b. The accumulation layer acts to create an n+ - n- junction

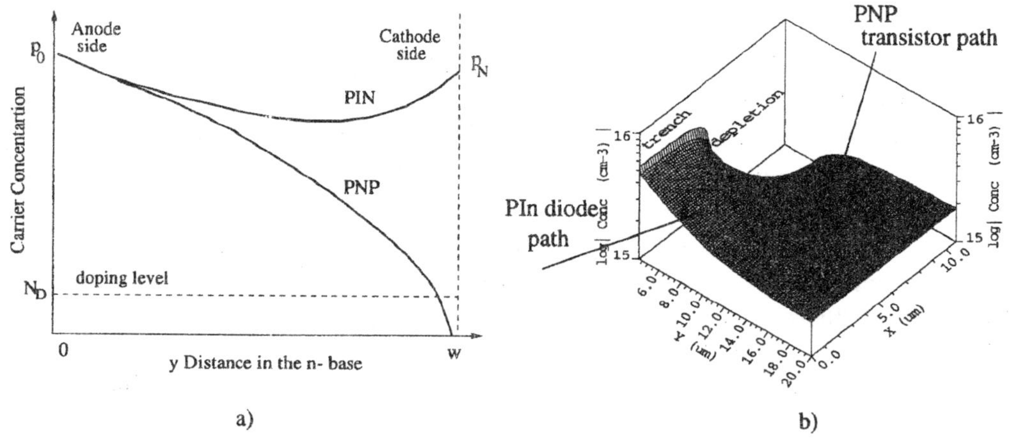

Figure 2: (a) The typical PNP transistor and PIN diode carrier distributions and (b) the two dimensional distribution of the carriers in a TIGBT at the cathode side. The distinctive PIN diode and PNP transistor paths can be clearly identified

(under gate control) which as an electron injector into the drift region, as is the case for a PIN diode [8,10]. Therefore in a TIGBT the accumulation layer acts as an injector, and the upper drift region has an excess of electrons and holes(required for neutrality) which contribute to a large recombination current component [11]. It has been experimentally verified that this carrier storage effect can be extremely well utilised in TIGBTs rated up to 4.5 kV [12]. A modified TIGBT structure, termed the Carrier Stored Trench-Gate Bipolar Transistor (CSTBT) which maximises the PIN diode-like carrier distribution has been recently proposed and verified [13], see Fig. 3a.

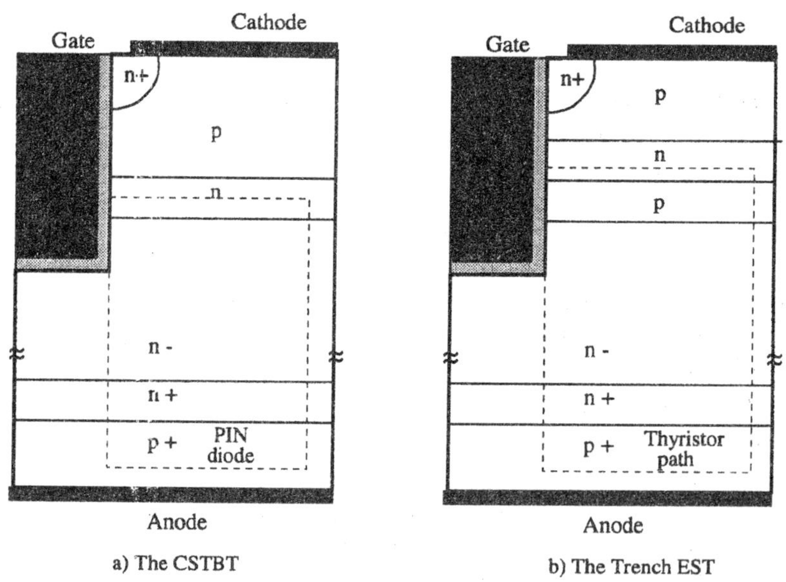

Figure 3: (a) The CSTBT structure. The device is comprised of a PIN diode in series with a MOSFET channel (b) The Trench EST (also TMT) structure. The device is comprised of a thyristor in series with a MOSFET channel

In this TIGBT structure an extra n-layer is placed between the upper drift region and the p-body region. The device operates as a PIN diode which is placed in series with a trench gate MOSFET. Parasitic latch up immunity of this device is also extremely high as hole current flow in the p-base is almost entirely suppressed. A specific on resistance of has been measured at a current density of 50 A cm^{-2} for a device rated at 1.7kV [13]. This is encouraging as it was only 30 % higher than that of a PIN diode made in the same process without the trench MOS gates. An important feature of the TIGBT structure with a PIN diode-like carrier distribution is that the lower on resistance through improved modulation is not achieved at the expense of greater turn off time. This is because the improved modulation is at the upper end of the drift region.

The n+ injector formed by the accumulation layer is physically removed when the gate is turned off and, due to the growth of the p-body/n-drift junction depletion layer as the device turns off, excess carriers are rapidly swept out of the upper drift region [11] (Fig. 4).

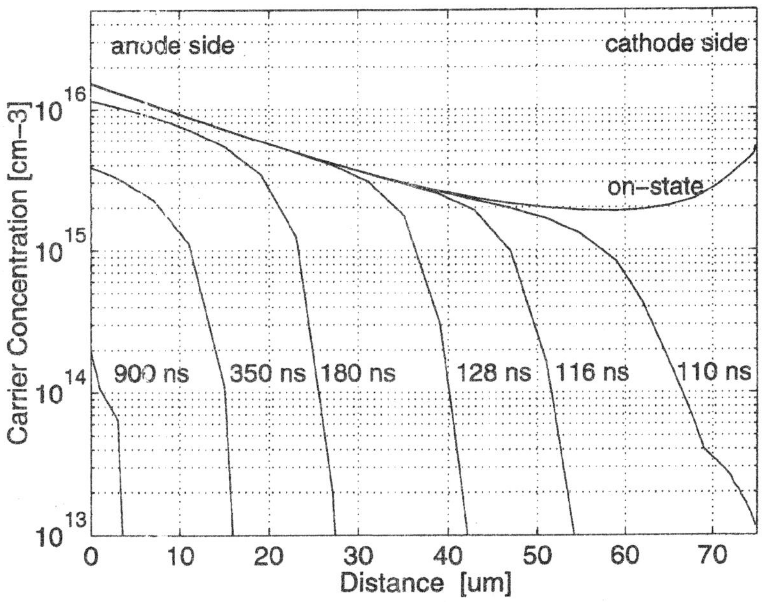

Figure 4: Electron distribution in the n- base of a 900 V Trench IGBT during turn-off. The gate is switched-off from 15 V to 0 V in 150 ns. A lifetime of 0.1 μs was considered.

Therefore, the TIGBT operating in PIN diode mode seems to offer an almost optimum solution to the problem of obtaining maximum modulation with maximum controllability.

3. Thyristors with trench MOS control

An alternative MCB concept which can achieve low on resistance is to utilise a thyristor structure to achieve a high degree of conductivity modulation in the n-drift region. The trench gate geometry is very amenable to the inclusion of additional MOS gates along the trench wall to control the device as there is no surface area overhead as in DMOS devices. Further, since the effective MOS channel length is defined vertically by implantation or epitaxy, it is possible to have narrow channel control devices without the requirement of fine lithography.

3.1 Trench gate MOS-thyristor (TMT)

A direct analogue to the TIGBT working in PIN mode is to have a trench MOS transistor placed in series with a thyristor. The TMT is such a structure, see Fig. 3b [14]. Instead of the n emitter being placed between the p-body and n-drift for a MOS channel in series with a PIN, it is placed within the p-body region. When a gate voltage is applied, two MOS inversion channel are formed on either side of the emitter, which has an accumulation channel in series with the two inversion channels. Once injection from the anode commences, the thyristor formed by the n-emitter, lower p-body, n-drift and the p+ anode is turned on. In the on state the device is operates as a trench gate emitter switched thyristor (TEST) [15]. The TEST structure (Fig. 3b) is considerably superior to its DMOS equivalent as it avoids to a large extent the problem of having parasitic thyristors inherent in its structure. When turning off the device, if the gate voltage is reduced to zero, then its operation is similar to that of an EST. However, if the gate voltage is taken negative to turn off the device, then with proper choice for the n-emitter doping, it will be possible to form a p-channel inversion layer in the emitter. The p-channel transistor will then act as a bypass for the hole current in the upper region of the device, and hence improve switching performance. In the turn-off mode therefore the device becomes the equivalent of the a trench gate MOS controlled thyristor (MCT)[16]. The limitation on the hole current that can be diverted in this way is the parasitic latch up of the n+ source junction.The TMT structure is a good example of how a hybrid device (EST and MCT) can be achieved with trench technology without any additional Si surface area consumption. The major technological development which will be required for use of the TEST in the next generation of high power devices is the requirement for a relatively deep trench gate technology 12μm. However, present results from research suggest that such a trench gate technology is viable.

3.2 Inversion layer emitter thyristor

In both the TIGBT-PIN and TEST structures a limitation to device performance in terms of on resistance is the resistance of the MOS channel placed in series with the main device. While it is possible to make this channel very narrow, its length is limited by the requirement that it must not punch through at the rated offset voltage. A way of overcoming this limitation has been found with the development of novel device concept. This is the use of a MOS inversion layer to form an emitter [17]. The new concept was recently demonstrated experimentally for lateral structures [18]. In a TIGBT structure it is possible to use the hole current which flows in parallel with the MOS channel, see 2.1 above, to actually open the inversion layer substrate junction [8,19]. When this occurs the effective MOS channel length is that up to the point from which electron injection form the channel into the p-body commences, see Fig. 5. A TIGBT which operates in this mode, with a very narrow effective channel due to a large part of the inversion layer acting as an electron injector is termed the inversion layer emitter thyristor (ILET) [19] (Fig. 5).

The thyristor is comprised of the injecting n+ inversion layer, p-body, n-drift and the p+ anode. A great advantage of this device is that when the gate voltage is removed to turn the device off, the emitter in the thyristor structure is also removed. The existence of such a dynamic emitter greatly improves the switching and turn-off

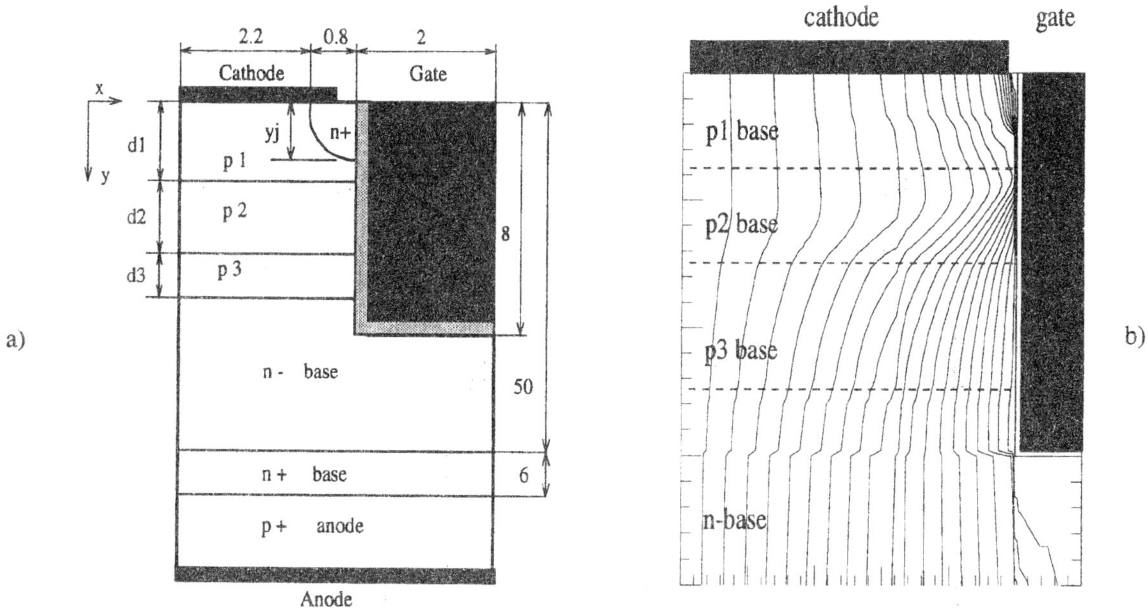

Figure 5: (a) Inversion Layer Emitter Thyristor (ILET) cross section and (b) simulated 2D flow-line contour showing inversion layer injection

performance of the ILET. In comparison to the TIGBT -PIN operation, the ILET requires a hole current in the p-body to enable the inversion channel to act as an injector. Therefore on first examination it may appear that the the carrier distribution in the drift region of the ILET is similar to that of wide base pnp transistor, and therefore not as well modulated until the injection from the inversion layer commences. However, it is possible to design the ILET so that there is a balance between PIN mode operation using the accumulation channel at the bottom of the trench and inversion layer injection. This can be seen from the unified model for the current transport in a TIGBT where both the PIN mode and pnp mode currents are taken into account in eqn (2) for the on-state voltage in TIGBT [11,20]:

$$V_{PIN} \simeq \frac{J\, sinh(w/L)}{q(\mu_n+\mu_p)} \int_0^w \frac{dy}{p_0 sinh(\frac{w-y}{L}) + p_w sinh(y/L)} + \frac{kT}{q}\left[\frac{D_n - D_p}{D_n + D_p} ln(\frac{p_0}{p_w}) + ln(\frac{p_0 p_w}{{n_i}^2})\right] \quad (2)$$

where p_0 and p_w are given by:

$$p_0 \simeq \frac{D}{2Lh_0}\left[-cth(w/L) + \sqrt{cth^2(w/L) + \frac{4b}{1+b}\frac{h_0\tau}{qD}J}\,\right] \quad (3)$$

$$p_w = \frac{D}{2Lh_w}\left[-cth(w/L) + \sqrt{cth^2(w/L) + \frac{4h_w\tau}{qD}(\frac{1}{1+b} - \alpha_{pnp})J + \frac{4h_w L p_0}{D sinh(w/L)}}\,\right] \quad (4)$$

In eqns. (2-4), w is the length of the n- base, L is the ambipolar diffusion length, D is the ambipolar diffusion constant, μ_n and μ_p are the mobilities of electrons and

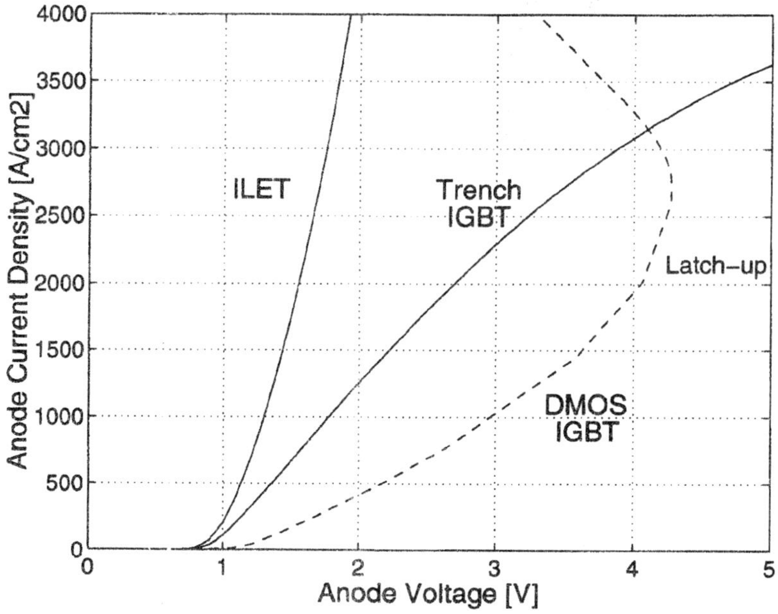

(a)

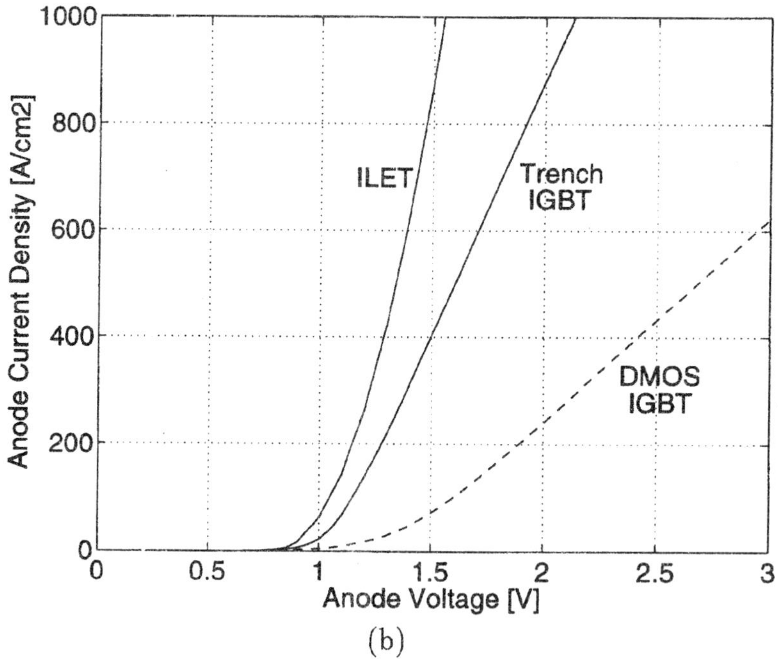

(b)

Figure 6: A comparison between the I-V characteristics of 600 V ILET, Trench IGBT and DMOS IGBT for (a) a lifetime of 10 μs and (b) a lifetime of 0.1 μs .

holes in the n- base, $b = \mu_n/\mu_p$, p_0 and p_w the peak carrier concentrations in the PIN region on the anode and cathode side respectively, J the anode current density, h_w, h_0 recombination parameters, α_{pnp} the on-state gain of the PNP transistor,τ is the ambipolar lifetime in the n- base.

In eq. 2 it can be seen that a higher static gain of the PNP transistor leads to a reduced value of the peak concentration p_w on the cathode side in the PIN region and as result a higher on-state voltage drop.

The total anode voltage drop in the on-state V_{on} is comprised of the voltage drop on the modified PIN diode V_{PIN} (which includes the PNP transistor effect in p_w) and the voltage drop across the MOSFET channel V_{MOS}. Eqns (1-4) could be combined to predict the on-state characteristics of the TIGBT.

The fact that the degree of injection from inversion layer can by controlled by adjusting the gate voltage (effectively controlling the concentration of the n+ emitter) is also advantageous in terms of safe operating area. The ILET can therefore be viewed as a device which embodies optimum operation through a hybrid TIGBT-PIN structure and a TMT structure. Numerical comparisons between an ILET, an optimised Trench IGBT and a conventional third generation DMOS IGBT rated for 600V for different values of the lifetime are shown in Fig. 6.

4. Conclusions

A range of key power device structures which are likely to form the next generation of solid state Si power switches have been presented. The core of all these structures in the trench gate MOS channel. By utilising the trench gate MOS geometry in an IGBT it is not only possible to improve device performance by maximising the MOS perimeter to cell area ratio (a geometric effect), but it also enables access to a number of novel semiconductor phenomena which can be utilised to improve device performance. From technology and device operation view points, it seems very likely that the TIGBT structure working in hybrid PIN-ILET mode will form the basis for next generation of power devices.

References

[1] J. M. Goldey, I. M. Makintosh and I. M. Ross, " Turn-off gain in p-n-p-n transistor triodes", *Solid State Electron.*, vol. 3, p. 119, 1961

[2] B.J. Baliga, M.S. Alder, R.P. Love, P.V. Gray, and N.D. Zommer, "The insulated gate transistor: A new three-terminal MOS-controlled power device", *IEEE Trans. Electron Dev.* vol ED-31, p. 821, 1984.

[3] J.P. Russel, A.M. Goodman, L.A. Goodman, and J.M.Neilson, "The COMFET-A new high conductance MOS-gated device", *IEEE Electron Devices Lett.*, vol. EDL-4, p. 63, 1983.

[4] Y. Takahashi, K. Yoshikawa, T. Koga, M. Soutome and Y. Seki " Experimental investigations of 2.5 KV-100 A PT type and NPT type IGBTs", in *Proc 7th Int. Symp. Power Semiconductor Devices and ICs,* p. 70, May 1995

[5] Y. Takahashi, K. Yoshikawa, M. Soutome, T. Fujii, M. Ickijyou and Y. Seki " 2.5 KV-1000 Power pack IGBT", in *Proc 8th Int. Symp. Power Semiconductor Devices and ICs,* p. 299, May 1996

[6] D. Ueda *et al.*, "A new injection suppression structure for conductivity modulated power MOSFET's", *Proc. 18th Int. Conf. Solid State Device and Materials,* p. 97,

1986.
[7] H.-R. Chang and B.J. Baliga, "500 V n-channel Insulated-Gate Bipolar Transistor with a trench gate structure", *IEEE Trans. Electron Device,* vol. ED-36, p. 1824, 1989.
[8] F. Udrea and G.A.J. Amaratunga, "Theoretical comparison between trench and DMOS technologies for Insulated Gate Bipolar Transistors", *IEEE Trans. Electron Device,* vol. 42, p. 1356, July 1995.
[9] F. Udrea, G.A.J. Amaratunga, and Q. Huang, "The effect of the hole current on the channel inversion in Trench Insulated Gate Bipolar Transistors (TIGBT)", *Solid State Electronics,* vol. 37, p. 507, 1994
[10] M. Harada, T. Minato, M. Takahashi, H. Nishihara, K. Inoue, I. Takata, "600 V Trench IGBT in comparison with planar IGBT", in *Proc 6th Int. Symp. Power Semiconductor Devices and ICs,* p. 411, May 1994.
[11] F. Udrea and G.A.J. Amaratunga, "A unified analytical model for the carrier dynamics in Trench Insulated Gate Bipolar Transistors", in *Proc 7th Int. Symp. Power Semiconductor Devices and ICs,* p. 190, May 1995.
[12] M. Kitagawa, I. Omura, S. Hasegawa, T. Inoue and A. Nakagawa, "A 4500 V Injection Enhanced Insulated Gate Bipolar Transistor (IEGT) operating in a mode similar to a thyristor", *IEDM Tech. Dig.*, p. 679, Dec. 1993
[13] H. Takahashi, H. Haraguchi, H. Hagino and T. Yamada, "Carrier Stored Trench-Gate Bipolar Transistor (CSTBT) - A novel power device for high voltage applications", in *Proc 8th Int. Symp. Power Semiconductor Devices and ICs,* p. 349, May 1996
[14] A.Q. Huang and G.A.J. Amaratunga *unpublished*
[15] M.S. Shekar, J. Korec and B.J. Baliga, "Trench Gate Emitter Switched Thyristor" in *Proc 6th Int. Symp. Power Semiconductor Devices and ICs,* p. 189, May 1994
[16] A.Q. Huang, G.A.J. Amaratunga and D.Y. Chen, "Analysis of 4500V double Trench MOS controlled thyristor" in *Proc 7th Int. Symp. Power Semiconductor Devices and ICs,* p. 153, May 1995
[17] F. Udrea and G.A.J. Amaratunga, "The Inversion Layer Emitter Thyristor - A novel power device concept", *Proc. 6th Int. Symp. Power Semiconductor Devices and ICs,* Davos, Switzerland, May 1994
[18] F. Udrea, G.A.J. Amaratunga, J. Humphrey, J. Clark and A. Evans, " The MOS inversion layer as a minority carrier injector", *IEEE Electron Device Lett.* vol. 17, No.9, Sept 1996
[19] F. Udrea and G.A.J. Amaratunga, "Analysis of a MOS-controllable Thyristor utilising an Inversion Layer Emitter", *Solid State Electronics,* vol. 37, p. 1999, 1994
[20] F. Udrea and G.A.J. Amaratunga, "An on-state analytical model for the Trench Insulated Gate Bipolar Transistor (TIGBT)" *submitted to Solid State Electronics* (1996)

INVITED PAPER

"Application Specific Devices: Transport and Performance of the Quasi-MODFET and the Graded Base Heterojunction Bipolar Transistor"

A. A. Iliadis, J. K. Zahurak*, and S. A. Tabatabaei

Electrical Engineering Department
University of Maryland
College Park, Maryland 20742, USA
Tel: (301) 405-3651/agis@eng.umd.edu

*Micron Tecnology Inc.
Boise, Idaho 83707, USA

Abstract

Two application-specific devices, a novel doped-channel Quasi-MODFET on AlInAs/InGaAs/InP, and an AlGaAs/GaAs graded-base heterojunction bipolar transistor, suitable for high current/gain applications in energy efficient mobile communication and optical communication and computer networks, have been developed. The study of transport and performance of the doped-channel MODFETs included the quantum well channel width, doping profile, and composition, and revealed that the position of the electron wavefunction peak can be engineered to shift in areas of minimum scattering, which resulted in the development of the 0.5 μm Q-MODFET with a g_m=703 mS/mm and I_{dsat}=800 to 940 mA/mm at V_g=0 to 0.4V, the highest g_m recorded for a doped-channel device.
In the HBT, the development of a rigorous two-dimensional model providing the optimum compositional grading of the base, the improvement of the collector using thinner layers and higher doping, and the emitter using compositional grading, resulted in the development of a high current heterojunction bipolar transistor appropriate for laser drivers.

Introduction

Application-specific devices (ASD) are designed to perform specialized tasks in an analog circuit, and as such their design must be optimized to provide the maximum possible capability for the specific task, at the expense of the other properties of the device. In this way, high current drive, high gain/high speed, or low noise, radiation hard, or high temperature devices, can be realized.
Modulation doped field effect transistors (MODFETs) take advantage of the spatial separation of the electrons from their parent atoms in the doped large band-gap semiconductor to form the two-dimensional electron gas (2DEG) in the undoped small band-gap semiconductor. This way transport in the 2DEG channel is improved and device performance is enhanced.
Such a structure is limited by the current that it can provide [1], and doped-channel FETs [2] are more suitable for higher current applications. The introduction, however, of dopants in the channel can degrade performance due to scattering, but the extend of degradation and the parameters of the device critical to counterbalance this performance trade-off, is not well understood.
Therefore, in the first part of this work, we have investigated two-dimensional electron transport in a variety of AlInAs/InGaAs/InP MODFET structures, where the InGaAs quantum well channel width was varied, uniformly doped or delta doped, and compositionally changed. Transport and FET

0-7803-3223-7/96/$5,00

performance in both micron (1.8 μm) and sub-micron (0.5 μm) gate devices, was studied. Our studies lead to the development of a novel device, the Quasi-MODFET, which yielded record performance for doped-channel devices.

In the second part of this work we present our studies to develop a graded base heterojunction bipolar transistor (HBT) [3] with high current driving capability, suitable for applications as a driver of semiconductor lasers in advanced optical communication and computing systems (OEICs). HBTs can provide the high current densities necessary to drive lasers in monolithic laser array systems, and, in principle, the higher the current capability the shorter the switching time of the device, since the capacitances (internal, parasitic, etc) can be charged faster at the higher current, to provide the high switching speeds necessary for the high bit rates targeted by advanced optical systems. In addition, current gain must be optimized to provide efficient operation of the transistor. In order to achieve an optimum high current performance, the critical parameters, that are competing in transistor operation, must be identified and optimized. For example, if the base width is increased to reduce base resistance (for the same base doping), the current increases, but the base transit time increases also, with subsequent reduction in the frequency (f_T) of operation of the device. On the other hand, a wider base is less susceptible to base neutral region modulation (Early effect), which is advantageous for high current operation. To counterbalance the base transit time increase [4], a compositional grading in the base of the AlGaAs/ GaAs HBT can be introduced. The grading provides an additional electric field in the base "neutral" region to facilitate carrier transport [5], and reduce base transit time. We have developed a rigorous two-dimensional model for optimizing the grading of the base, and fabricated HBTs grown by MBE on GaAs SI substrates according to the model. The fabricated HBTs demonstrated high current capabilities with good current gain.

Experimental Details

All MODFET structures were grown on (100) SI InP substrates using gas-source molecular beam epitaxy (MBE). The structures are grown beginning with a buffer layer of 400 Å of $In_{0.52}Al_{0.48}As$. This layer thickness was selected to provide good carrier confinement while minimizing layer thickness. Next, the quantum well channel layer was grown consisting of a layer of $In_{0.53}Ga_{0.47}As$ doped with Si. The parameters examined here were, the well width (40 to 300 Å), the doping profile (uniformly-doped or delta-doped), and the compositional profile (excess In) of the well (channel). The uniform doping level was at $8\times10^{17}\ cm^{-3}$, resulting in an electron sheet carrier concentration of $2.4\times10^{12} cm^{-2}$. The channel layer was followed by a 250 Å $In_{0.51}Al_{0.49}As$ undoped gate layer, followed by a 50 Å of heavily doped n^+ $In_{0.53}Ga_{0.47}As$ for ohmic contact formation.

Electron transport was evaluated by Van der Pauw low-field mobility measurements and high-field microwave-heated-carrier velocity measurements [6]. The microwave measurement technique, provides accurate electron velocity measurements up to 10 KV/cm without domain formation [7]. Devices with 1.8 μm and 0.5 μm gate lengths, were fabricated using standard photolithography and E-beam lithography respectively, under uniform conditions. Au-Ge based source-drain ohmic contacts were deposited and annealed at moderate temperatures to avoid lateral diffusion into the channel [8]. Ti/Au gates were deposited on $In_{0.51}Al_{0.49}As$ through the gate-recess.

The graded-base HBTs were grown on (100) SI GaAs substrates by solid-source MBE. The structures were formed by first growing an undoped GaAs buffer layer to provide the smooth surface morphology needed for the subsequent layers of the subcollector, collector, base and emitter. Then, a 8000 Å GaAs subcollector layer doped n-type (Si) at $5x10^{18}$ cm^{-3} was grown, followed by 4000 Å GaAs collector layer at $1x10^{17}$ cm^{-3}. This collector design deviates significantly from conventional collector design (6000 Å, $5x10^{16}$ cm^{-3}), in both the thickness and the doping level. We use here a thinner collector to reduce transit time, and higher doping to increase the maximum collector current. The base was grown compositionally graded from GaAs at the collector to $Al_xGa_{(1-X)}As$ at the emitter, where an optimum Al mole fraction x was determined by our two dimensional modeling. Base thicknesses and Al mole fraction were varied between 500 Å and 2000 Å and 0.1 to 0.2 respectively, and a non-graded structure was also fabricated for comparison. Next was a 50 Å undoped GaAs layer, followed by a graded 300 Å n-type $Al_xGa_{(1-X)}As$ emitter layer with x=0 to 0.3. This layer reduces the barrier between the emitter and base created by the energy band discontinuity at the conduction band of the $Al_{0.3}Ga_{0.7}As$ EB junction. A further 1200 Å n-type ($5x10^{17}$ cm^{-3}) $Al_{0.3}Ga_{0.7}As$ layer and a final GaAs heavily doped n^+ layer constitute the emitter and emitter contact layer of the transistor, respectively. Standard photolithographic processes were employed to fabricate HBTs out of the various structures, and evaluate their properties and performance.

Experimental Results and Discussion

I. Doped-Channel MODFETs.

(a). Quantum well channel width study.

In order to evaluate transport and device performance related to the width of the well, MODFET structures with well width of 300 , 150 , 75 , and 40 Å were grown. The center of each well was delta-doped to $2.4x10^{12}$ cm^{-2} Si atoms. The upper limit of 300 Å is established in order to maintain a high aspect ratio of gate-length to channel thickness, for efficient modulation of the channel [9]. Fig. 1 shows a summary of the mobility, μ, peak velocity, v(peak), and the transconductance g_m of devices with 1.8 μm gates. The degradation of all parameters with the reduction of the well width is clearly evident. Interestingly the peak velocity and transconductance are varying in exactly the same manner. The degradation of the transport properties with well width reduction is explained by regarding the electron wavefunction (EWF) and its increased overlap with the Si doping ions due to higher confinement in the thinner wells. This produces increased scattering and reduces transport. In addition, more scattering is produced at the outer walls of the well due to interface roughness. Most importantly, the energy difference between the ground state in the well and the adjacent AlInAs barrier layer Γ-valley minimum, is reduced, which increases the probability for real-space transfer of electrons to that valley, thus degrading transport in the channel. On the basis of these results and the high aspect ratio needed for good gate control, it was concluded that the 300 Å well width was the optimum thickness for the structure, and was adopted for all subsequent device structures.

(b). Channel doping profile.

To evaluate and optimize the doping profile parameter, four structures were grown: uniformly doped well, center delta-doped, top delta-doped (close to gate), and back delta-doped (close to buffer layer). A standard MODFET structure was also grown with the same doping in the top AlInAs layer. The uniformly doped channel structure had a doping of $8x10^{17}$ cm^{-3} Si atoms, while the delta doped structures had a doping dose equivalent to this, but as a sheet of charge in a 2-dimensional plane. Each structure with doped channel resulted in a 2-DEG concentration of ~$2.4x10^{12}$ cm^{-2}. The standard MODFET structure resulted in a 2-DEG concentration of $1.84x10^{12}$ cm^{-2}, implying that only 77%

of the dopants contributed electrons to the channel.

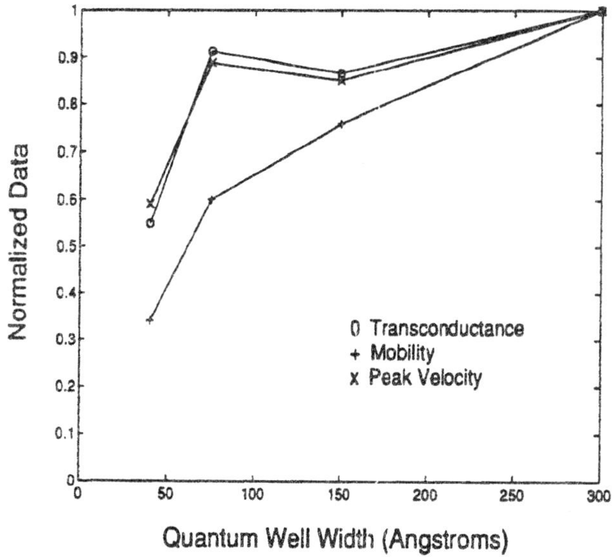

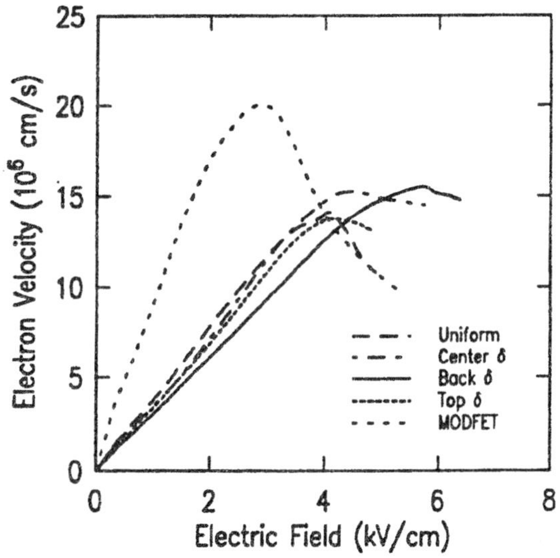

Fig. 1. Well width effects on μ, v(peak), g_m. **Fig. 2.** Velocity vs field for all 300 Å structures.

The evaluation of the structures revealed that the uniformly doped samples had the highest mobility, but v(peak) was higher in the center and back δ-doped. The lower mobility in the case of the center delta-doped structure is attributed here to increased scattering from the sheet of ionized impurities that coincides with the EWF peak for the ground state, while the top and back delta structures had enhanced scattering due to interface roughness. Fig. 2, shows the measured velocity-field curves. The standard MODFET structure reached a peak velocity of $2x10^7$ cm/s at 2 KV/cm, which was considerably less than the $2.6x10^7$ cm/s value measured here in undoped bulk InGaAs. This indicates

Doping Profile	1.8μm g_m^{ext} (mS/mm)	1.8μm g_m^{int} (mS/mm)	1.8μm I_{DSsat} (mA/mm)	0.5μm g_m^{ext} (mS/mm)	0.5μm I_{DSsat} (mA/mm)
Uniform	267	298	128.4	315	400
Top δ	214	232	168.0	435	539
Center δ	230	267	240.0	446	593
Back δ	200	211	255.2	310	517.5
MODFET	347	388	240	588	487.5

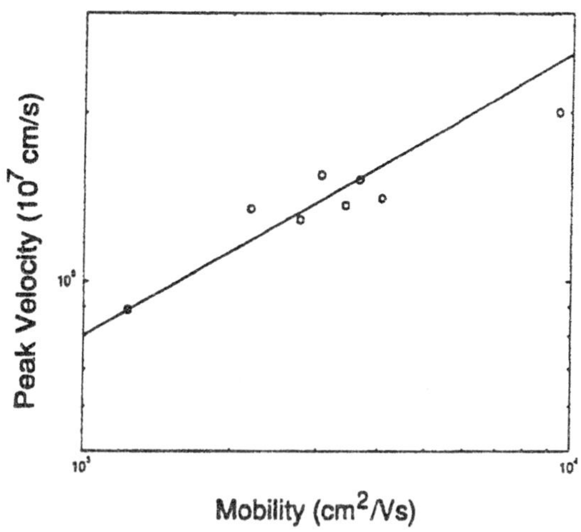

Table I. MODFET performance with doping profile. **Fig. 3.** v(peak) dependence on $\mu^{1/2}$.

that in addition to non-parabolicity, a main factor limiting electron velocity in MODFETs is real-space transfer from the Γ-valley of InGaAs to Γ-valley of AlInAs.

Table I summarizes the transistor performance results. As can be seen, the standard MODFET showed the highest transconductance. However, the magnitude of the increase in transconductance over the doped-channel devices, is considerably less than that expected from the difference in mobilities: while the mobility increased by 170%, the transconductance increased only by 32%, which corresponds to the observed 32% increase in the peak velocity of this structure. Nevertheless, in the 0.5 μm devices, the highest drain current values are observed in the doped-channel devices and in particular in the center delta-doped device.

A qualitative relation between μ, v(peak) and critical field has been derived previously for bulk GaAs [6], predicting a mobility $\mu^{1/2}$ and $\mu^{-1/2}$ dependence of peak velocity and critical field respectively. Fig. 3 is a plot of v(peak) vs μ, confirming the ½ power dependence on mobility, in the case of doped channel FETs.

(c). Channel compositional profile: Excess In in the channel.

The well width and doping profile studies, indicate that additional and significant improvements in transport and device performance, must address the issue of positioning the EWF in the well. It is evident that the EWF peak must be shifted in areas of the well removed from interface and doping charges. Furthermore, the introduction of a delta-doping in the center of the well where the EWP is maximum, gave increased drain currents I_{dsat}, in submicron devices, but at the expense of gain (g_m). It is important therefore that we introduce an additional parameter to improve gain (g_m) and further enhance I_{dsat}. To achieve such task, we introduced excess In in the channel in the form of an additional undoped layer in the channel with the composition $In_yGa_{(1-y)}As$ with y=0.65, 0.75 and 1.00. The thickness of the layers for y=0.65, and 0.75 was 75 Å, while for y=1.00 (InAs) was 30 Å. The delta-doping was placed in the center and the excess-In layers were placed 75 Å away from that towards the top (close to gate) of the structure, to avoid scattering and shift the peak of the electron wavefunction in the excess-In layer. Increasing the In concentration in InGaAs channels has been reported to improve mobility in the channel [10], and pseudomorphic MODFETs having significantly higher In concentration than the lattice matched InGaAs showed higher g_m and f_T (2). Fig. 4 shows the measured mobility versus temperature for each one of the structures. The control structure, y=0.53, with no excess In, shows no significant increase in mobility at low temperatures, indicating no separation of the electrons from the dopant atoms in the channel, while the mobility enhancement is observed at low temperatures in the excess-In structures, and in particular in the y=1.00 structure, where the enhancement is substantial. The mobility enhancement indicates the separation of the EWF peak from the ionized dopant atoms, in much the same way as a standard MODFET, but in this case, within the active channel of the structure. Hence, this novel type of device structure introduced here, is called the Quasi-MODFET.

Fig. 5 shows the transfer characteristics of the y=1.00 top layer structure, which for the sub-micron (0.5 μm) gate devices produced the highest g_m=537 mS/mm and a drain current of ~600 mA/mm. We have demonstrated that the EWF can be shifted to peak away from the ionized delta-doped impurity region to achieve enhanced transport in doped channel MODFETs. We can further utilize the natural tendency of the wavefunction to peak at the center of the quantum well channel, and strategically place an undoped InAs layer in the center, thereby further improving transport and

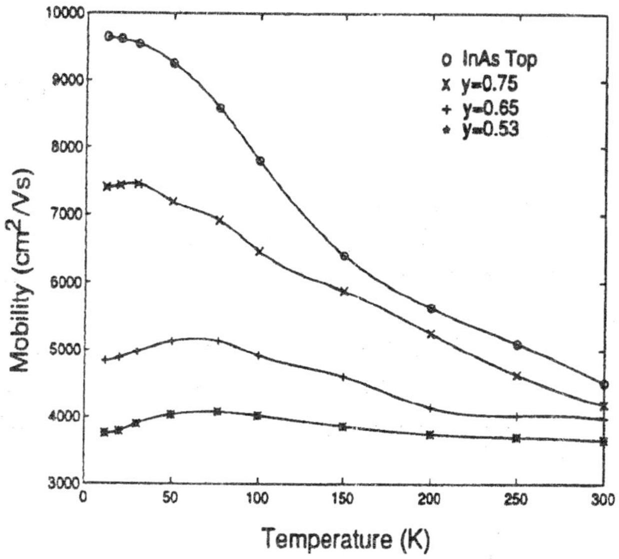

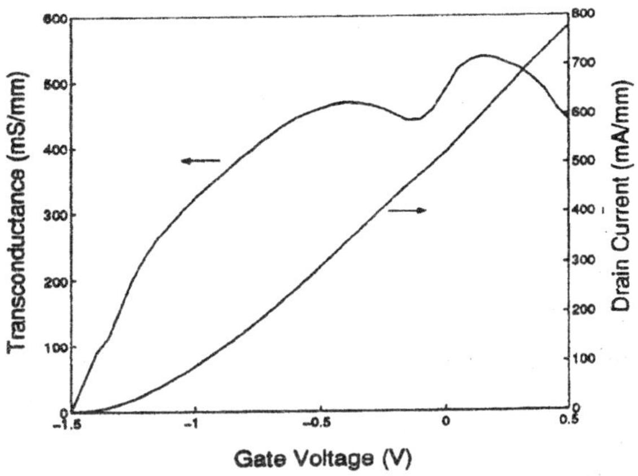

Fig. 4. μ vs T with excess In mole fraction y. **Fig. 5.** 0.5 μm device center δ-doped/top 30Å InAs.

device performance. In this case the channel is uniformly doped, and the InAs is undoped. This layer is varied from 0 to 40 Å, in 10 Å increments, to examine optimum performance. InAs thicknesses above 35 Å are beyond the critical thickness for dislocation formation due to strain relaxation from the lattice mismatch [11], and as such, transport in the layer deteriorates due to dislocations.
Fig. 6 shows the variation of mobility with temperature for these structures. The room temperature mobility of the 30 Å InAs sample is now 5330 cm^2/Vs, an increase of 31% over the control sample. The 40 Å sample mobility (omitted for clarity), decreases to 3450 cm^2/Vs, suggesting that dislocation formation has already started at 40 Å of InAs. The v(peak) also increases with InAs thickness, to become 1.63x10^7 cm/s, a 16% increase over the control sample (without exess In). Fig. 7 shows the

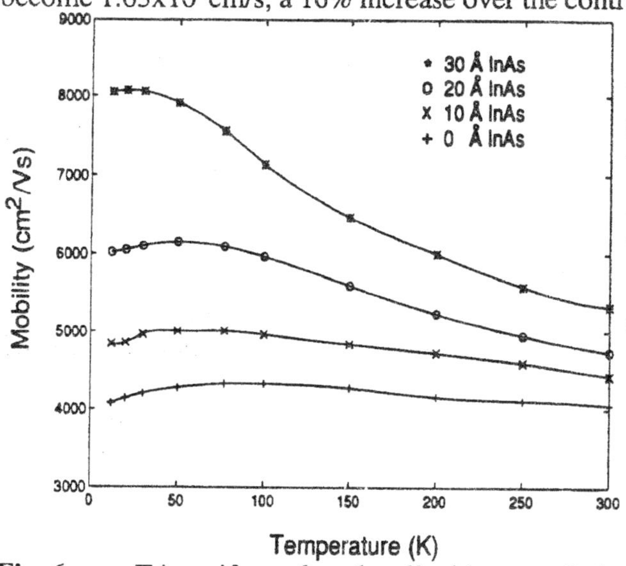

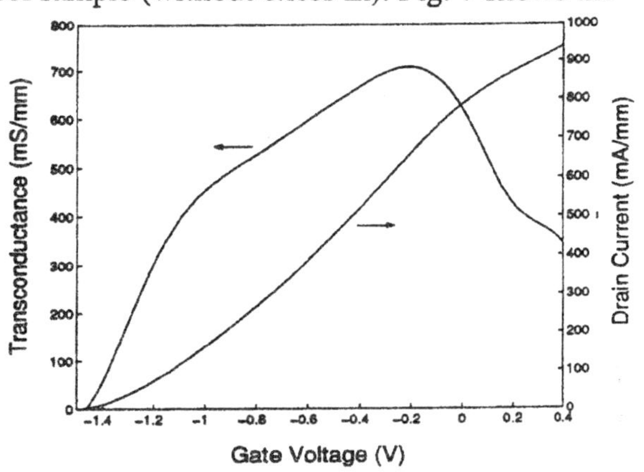

Fig. 6. μ vs T in uniform-doped well with center InAs. **Fig. 7.** 0.5μm Q-MODFET with center InAs.

transfer characteristics of the uniform-doped/center (30 Å) InAs structure. This 0.5 μm FET gave record g_m over 700 mS/mm and drain currents approaching 1 A/mm at small positive gate biases. Table II provides the transport results on the uniform-doped channel device structures, with various thicknesses of undoped InAs in the center, and Fig. 8 summarizes the mobility, v(peak), and g_m for these devices normalized against the control sample. Optimum values are observed for the 30 Å InAs Quasi-MODFET.

InAs Thickness	N_{2-DEG} (cm^2)	μ (cm^2/Vs)	v_{peak} (cm/s)	$E_{critical}$ (kV/cm)
0 Å	2.44×10^{12}	4059	1.41×10^7	4.11
10 Å	2.56×10^{12}	4434	1.42×10^7	3.70
20 Å	2.44×10^{12}	4725	1.47×10^7	3.66
30 Å	2.44×10^{12}	5329	1.63×10^7	3.30
40 Å	2.11×10^{12}	3452	1.35×10^7	4.12

Table II. Uniform-doped channel/center InAs transport data.

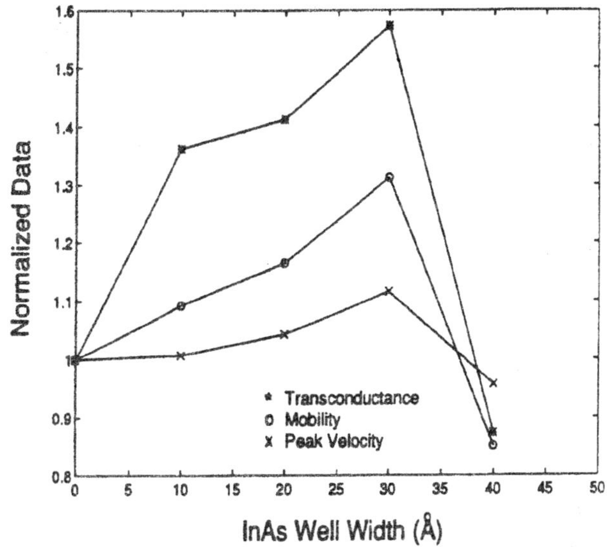

Fig. 8. μ, v,and g_m vs InAs width.

Break down voltages BV_{DS} under pinch-off conditions were measured and found to be between 9.5 and 14 V, with the 30 Å InAs optimum structure having a BV_{DS}=12.2 V. Although it is difficult to make direct comparisons of BV_{DS} with literature due to different geometries, our values are much greater than those of InAsSb based FETs and comparable to those reported for 2 μm gate InAlAs/InGaAs pseudomorphic FETs [2] and 0.25 μm GaAs based FETs [12].

II. Graded Base Heterojunction Bipolar Transistors (HBTs).

The design and optimization of the HBT structure entailed developing the emitter, base and collector regions, by identifying critical parameters relevant to the high current laser driver application. In such an application, typical threshold currents for the lasers are I_{TH}=50 mA, requiring the HBT driver to provide a maximum output current at twice that level, ie $I_c$max=100 mA. Furthermore, a reasonable dc current gain β of 20 to 50 is necessary for efficient operation.

The design of the base has a strong effect on β, either by compositional or doping variation. For example a thicker base would provide less spreading resistance for the same doping level than a thinner base, thus improving current levels, but will also increase the base transport time, thus reducing the speed of operation. The introducion of a compositional grading provides an additional electric field in the "neutral" region of the base, to help the injected minority carriers traverse the base faster, thus reducing transport time. However, grading reduces the valence band discontinuity barrier for holes at the BE junction of the HBT, thus increasing carrier injection into the emitter and reducing emitter efficiency, which in turn reduces β.

Therefore, the antagonistic nature of these parameters require strong optimization processes depending on application. In order to design the base grading correctly, we have developed a rigorous self-consistent two-dimensional drift-diffusion and energy-balance simulation of Npn AlGaAs/GaAs HBT [13] to study the effect of compositional grading of the base, on performance.

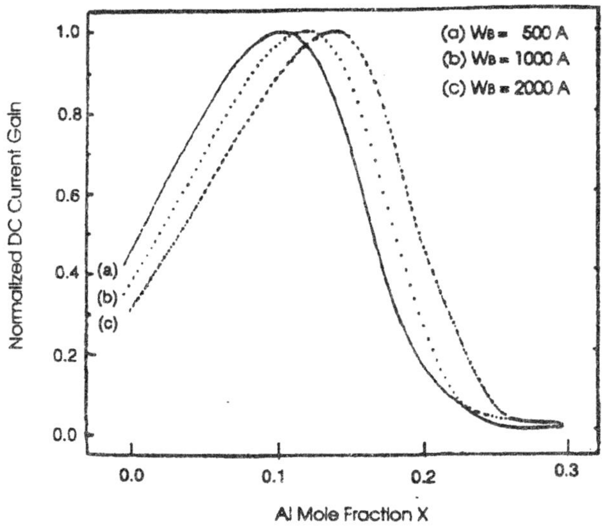

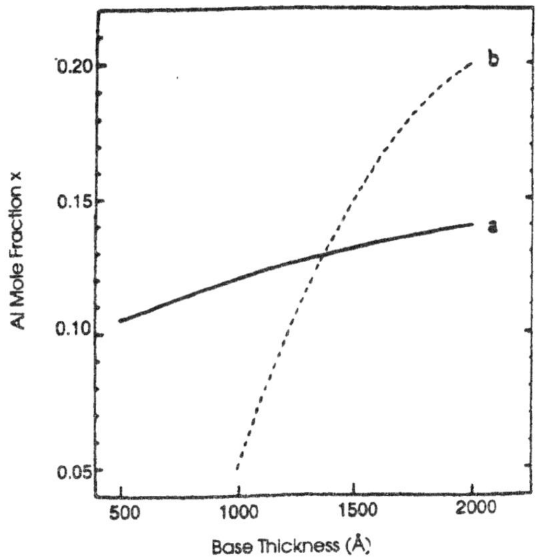

Fig. 9. Normalized β vs Al fraction x of graded-base. **Fig. 10.** (a) Optimized β and (b) transit time.

Fig. 9 shows the results of the simulation of the normalized dc gain β with the Al mole fraction x of the base grading for three different base thicknesses, w_B=500 , 1000 and 2000 Å. As can be seen, the range of Al mole fraction x in the $Al_xGa_{(1-x)}As/GaAs$ graded base where β is maximum, is relatively narrow (between 0.10 and 0.15), the thicker base affording more Al grading than the thinner base transistors.

Fig. 10 shows the maximum β with base thickness measured here, and the optimum (minimum) electron transit time from ref. [14] for comparison. As can be seen, the intersection of the two curves provides the optimum x and w_B for current gain and transit time. Any values below or over would enhance one parameter and degrade the other.

Fig. 11 shows the I_C-V_{CE} output characteristics of a graded base HBT with x=0.1 and w_B=2000 Å. It is evident that the high current capability required is not reached in this transistor.

Collector design is critical to the current capability of the HBT, since a long collector (w_C large) has reduced resistance for the same doping level, but collector transit time increases, while a shorter collector will require higher doping to carry the high current, thus reducing break down voltages in the transistor. For the collector, usual parameters are w_C=6000 Å and $N_D=5x10^{16}\,cm^{-3}$. However, in our case, we increased the doping of the collector substantially to $N_D=1x10^{17}\,cm^{-3}$, and reduced w_C to ~4000 Å. This in turn will increase the $I_c$max of the transistor significantly, but will decrease the break-down voltage BV_{CE} as well. However the reduction in BV_{CE} is not a significant factor in this application of a laser driver. This is because laser drivers operate under high current and low voltage (3 to 4 V) conditions in contrast to microwave power amplifiers that operate under low

current/high voltage conditions, and hence the requirement for high BV_{CE}. A subcollector layer of 8000 Å and $N_D=5x10^{18}$ cm^{-3} is also incorporated to facilitate with current flow. In addition to the collector design, the EB junction is optimally graded to reduce the conduction band discontinuity barrier for electrons, and the emitter metal contact junction is graded from AlGaAs to GaAs to facilitate ohmic contact operation at high currents.
Fig. 12 shows the I_C-V_{CE} output characteristics of a graded base, optimized collector HBT. Clearly the high current target of $I_c max=2I_{TH}$ is satisfied. A current gain β of over 30 is measured, which is within the targeted range. The negative output conductance observed in the characteristics is due to thermal effects under high current conditions, possibly also related to ohmic contact problems as well. Thermal effects are reached in this case, before the Kirk effect, because of increased collector doping, which implies no collector push-out due to high injection levels of electrons from the base. Further work, however, is necessary to address the thermal effects, optimize the geometry of the HBTs, reduce surface recombination by surface passivation, and improve ohmic contact formation.

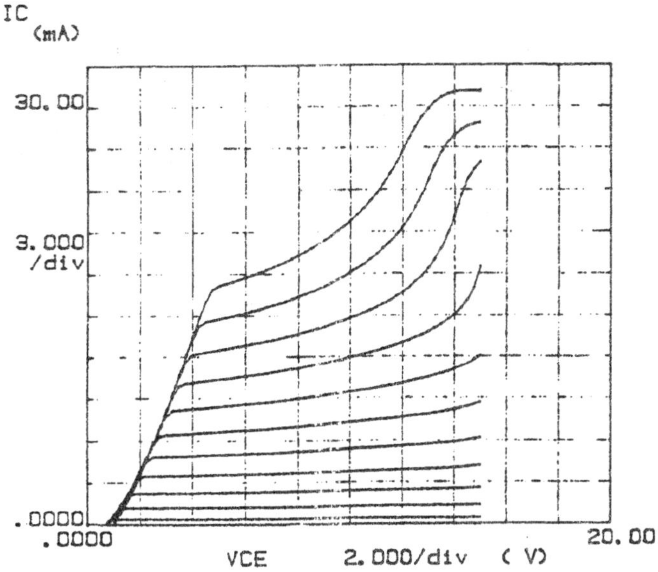

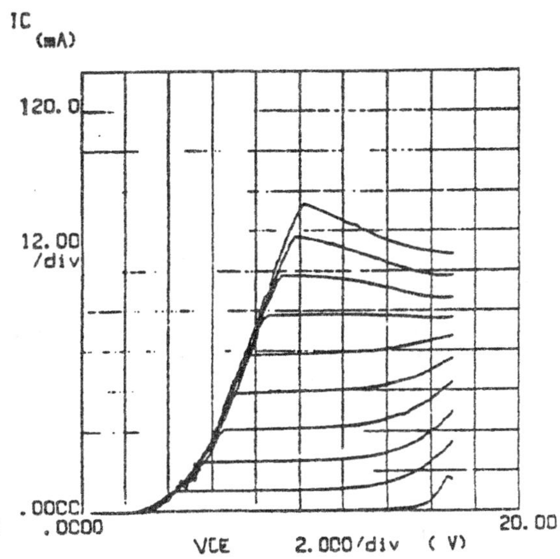

Fig. 11. I-V characteristics for optimized base-grading. **Fig. 12.** Optimized base and collector.

Summary.

Application specific devices with high current/gain capabilities for applications in mobile communication and optical communication networks, have been developed. The transport and device characteristics of doped channel MODFETs have been investigated to show that engineering the position of the electron wavefunction in the channel, using doping and compositional profiles, is an appropriate way of achieving high gain/current doped channel devices. This approach resulted in the development of a novel record high performance device, the Quasi-MODFET.
The compositional grading of the base of an AlGaAs/GaAs HBT, has been modeled by a rigorous two-dimensional drift-diffusion and energy-balance analysis. The model was used optimize the base of HBTs and fabricate transistors with varying Al mole fraction in the base designed for high current laser driving applications. HBTs fabricated with optimized graded base, and collector, satisfied the

current/gain requirements for the driver, but further work is needed to fully optimize performance.

Acknowledgements: The authors wish to acknowledge Drs W. T. Masselink, S. A. Rishton, C. E. C. Wood, D. Stone, and the partial support by the Joint Program for Advanced Electronic Materials.

References.

[1] L. D. Nguyen, A. S. Brown, M. A. Thompson, L. M. Jelloian, IEEE Trans. ED 39, 2007, 1992.
[2] S. R. Bahl, B. R. Bennett, J. A. Del Alamo, IEEE Elect. Dev. Lett. 14, 22, 1993.
[3] D. L. Miller, R. M. Asbeck, R. J. Anderson, F. H. Eisen, Elect. Lett. 19, 367, 1983.
[4] C. M. Maziar, M. S. Lundstrom, IEEE Elect. Dev. Lett. 8, 90, 1987.
[5] H. Ito, T. Ishibashi, T. Sugeta, Jap. J. Appl. Phys. 24, L241, 1985.
[6] W. T. Masselink, Semicond. Sci. Tech., vol 4, pp 503, 1989.
[7] J. K. Zahurak, A. A. Iliadis, S. A. Rishton, and W. T. Masselink, IEEE/Cornell Conf., IEEE Proc., New York, pp. 270-278, 1993.
[8] J. K. Zahurak, A. A. Iliadis, S. A. Rishton, W. T. Masselink, and T. Neal, Proc., 1995 Intern. Semiconductor Device Research Symposium, Charlottesville, Virginia, pp. 571, 1995.
[9] Y. Awano, M. Kosugi, K. Kosemura, T. Mimura, M. Abe, IEEE Trans. ED 36, 2260, 1989.
[10] G. I. Ng, D. Pavlidis, M. Quillec, Y. J. Chan, M. D. Jaffe, J. Singh, IEEE Elect. Dev. Lett. 52, 728, 1988.
[11] J. Matthews, A. Blakeslee, J. Crystal Growth 29, 273, 1975.
[12] T. Hwang, M. Feng, IEEE Elect. Dev. Lett. 13, 445, 1992.
[13] S. A. Tabatabaei, A. A. Iliadis, C. E. C. Wood, D. Stone, 1995 Intern. Semiconductor Device Research Symposium, Charlottesville, Virginia, pp. 681, 1995.
[14] X. Zhou, IEEE Trans. ED 41, 484, 1994.

Factors influencing the electric conductance of SnO_2 gas sensors

R. Ionescu, Ana Vancu
Institute of Physics and Technology of Materials, Bucharest

Abstract

The ever increasing concern about air pollution, the need to monitor potentially hazardous gases and to improve the efficiency of combustion processes has stimulated during the past three decades an extensive research and development of gas sensors.

Among various types of gas sensing devices, most popular became the chemoresistive sensors based on the surface characteristics of semiconducting oxides, most commonly termed semiconductor gas sensors SGS. They offer the advantages of small size, simple operation, high sensitivity and simple associated electronics. They detect small concentrations (up to a few ppm) of reducing gases (H_2, CO, NO_x, hydrocarbons, alcohol vapours) in oxygen containing atmospheres (usually air) by variations in their electrical conductance.

Besides the atmospheric composition, however, and depending on the conditions of preparation of the semiconducting oxide (heat treatment, doping), other important factors which produce conductance variations are the temperature of operation and the electrical phenomena at the electrodes.

In our laboratory we concentrated on the study of sintered SnO_2 thick-film samples, by means of electrical conductance measurements.

In the following we shall try to point out some ways of exploiting such measurements for the explanation of some peculiarities observed in the operation of SnO_2 SGS.

I. Introduction. Generalities about SGS

SGS typically consist in a semiconducting oxide film deposited on an insulating substrate, provided with two noble metal electrodes and a heating system yielding operation temperatures of 200-400 °C.

The semiconducting oxides used as gas-sensing materials have wide band gaps typical for insulators (3.6 eV for SnO_2), but their conductivity enters the range of semiconductors due to point defects in the crystal structures (oxygen vacancies in SnO_2, acting as donors).

The determinant element in the interaction with the surrounding atmosphere is the surface. Changes in conductance are mainly due to changes in the free electron concentration occurring by charge exchange with adsorbed species from the gas phase, in a thin region below the surface. For a high sensitivity of gas detection, a specific area as high as possible is needed. This is realised in practice by using thick, porous layers of partially sintered SnO_2 material, consisting of assemblies of fine oxide particles.

With such samples, operating at atmospheric pressure in air with a normal concentration of water vapours of 20-30000 ppm we are far from the "ideal" conditions and the "direct" conclusions of conventional methods of surface analysis. The study of SGS is situated somewhere at the border between engineering, chemistry and surface physics. The complexity of the phenomena involved has till now left scientific interpretations far behind the technological performances attained.

0-7803-3223-7/96/$5,00

For SnO_2 SGS, the mechanism most often quoted as dominant is adsorption/desorption. It is generally accepted that in air, at operating temperatures of 200-400 °C, oxygen is chemisorbed on ionised donor sites (lattice oxygen vacancies), trapping free electrons. The chemisorption of oxygen species O_2^-, O^-, O^{2-} produces at the surface of each crystalline grain a surface potential which can be evaluated by solving the Poisson equation. In each grain a depletion layer is formed between bulk and surface, and potential barriers arise between neighbouring grains (Fig. 1).

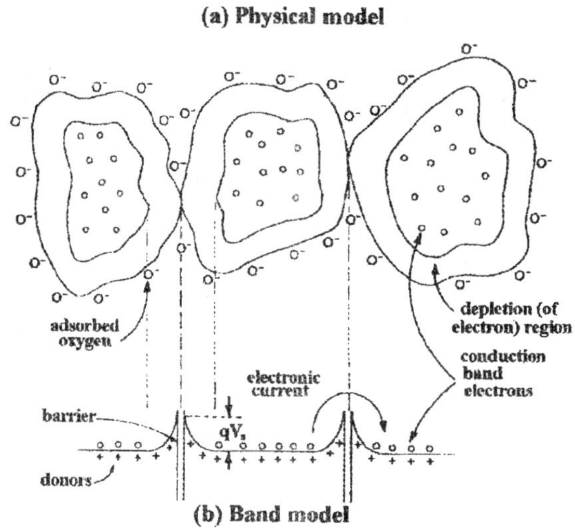

Figure 1. Physical and band model for SnO_2 sintered powder in air

The barrier height eV_S is given by:

$$eV_S = \frac{Q_S}{2en_D\varepsilon\varepsilon_0} \quad (1)$$

where Q_S is the total localised surface charge, resulting from the coverages with chemosorbed oxygen species.

The electrical conductance will be proportional to the electron probability of jumping over the potential barriers:

$$G \sim \exp\left(-\frac{eV_S}{kT}\right) \quad (2)$$

When a reducing gas R is present in the atmosphere, surface reactions between R and chemosorbed oxygen produce free electrons, thus lowering the potential barriers and increasing the conductance (the sensing effect).

The surface processes are evidently dynamic, resulting from a competition between oxygen chemisorption (rate k_A), desorption of chemisorbed oxygen (rate k_D) and oxygen reaction with R (rate k_R). They are governed by a classical kinetic equation (simplified by considering a single chemisorbed oxygen species):

$$\frac{d\theta}{dt} = k_A(1-\theta) - k_D\theta - k_R\theta \tag{3}$$

where θ is the surface coverage by chemisorbed oxygen

n is the concentration of free electrons ($n \sim \exp\left(-\frac{eV_s}{kT}\right)$).

The first term in (3) describes oxygen chemisorption, proportional to the concentration of unoccupied sites 1-θ and to the surface concentration n of free electrons (partners in chemisorption). Evidently $k_A \sim p_{O_2}$ and $k_R \sim p_R$.

Neglecting the desorbtion term an taking $1-\theta \sim n$, we get at equilibrium:

$$n \sim G \sim p_R^{1/2} \tag{4}$$

Eq. (4) describes satisfactorily the experimental dependence (Fig. 2), but evidently contradicts eq. (2). We discussed this contradiction in [1] by analysing the variation of G in time after steep transitions between oxygen and nitrogen and came to the rather surprising conclusion that the SnO_2 surface behaves apparently like a n-type semiconductor electrically influenced by the ambient atmosphere, but isolated from the bulk by a highly resistive, "intrinsic" SnO_2 region.

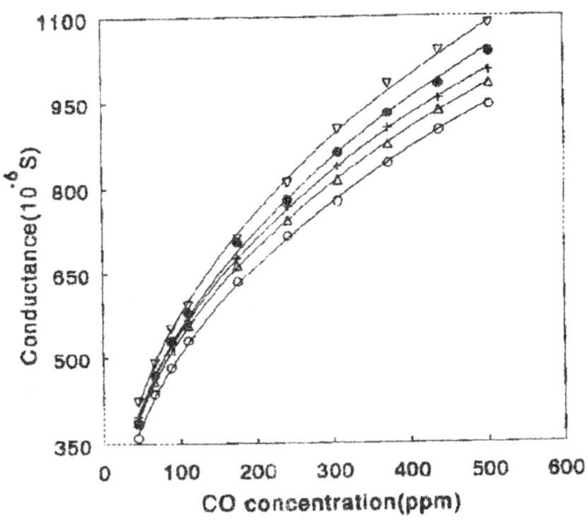

Figure 2. The dependence of conductance on CO concentration; different curves correspond to measurements performed in five consecutive days: (+) first day; (Δ) second day; (o) third day; (∇) fourth day; (●) fifth day.

One of the major problems in exploiting SGS is the influence of ambient humidity upon conductance and gas sensitivity. Starting from the "surface defect mechanism" proposed by Heiland and Kohl [2] we have studied the role of water vapours in the detection of CO with SnO_2 SGS [3]. We proposed in this paper a model in which interaction of water vapours with the surface of SnO_2 creates a supplementary number of lattice oxygen vacancies, which represent a supplementary number of sites for oxygen chemisorption.

II. Why sensors and not transducers ?

A transducer relates two physical entities by a univocal mathematical relation; the influence of different external parameters upon this relation is also univocal.

In the case of SGS, the conductance varies in time even under stable external conditions (ageing). Therefore neither the relation between conductance and concentration of the reducing gas, nor the relations between conductance and other external parameters (temperature, humidity, oxygen partial pressure, applied voltage) are univocal. They all show strong hystereses.

In the following we describe experiments performed in our laboratory, by which we try to explain this behaviour.

A statistical analysis has shown that in ~10% of our samples after applying the d.c. voltage, the conductance G decreases in time (Fig. 3), and after interrupting the voltage the samples generate currents of the order of nA, with the polarity of the d.c. voltage formerly applied.

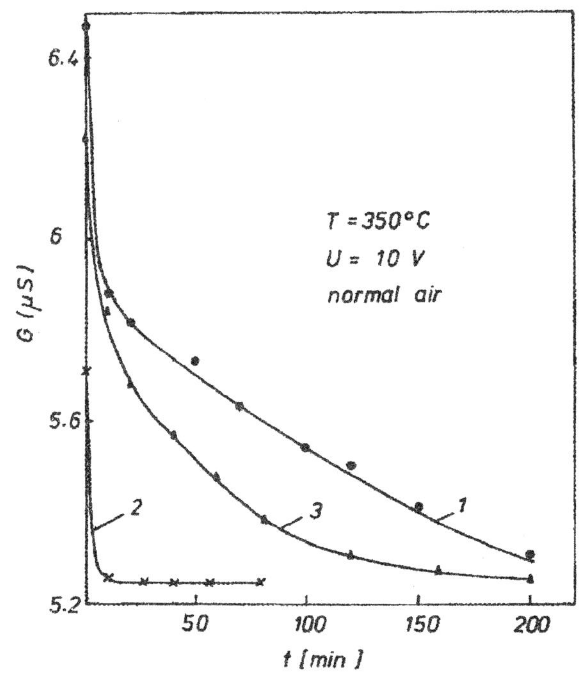

Figure 3. Time dependence of the conductance in usual operating conditions: (1) after having kept the sensor for 3 days at 350 °C without any applied voltage U; (2) after 15 min at 350 °C without U; (3) after 24h at 350 °C without U.

With the same samples, when registering the conductance evolution with constant-rate increasing temperature, we observed that the characteristic maximum - usually situated at ~250 °C and attributed to the competition between donor excitation and oxygen chemisorption in SnO_2 - is situated at temperatures varying from 200 to 350 °C for different samples.

Taken together, these three experimental observations lead us to the idea that near one of the noble metal electrodes some phenomena occur, which influence the conductance.

When heating these samples up to 700 °C and cooling them under applied d.c. voltages of 10÷20V down to usual work temperatures of ~300 °C, higher generated currents (tens of nA) and asymmetric I-V characteristics (in a factor of 3÷5) are obtained, which maintain themselves during several days.

In order to erase these currents we repeatedly heated and cooled the samples between 300 and 700 °C without any applied voltage.

After each cooling, the generated currents increase exponentially with temperature, with activation energies which increase with decreasing current intensity (Fig. 4) and with attenuating asymmetry of the I-V characteristics.

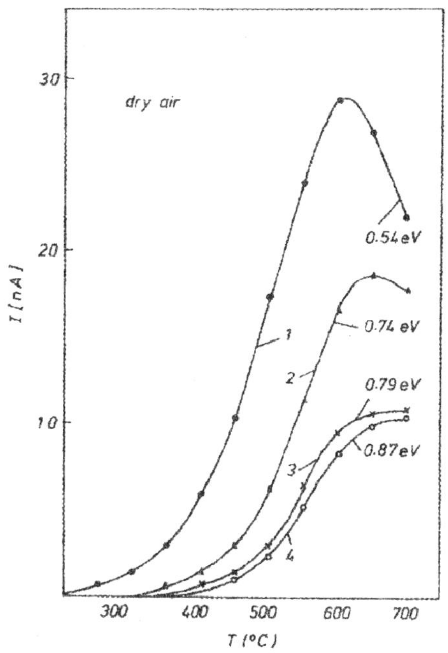

Figure 4. Temperature dependence of the currents generated during repeated heatings without any applied voltage.

These results can be understood if we consider that in the forbidden gap of SnO_2 local electronic traps are present in a concentration comparable with that of the donors, distributed over a broad energy range (of the order of one eV below the edge of the conduction band) [4].

In order to verify that the manifestation of the deep traps was due to accidentally high electron densities injected at the cathode, we deliberately prepared samples with electrodes differing as much as possible by their contact area with SnO_2.

Under a.c. voltages (of the order of ~100Hz) their I-V characteristics is almost symmetrical.

Under d.c. voltages, however, after a strong (in a factor of 50) and rapid (1 min) decrease of the current in one direction (reverse bias), a pronounced asymmetry is installed in the I-V characteristics (in a factor of ~100). Samples with different contact area of the electrodes behave therefore like diodes.

Under direct voltages, the behaviour of diode-like samples is identical with that of usual SnO_2 SGS. Under reverse voltages, however, different behaviours are observed:

- after interrupting the voltage applied, the currents generated are very high (of the order of μA); their decrease in time can be described by a combination of exponentials with time constants ranging from seconds to hours;

- conductance versus temperature curves show no maximum, but rather Arrhenius dependencies (Fig.5).

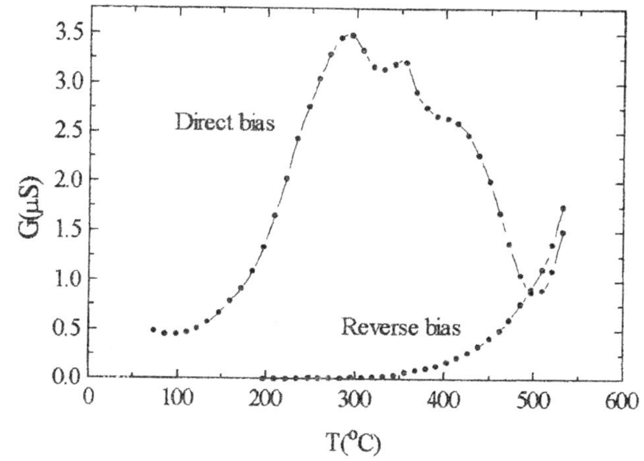

Figure 5. Conductance versus temperature dependence (by constant-rate heating) for diode-like samples: (a) direct voltage; (b) reverse voltage.

With increasing reverse voltage the sensitivity for gas detection decreases and at 5÷10V the device becomes quite insensitive to ambient atmospheric composition (oxygen, humidity, reducing gases).

When the contact areas are not very different, the decrease in time of the (reverse) currents are slower and less important, the asymmetry in the d.c. I-V characteristics is less pronounced (in a factor of 5÷10 only) and the gas sensitivity under reverse bias is in an order of magnitude higher than under direct bias, in contrast with highly asymmetric diode-like samples.

These experimental results can be understood by remembering the former hypothesis of a complex, broad distribution of high density electron traps in the forbidden gap at the SnO_2 surface, competing with donors and states of oxygen chemisorption and manifesting itself in the phenomena of ageing and of conductance hystereses. In diode-like samples, the large area electrode acts as an "ohmic" contact, with a relatively low current density. At the low-area electrode, however, the Schottky metal-SnO_2 barrier empties deep electron levels on which many electrons are trapped (high density of injected electrons), progressively forming a localised space charge which further increases the Schottky contact barrier. This supplementary bending of the energy bands may result in bringing the Fermi level below the deepest levels of oxygen chemisorption, and thus in the disappearing of any gas sensing effect.

Our experiments have shown that diode-like samples "memorise" the effects of the electrical fields applied. These effects and their memory are the more important, the higher the voltage and the higher the product between the temperature and the time of applying the voltage. Memory effects manifest themselves in conductance (changes in the shape of the conductance-versus-temperature curves, strong conductance hystereses) and in generated currents (current intensity, times needed for erasing the currents).

The "history" or "memory" parameter, containing applied voltage and the product of temperature and time, reflects the activation of more and more deep (and slow) electron traps and therefore the downward shift of the Fermi level.

In any simple experiment of measuring conductance during a heating-cooling sequence, when the temperature increases, electrons are progressively freed from deeper and deeper traps (depending on temperature and time spent at each temperature); when the temperature decreases, electron re-trapping on deep traps is so slow that the conductance values are lower than during heating, and hysteresis results.

Such hystereses were repeatedly observed even under moderate applied voltages. This shows that the redistribution of the occupancy of shallow, intermediate and deep traps, i.e. the stabilisation of the Fermi level (of the thermodynamic equilibrium) is slow even with weak electric fields.

When a higher electric field is applied at one of the electrodes, the effects of occupancy redistribution are more easily evidenced, because the Fermi level lies in the range of deep traps.

Experimental results point out that the conductance is determined not only by the levels of donors and of oxygen chemisorption, but by the total assembly of chemical and electronic states.

III. Conclusions

SnO_2 SGS can be described by two reversely polarised diodes (the electric contacts) bound by a gas-sensitive resistor (SnO_2); the Schottky barriers at the electrode contacts reflect every shift in the Fermi level of SnO_2. Usually the contribution of the electric contacts to the conductance of the device is weak, but it becomes dominant in diode-like samples, evidentiating the deep electron traps in the forbidden gap of the SnO_2 surface by effects of ageing, conductance hystereses, generated currents.

In order to transform SnO_2 SGS into transducers, we consider that technologic efforts should be directed towards the elimination (annihilation) of the deep electron traps.

Diode-like SnO_2 samples are an instrument for the investigation of electron traps distribution, but also an interesting new type of gas sensing device.

IV. References

[1] R.Ionescu, C.Moise and Ana Vancu, Are modulations of the Schottky surface barrier the only explanation for the gas-sensing effects in sintered SnO_2? Applied Surface Science 84 (1995) 291-297

[2] G.Heiland and D.Kohl, in T.Seiyama (ed.), Chemical Sensor Technology, Vol.1, Kodansha, Tokyo, 1988, Ch.2, pp.15-38

[3] N.Bârsan and R.Ionescu, The mechanism of the interaction between CO and the SnO_2 surface - the role of water vapour, Sensors and Actuators B, 12 (1993) 71-75

[4] R.Ionescu and Ana Vancu, Polarisation Effects in SnO_2 Thick-Film Sensors, Physica Status Solidi (a) 151 (1995) 135-142

INVITED PAPER

APPLICATION OF CATHODOLUMINESCENCE MICROSCOPY TO THE STUDY OF NATIVE ACCEPTORS IN GALLIUM ANTIMONIDE

J. Piqueras[a], B. Mendez[a], G.N. Panin[a,*], P.S. Dutta[b,+] and E. Dieguez[b]

[a]Departamento de Fisica de Materiales, Facultad de Fisica Universidad Complutense, 28040, Madrid, Spain.

[*]Institute of Microelectronics Technology and High Purity Materials, Russian Academy of Sciences, Chernogolovka, Moscow district 142432, Russia.

[b]Instituto Nicolas Cabrera and Departamento de Fisica de Materiales,C-IV, Universidad Autonoma, 28049, Madrid, Spain.

[+]Department of Physics, Indian Institute of Science, Bagalore-560 012, India.

Summary

Cathodoluminescence in the scanning electron microscope is used to ivestigate growth and prosess induced defects in GaSb crystals. In particular, luminescence emission has been used to study the nature of acceptor defects present after different annealing and irradiation treatments.

1.Introduction

In recent years, Gallium Antimonide has generated increasing interest as a potencial substrate material for various lattice-matched devices in the bandgap range 0.3eV to 1.58eV (1). However, significant progress has to be made both in materials and processing aspects before it can be employed for device applications. It is well known that the quality of epitaxial layers and the yield of devices made from them depend on the structural perfection of the substrate. Hence, characterization of defects is of great importance for reliable device perfomance. The interest of defect studies in GaSb refers notonly to the as-grown substrate material but to the effect of different processes as doping,ion implantation,etc. For

0-7803-3223-7/96/$5,00

defect characterization, the cathodoluminescence (CL) technique has been found to be highly sensitive and is extensively used for the spatial defect mapping of various elemental and compound semiconductors. However, CL microscopy has been seldom applied to defects study in GaSb (2) (3) (4). In this paper, we present the results of CL investigation of defects in bulk GaSb generated during growth, post growth annealing treatments, doping and ion beam irradiation.

Undoped GaSb is always p-type in nature with the acceptor concentration of approximately $10^{17}cm^{-3}$ at room temperature. The acceptor is intrinsic and is due to vacant gallium site (V_{Ga}) and gallium antisite (Ga_{Sb}). The acceptor concentration can be reduced either by non-stoichiometric melt growth or by employing low temperature growth techniques. In this work we have perfomed post growth annealing in vacuum, gallium and antimony atmospheres to examine the evolution and nature of the native acceptors. Due to the presence of high concentration of native acceptors, impurity doping to obtain n-GaSb always leads to compensation (5) (6). As a result, the mobility of electrons reduces and controlled doping to very low levels poses problems. Dopant incorporation by diffusion has been given limited attention till now (1). In this paper we present results concerning the nature of luminescent centers in Te-diffused GaSb. Such a study of diffusion doping is of interest for basic p-n junction device structures. From the device fabrication point of view ,the ion milling technique is of utmost importance and has been widely used in other materials. More limited attention has been devoted to the problems of ion implantation and related studies in GaSb (7) (8) (9) (10). In the last part of this work we have carried out ion milling of p-GaSb by Ar^+ ions and studied its effect on the electronic properties by CL. Electron beam induced current (EBIC) measurements in the scanning electron microscope have been also made to complement the CL measurements after ion milling.

2.Experimental Method

The samples used in this study were vertical Bridgman grown GaSb single crystals (11). For the annealing treatments the wafers were prepared by conventional chemo-mechanical polishing. Prior to annealing, the samples were etched in CH_3 COOH: HNO_3: HF (20:9:1) to remove any damaged layer left behind after polishing, dipped in HCl and rinsed in methanol. Thermal annealing of the samples was carried out by placing the wafers in quartz ampoule under vacuum. For the gallium annealing, 6N pure gallium was spread on the wafer surface at room temperature. After annealing, the gallium sticking to the surface was removed by rinsing in HCl. For annealing in antimony atmosphere, 6N pure antimony balls were kept along with the wafer. Other wafers were annealed under vacuum of 10^{-6} torr. The annealing temperature and time was kept constant at 500°C and 12 hours respectively for all the cases. The Te-diffusion experiments were carried out in evacuated and sealed quartz ampoules at 500°C for 2,15 and 24 hours. Prior to diffusion, the quartz ampoules were prebacked at 800°C under 10^{-6} torr to remove any volatile contaminants. The volume of sealed ampoules was about 6 cm^3. The diffusion source mass amount was approximately 100 mg. During the diffusion experiments, the wafers of GaSb were placed along with Te. Pieces of Sb were also kept to create an overpressure of Sb during the diffusion cycle. Diffusion was simultaneously carried out on several samples under identical conditions. At the end of diffusion, the ampoule was withdrawn from the furnace and cooled rapidly in liquid nitrogen to condense the Te vapour on the ampoule wall and thus prevent any metallic condensation on the sample surface. The wafers were rinsed in organic solvents and HCl before any measurements. No visible degradation of the sample surfaces occurred even after 24 hours of heat treatment. The depth profiles of the net donor concentration of varius samples were evaluated from the Capacitance-Voltage (C-V) measurements at room temperature.

The ion beam milling treatment was perfomed in an ion beam milling unit with 4kV Ar^+ ions for 30 min at room temperature. The samples,rotating during the ion milling process,were bombarded at an angle of 45° to the surface. The ion beam current density was 70 $\mu A/cm^2$.

CL measurements were carried out using a Hitachi S-2500 scannig electron microscope in the emissive and CL modes at 77K. A liquid nitrogen cooled North Coast EO-817 germanium detector was used for the signal detection. The details of the experimental set-up for spectral and panchromatic CL measurements are presented elsewhere (12). The EBIC measurements in the electron microscope were carried out at liquid nitrogen temperature with electron beam current and voltage of 10^{-9} A an 25 kV respectively. Electrical contacts for the EBIC measurements were provided by silver paste with gold wires to the treated (ion milled) region and back side of the sample.

3. RESULTS and Discusion

3.1 Annealed Samples

Fig.1 shows the CL spectrum at 77K of an as-grown sample. The main emission bands appear at 796 meV and 775 meV. Weak transitions at 756 and 830 meV can also be seen. The 796 meV corresponds to the band-band transition and the 775 meV (commonly known as band A) to a transition from the conduction band to the neutral state of the native acceptor level V_{Ga} Ga^{Sb} (13)(14)(15). The peak at 756 meV (designated as band B) has been previously attributed to a transition from the conduction band to to a triple native acceptor V_{Ga} Ga_{Sb} V_{Ga} , arising from excess of Ga-vacancies (16). However, the origin of this transition has not been clearly resolved up till now. The annealed samples exhibit only the 775 and 756 meV transitions, but with difference in relative intensities depending on the annealing conditions

(Fig.2a). In general, the relative intensity of the 775 meV with respect to the 756 meV

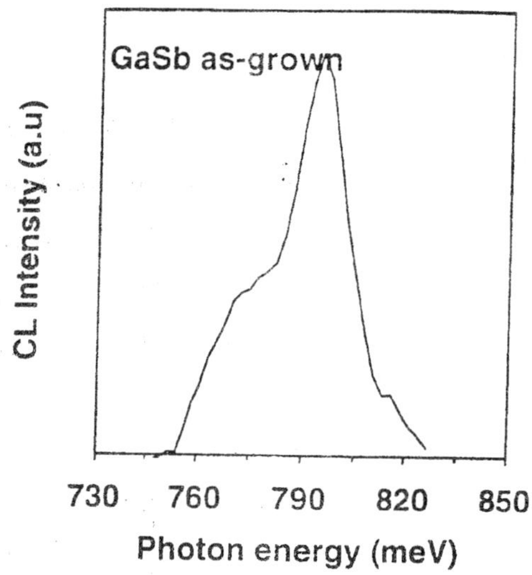

Fig. 1. Cl spectrum of as-grown GaSb

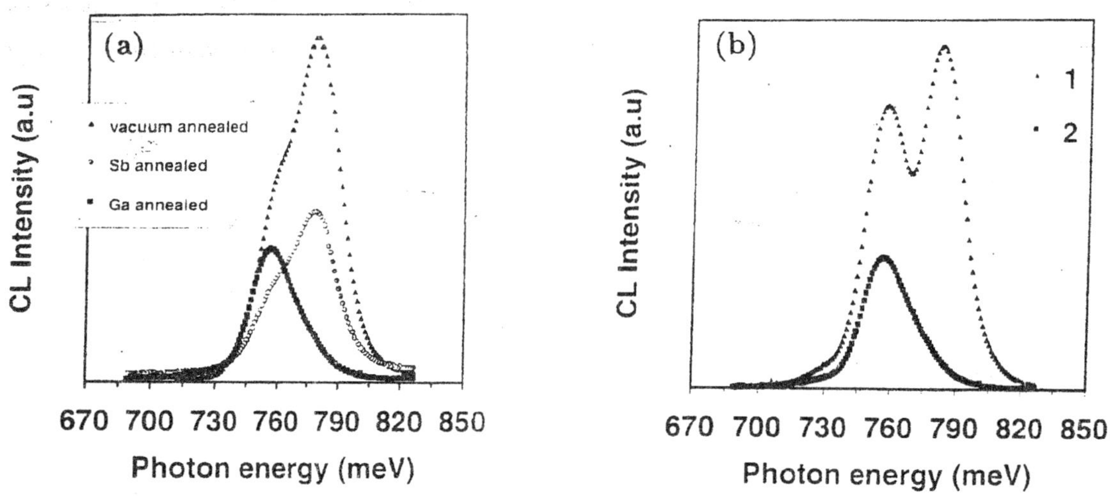

Fig. 2. Spectra of GaSb: a) after annealing in different atmospheres for 12 hours at 500 oC, b) in different regions of a Ga-annealed sample.

transition decreases after annealing. The vacuum and Sb- annealed samples show similar CL spectrum. Moreover, the relative intensity of the 775 meV with respect to the 756 meV transition is less in the Sb-rich samples. In the Ga-annealed sample, the 756 meV transition enhances drastically andat certain locations of the sample, it is even more intense than the 775 meV peak as

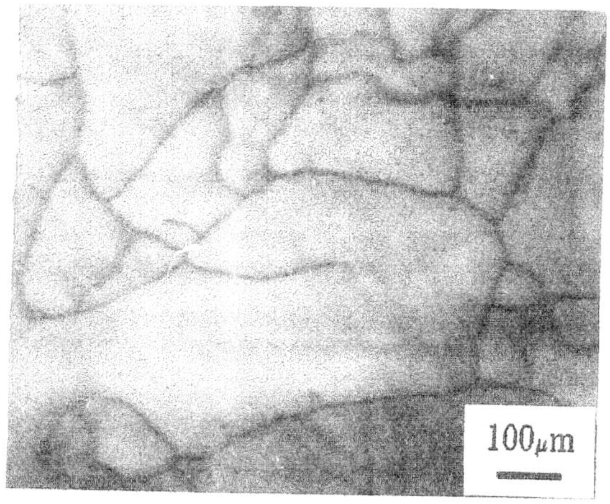

Fig. 3. CL image of as-grown GaSb

shown in Fig. 2b. Thus the band B transition is probably associated with a defect involving excess gallium and can be due to a gallium antisite or a related complex. Hence the decrease in relative intensity of the 775 meV peak compared to the 756 meV in the Sb and vacuum annealed samples can be explained by considering the reduction of the native acceptors on annealing, which shifts the Fermi level position towards the conduction band and hence the intensity of the 756 meV transition increases. The CL image of as-grown GaSb shows dark features related to the presence of sub-boundaries in the material (Fig.3). The CL images of the Sb-annealed sample revealed dark precipitate like defects which contribute to the reduction of luminescence emission (Fig.4a). These defects are not seen in as-grown or vacuum annealed samples. The defect structure in the vacuum annealed sample is similar to that of the untreated ones. The Ga-annealed sample exhibits a uniform CL image (see Fig.4b) without any contrast inhomogeneities(sub-boundaries etc.) as seen in the other samples. This observation suggests that post-growth annealing in Ga-bath can be technologically important for preparing GaSb wafers with uniform properties for opto-electronic applications. On the other hand, although annealing in Sb atmosphere reduces the native defect concentration, precipitate formation takes place modifying the potential applications of the GaSb crystals as substrates.

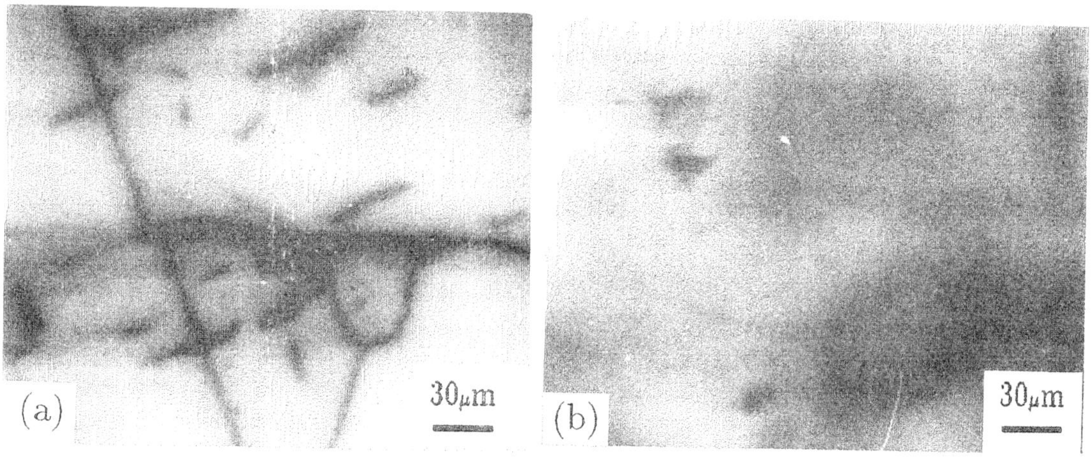

Fig. 4. CL images of GaSb annealed in Sb (a) and in Ga (b).

3.2.Te-diffused samples

CL images of Te-diffused samples are shown in Fig.5. The sample for which the diffusion was carried out for 2 hours appeared to be almost precipitate free (see Fig.5a), but small dark dots of 1-2 um size are clearly revealed. This contrast could be due to the decoration of dislocations associated with the presence of tellurium as occurs in other III-V compounds. As described above in undoped GaSb, Cl images show dark contrast related to the presence of subboundaries in the material and the inner portion of the grains appears uniformly bright. With increase in diffusion time to 15 hours, precipitates start appearing throughout the sample. The size and concentration of the precipitates increase with further increase in diffusion time. For the sample with 24 hours of diffusion, large clusters and precipitates could be seen (Fig.5b). Characterization of extended defects in heavily Te- doped CZ grown GaSb has been previously carried out by transmission electron microscopy (17) and planar defects have been identified as $Ga_2 Te_3$ precipitates. In our case, the precipitates can be either Te or $Ga_2 Te_3$. In order to get a more accurate information CL spectra obtained from different Te-diffused samples were analyzed. The sample with shortest diffusion time exhibits luminescence peaks at 744, 775 and 796 meV. The 744 meV peak is intrinsic to Te and has been attributed to transition from conduction band to an

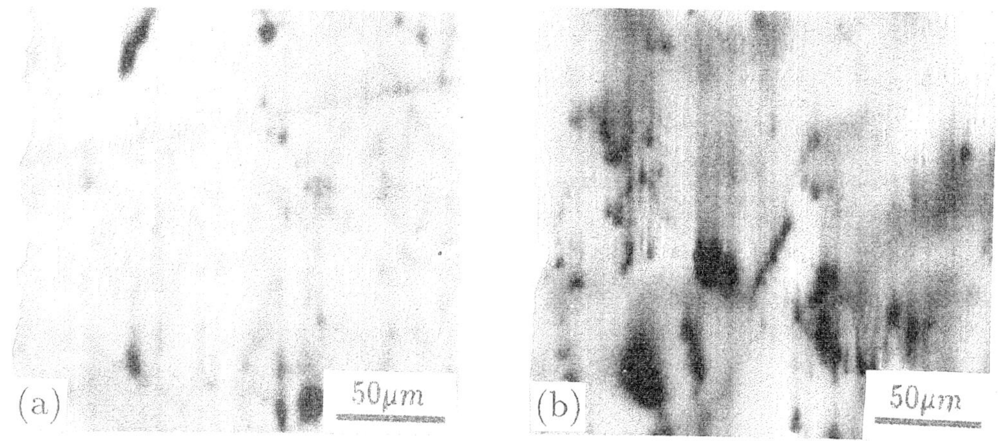

Fig. 5. CL images of Te-diffused sample after diffusion for 2 hours (a) and 24 hours (b).

acceptor state V_{Ga} Ga_{Sb} Te_{Sb}. The spatial distribution of this complex seems to be connected to dislocations as is shown in Fig.5a. With increase in Te concentration, the intensity of 744 meV peak increases (Fig.6), indicating the enhancement of

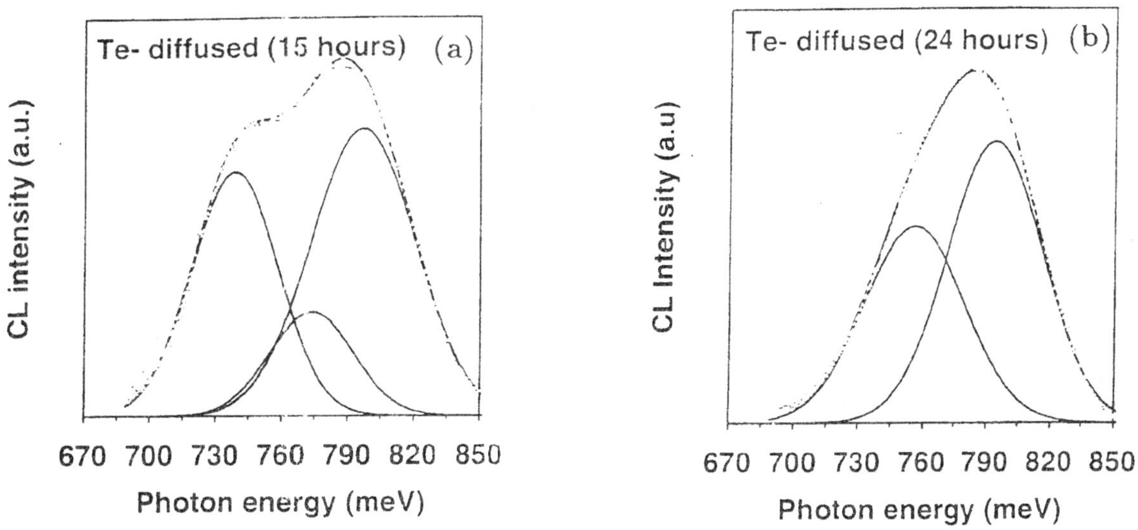

Fig. 6. CL spectra of Te-diffused samples after diffusion for 15 hours (a) and 24 hours (b).

the formation of the related acceptor centre. In the CL spectrum of the sample with highest Te concentration it is observed that the dominant transition occurs at 756 meV while the 744 meV emission is almost disappeared. The fact that the intensity of the 756 meV peak increases after long diffusion times can be explained by an increase in volatilization of Sb. This in turns results in excess of Ga and is similar to the situation of annealing under Ga- rich conditions. The probability of formation of Ga_2Te_3 precipitates and the increase in concentration and size of precipitates with increasing diffusion time can also be explained by the presence of excess Ga. It appears from the above results that the dominant compensating acceptor varies with increase in diffusion time.

3.3. Ion irradiated samples

CL images show that ion milling causes an enhancement of the panchromatic CL emission in the irradiated area and its surroundings as shown in Fig.7a. CL intensity is higher at the center of the treated region and decreases radially. Apart from the increased luminescence intensity in the irradiated regions, uniformly distributed precipitates could also be seen under higher magnification as depicted in Fig.7b. At exactly the location of a dark precipitate like in the CL image the secondary electron image exhibits precipitates like topographic features. These extended defects were absent on the

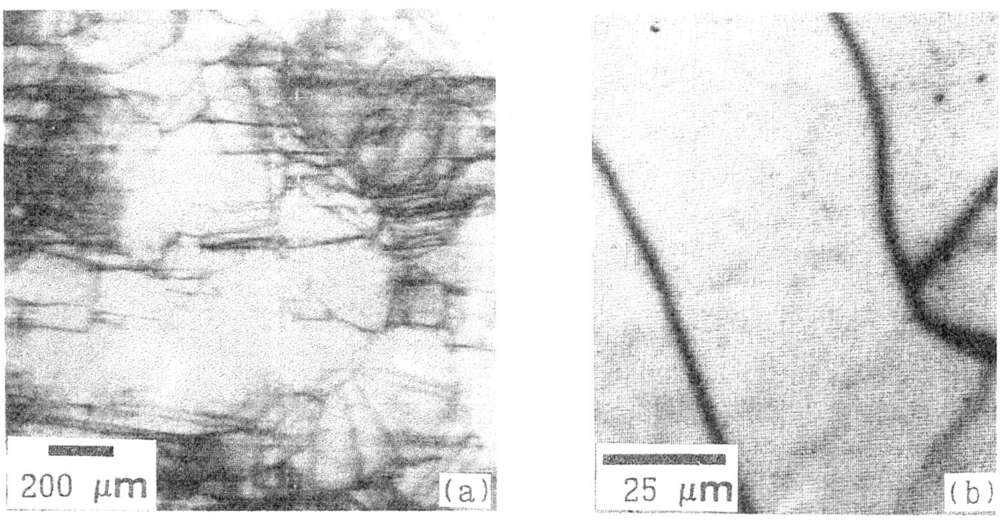

Fig. 7. CL images at the centre of the irradiated region at two magnifications.

surface of the as-grown samples and are created during the ion milling treatment. Apart from the increased luminescence no differences in the relative band intensities or spectral features of the treated and untreated samples could be observed. Spectra show the bands at 775 and 756 meV associated with the native acceptors V_{Ga} and Ga_{Sb}. The increase in CL intensity after ion milling can be explained either by a bombardment induced increase in acceptor concentration or by the reduction of non-radiative recombination centers at grain boundaries or dislocations. In order to study these possibilities and in general to investigate the effect of ion milling on the electronic properties of GaSb, the spatial resolved EBIC technique in the scanning elctron microscope was used. EBIC measurements with metal contact on p- GaSb cannot be carried out due to the low barrier height produced and therefore to the low efficiency of excess charge carriers collection. However, if as in other semiconductors, ion milling causes local p to n- type conversion EBIC measurements would be possible. Fig.8 shows a typical EBIC image of the ion irradiated

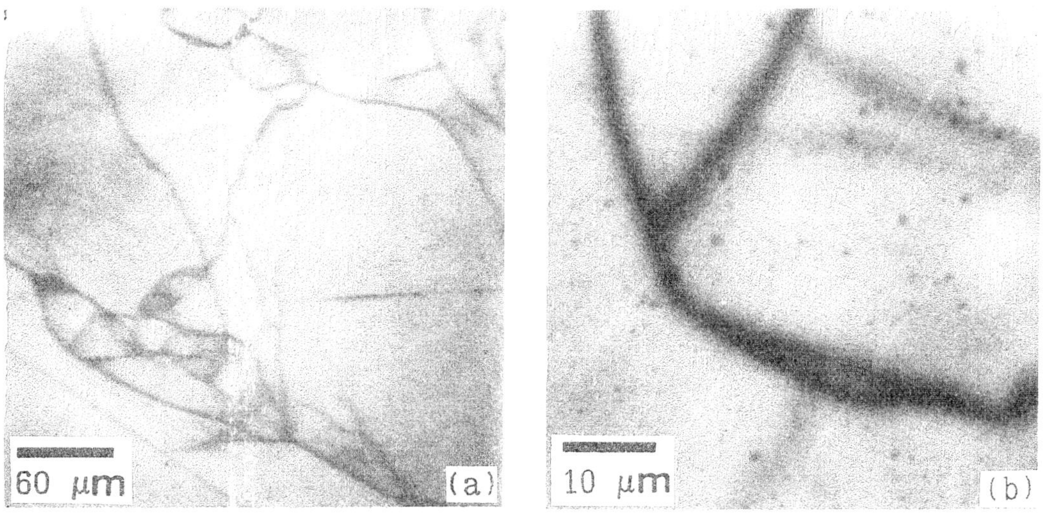

Fig. 8. EBIC images of GaSb after irradiation at two magnifications.

sample. The detection of EBIC contrast implies a type conversion caused by the ion milling treatment. The untreated samples did not show any EBIC signal.The polarity of the EBIC signal corresponds to the formation of a n-type surface layer.This result enables to rule out the possibility that the CL spectral changes after ion milling are due to the increase of acceptor concentration because such increase would not lead to an n-type surface region. The reduction of non-radiative recombination centers during ion milling seems to be a probable cause of the observed CL intensity increase.

4.Conclusions

CL mode in the scanning electron microscope is a suitable technique that enables to study the morphology and nature of growth and process induced defects in GaSb with high spatial resolution.CL images reveal that defect distribution in as-grown crystals is inhomogenous and annealing in vacuum ,Sb and Ga atmospheres causes an increase in homogeneity. Ga-annealed samples exhibit featureless CL images suggesting an uniform nature of the luminescent defect distribution. In these samples it has been found that a luminescence band at 756 meV,with an unclear origin till now, is related to excess Ga in the sample. Te diffusion provides new information about defects in GaSb. During Te-difussion the type of defects formed in the samples is a function of diffusion time. CL images show that this is a consequence of precipitation processes. Ion beam milling causes also an increase of the luminescence in GaSb as well as the type conversion from p to n of the subsurface layer as revealed by EBIC measurements.

References

1) A.G. Milnes and A.Y. Polyakov, Solid State Electronics 36, 803 (1993)

2) B. Méndez, P.S. Dutta, J. Piqueras and E. Dieguez, Appl. Phys. Lett. 67, 2648 (1995)

3) G.N. Panin, P.S. Dutta, J. Piqueras and E. Dieguez, Appl. Phys. Lett. 67, 3584 (1995)

4) P.S. Dutta, A.K. Sreedhar, H.L. Bhat, G.C. Duber, V. Kumar, E. Dieguez, U. Pal and J. Piqueras, Accepted in Journal of Applied Physics

5) K. Nakashima, Jpn. J. Appl. Phys. 20, 1085 (1981)

6) A. I. Lebedev and I.A. Strel'nikova, Sov. Phys. Semicond. 13, 229 (1979)

7) R.Callec, P.N. Favennec, M.Salvi, H.L. Haridon, and M.Gauneau, Appl.Phys.Lett.59,1872 (1991)

8) S.J. Pearton, A.R. Von Neida, J. M. Brown, K.T. Short, L. J. Oster, and U.K. Chakrabarti, J. Appl. Phys. 64, 629 (1988)

9) M. Perotin, L. Gouskov, H. Luquet, P. Abiale Abi, A. Sabir, and A. Pérez, J. Appl. Phys. 68, 3756 (1990)

10) Y. K. Su, K.J. Gan, J.S. Hwang, and S.L. Tyan, J. Appl. Phys. 68, 5584 (1990)

11) P.S. Dutta, K.S. Sangunni, H.L. Bhat and V. Kumar, J. Cryst. Growth. 141, 44 (1994)

12) B. Méndez and J. Piqueras, J. Appl. Phys. 69, 2776 (1991)

13) W. Jakowetz, W. Ruhle, K.Breuninger, and M. Pilkuhn, Phys. Stat. Sol.(a) 12, 169 (1972)

14) M. Lee, D.J. Nicholas, K.E.Singer, and B. Hamilton, J. Appl. Phys. 59, 2895 (1986)

15) M.C. Wu, and C.C. Chen, J.Appl. Phys. 72,4275 (1992)

16) D. Weiler and H. Mehrer, Philos. Mag. A 49, 309 (1984)

17) J. Doerschel and U. Geissler, J. Crystal Growth. 121, 781 (1992)

INVITED PAPER

NONLINEAR OPTICAL PROCESSES IN NONCRYSTALLINE SEMICONDUCTORS

Valentin Chumash, Ion Cojocaru

Center of Optoelectronics, Academy of Sciences of Moldova, Academiei str. 1, Chisinau, 2028, MOLDOVA

ABSTRACT

The paper summarizes the peculiarities of the nonlinear interaction of short laser pulses with the noncrystalline semiconductors (chalcogenide glasses, a-Si:H), showing that noncrystalline semiconductors (NS) are an interesting material for high speed optoelectronics. The physical mechanisms, which can contribute to the nonlinear light absorption in NS under pulsed excitation, are critically reviewed.

1. Introduction.

Nonlinear optical effects in semiconductors are studied to a much less extent in the noncrystalline than in the crystalline state. The interest to the study of the nonlinear propagation of laser radiation in NS is determined not only by the new fundamental physical mechanisms present in these materials, but also by a wide spectrum of possible applications in optoelectronic devices and photonic switching. NS are attractive candidates for the fabrication of all-optical passive and active devices in the same medium. In recent years a variety of both passive (fibers, planar waveguides, lenses, gratings) and active (nonlinear devices mainly based on Fabry-Perot interference, optical bistability and optical hysteresis) elements have been demonstrated [1, 2].

Special attention is paid to the phenomena that can be observed at relatively low levels of light intensity, in connection with their possible use in active elements of integrated optics and optoelectronics. As it will be shown the nonlinear change of the optical constants may be reached in the NS at much lower irradiation intensity than in the crystalline ones, when interband transitions with the participation of nonequilibrium phonons and nonequilibrium localized vibrational excitations play an important role. Structure inhomogeneities in NS lead to some spatial localization of the vibrational excitations, that is, they may be considered as quasiresonators for the vibration energy [3, 4], which can be similar to the electron localization in quantum well structures. The capture of nonequilibrium phonons by such quasiresonators leads to the increase of their effective lifetime [4, 5]. Promising effects are optical bistability and optical hysteresis, on the basis of which it is possible to build fast-response, all-optical switching and logical elements [6, 7].

In a number of works some of the light induced reversible changes of the NS optical constants were studied considering the nonlinear transmission of focused CW laser radiation through thin film samples. Three types of nonlinear interaction of CW laser radiation with NS were observed: the nonlinear change of the light transmission of NS samples; the periodic oscillations in time of the photocurrent, of the transmission and of the reflection coefficients of NS films; and the light transmission bistability. In spite of a considerable number of works up to now there are no reliable data about the physical processes, describing adequately the nonlinear interactions of the CW laser radiation with noncrystalline semiconductors. The peculiarities of the nonlinear transmission of CW and pulsed laser radiation through thin NS films are reviewed in [1, 2].

0-7803-3223-7/96/$5,00

The aim of this work is to illustrate the present state of knowledge and some unresolved problems of the nonlinear interaction of laser pulses with NS (chalcogenide glasses: As_2S_3, As_2Se_3, AsSe, $GeSe_2$, $As_{22}S_{33}Ge_{45}$, and a-Si:H). Attention is primary given to those nonlinear phenomena that are characteristic of the NS and, as a rule, have no analogous in crystalline phase. The investigated peculiarities of the optical hysteresis and nonlinear interaction of short laser pulses with the NS, are briefly discussed in section 2. The physical mechanisms, which can contribute to the nonlinear light absorption in NS under pulsed excitation, are discussed in the next section. A model of the physics of the nonlinear interaction of powerful light with NS, taking into account the light interaction with nonequilibrium phonons and localized vibrational excitations, is considered for the explanation of the experimental results.

2. Optical hysteresis and nonlinear absorption of laser pulses in NS.

The results of a systematic investigation of the optical hysteresis and nonlinear absorption at interband excitation of thin film samples (0.2 - 5 μm thick) of a-Si:H, a-SiC:H, As_2S_3, As_2Se_3, AsSe, $GeSe_2$, $As_{22}S_{33}Ge_{45}$ with short laser pulses (300 K, 77 K) are summarized in this section.

When the input light (with $h\nu \geq E_g$) pulse intensity was relatively low, the transmission of the NS films decreases with the thickness according to the usual Beer law (Fig. 1). However, increasing the incident light intensity over some threshold value (I_t) leads to a nonlinear light transmission by the NS films [1, 2, 8, 9]. The characteristic value of the threshold light intensity depends on the NS film composition, wavelength of excitation, temperature and laser pulse duration. For example, for As_2Se_3 , at 570 nm, the I_t values were equal: ~ 2 MW/cm^2 for laser pulses with the 7 ns duration and ~ 4 kW/cm^2 for laser pulses with microsecond duration (300 K). As a result of the nonlinear light absorption a change of the time profile of the laser pulses was recorded at the output of the amorphous semiconductors. The change of the time profile of the laser pulses leads to hysteresis-like dependencies of the output light intensity (passed through the sample) on the corresponding value of the input (Fig. 2). The extension of the optical hysteresis loop increases (changing its shape) with the input light intensity up to the damage of the sample surface.

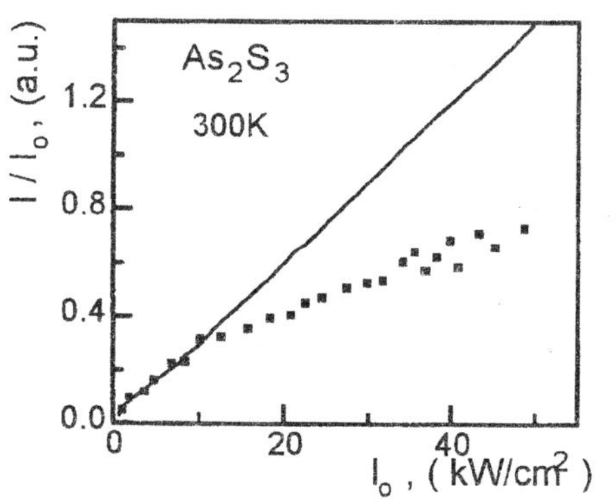

Fig. 1. Full line: continuation of low intensity light transmission. Symbols: nonlinear light transmission of a-As_2S_3 film (thickness 2.6 μm) versus input pulse intensity; $\hbar\omega = 2.58$ eV.

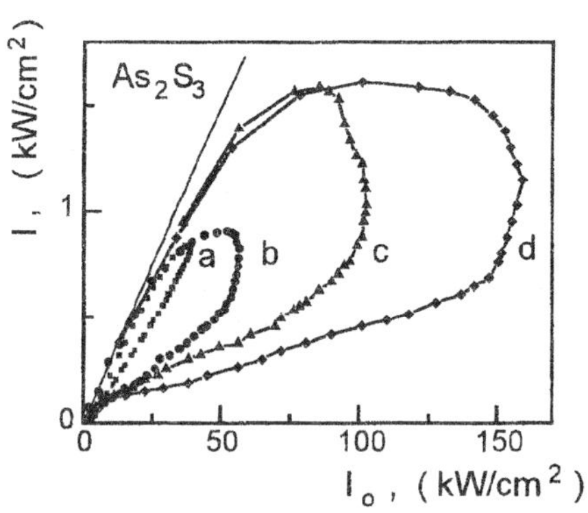

Fig. 2. Hysteresis dependencies of output light intensity versus input one. Incident pulse peak intensity (I_o): 40 (a), 55 (b), 105 (c), and 160 kW/cm^2 (d). Film thickness - 2.6 μm, $\hbar\omega = 2.58$ eV.

The kinetics of the NS films induced darkening in the field of laser pulses has been determined. Fig. 3 presents the curves of the transmission change (in per cents) during the NS film irradiation by a laser pulse of microsecond duration. The characteristic times of establishing the new NS light absorption quasiequilibrium state (corresponding to the horizontal part in Fig. 3) in the field of laser pulses are less than the duration of the laser pulses.

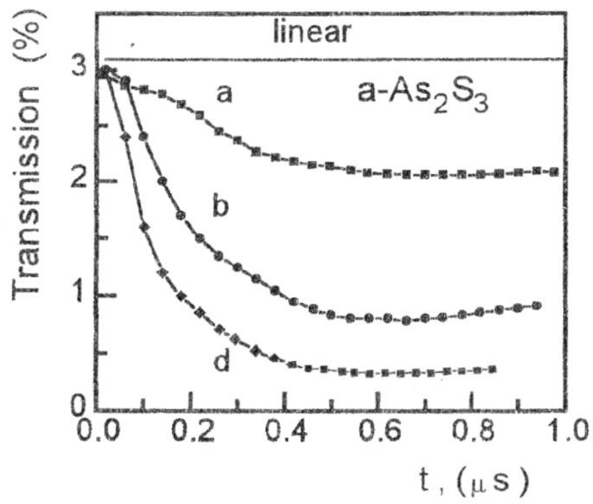

Fig. 3. Kinetics of the a-As_2S_3 darkening during the pulse irradiation. The parameter values - the same as in the Fig. 2.

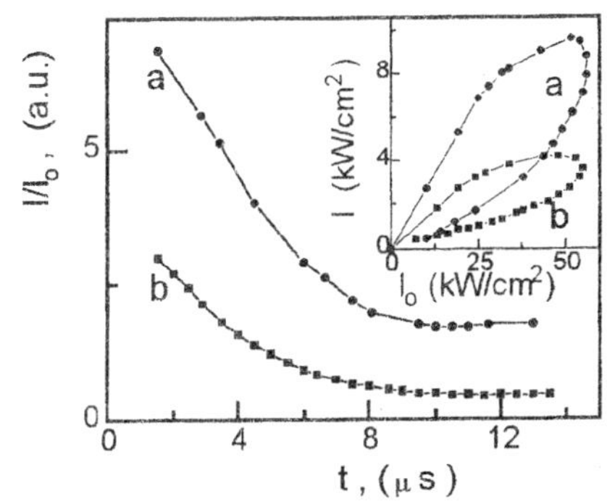

Fig. 4. Kinetics of the As_2Se_3 photodarkening during 7 ns pulse irradiation at 77K (a) and 300K. Insertion: Hysteresis loops at 77K (a) and 300K.

The photoinduced increase of the NS light absorption coefficient in the field of the laser pulses has a reversible character, i.e. the medium fully restores to its initial transmission state after the ending of the laser pulse, and if the same place of the sample is irradiated with another laser pulse the effect can be completely repeated. The restoration of the NS initial absorption state was measured by a low intensity probing light at the exciting light wavelength. The time values after which the light transmission completely restored its initial value lie in the interval from several microseconds up to several dozens of microseconds and depend on the NS film composition, its thickness, the substrate material, and the excitation intensity [10].

The cooling of the NS samples leads to a considerable increase of the surface surrounded by the optical hysteresis loop (Fig. 4). It was ascertained that by cooling the samples from 300 K to 77 K the light intensity threshold values I_t does not change. The temperature behavior of the linear and nonlinear optical absorption in amorphous semiconductors points to the important role which phonons play in the process of the interband light absorption. In fact the characteristics times of the NS photoinduced transition into the new quasi-equilibrium state (corresponding to the horizontal part of the curves in Figs. 3 and 4) do not change during the sample cooling, from 300 K to 77 K. Thus, we can come to the conclusion, that the quasi-stationary value of the light absorption coefficient in the laser pulse field is determined first of all by the light intensity in the medium.

The dependence of the NS nonlinear light absorption characteristics on the laser radiation wavelength (higher than E_g up to 0.4 eV and lower than E_g up to 0.2 eV) were investigated. It was established that in this spectral range the nonlinear absorption characteristics were maintained. However, the pulse intensity threshold values I_t change with the wavelength inversely as the change of the linear absorption coefficient of the medium.

By reducing the laser pulse duration down to picoseconds or femtoseconds, it is possible to exclude or to lower the influence of some processes, for example, the laser heating of the sample, and to investigate the mechanism of the nonlinear or nonstationary interaction of light with the NS. Nonlinear light absorption of 100 femtosecond pulses in the case of inter- and intraband excitation of As_2S_3 and As_2Se_3 thin films with the pump-probe technique was measured in [11, 12]. Different behavior of time evolution of induced nonlinear absorption in As_2S_3 and As_2Se_3 was revealed, that can be explained by different mechanisms of interaction of powerful light with amorphous semiconductors in these two cases.

3. Physical mechanisms of the nonlinear light absorption in NS.

The main peculiarities of the nonlinear light transmission in NS thin films are the following: a nonlinear increase of the light absorption coefficient; a nonlinear refractive index change, whose magnitude depends strongly on the excitation intensity; the threshold character of the nonlinear transmission; the change of the time profile of the laser pulses, leading to a hysteresis transmission dependence; the characteristic time constants of the sample photoinduced darkening, which does not exceed the laser pulse duration; the independence of the nonlinear transmission character on the light beam diameter, on the film thickness, and on the substrate material; the weak dependence from the laser pulse wavelength; the successive step character of restoration of light transmission initial state; the full restoration of light transmission initial state in a microsecond time interval; the temperature behavior of the parameters of the nonlinear interaction (the invariance of the light intensity threshold values and of the transient time into the nonlinear light absorption state); the correlation with the photostructural transformations.

The different viewpoints of the physical mechanisms that can explain the peculiarities of the nonlinear light absorption in NS are discussed briefly in [1, 2, 9, 12].

In the case of the chalcogenide glasses (ChG) photoinduced darkening under laser pulses is accompanied by photostructural transformations of the medium. These photostructural transformations (with the E_g shift) are noticed only in freshly prepared (previously not illuminated) samples and are irreversible after the ending of the exciting pulse. The photoinduced ChG darkening in the laser field, which was described previously have another physical nature than the ChG photostructural transformations, because, first of all, it is reversible (the initial ChG light transmission sets completely in during 1-100 μs) and, then, the photoinduced darkening has been found in materials, which, as a result of the photostructural transformations, have a different sign of the E_g shift, that is, give bleaching ($GeSe_2$, $As_{22}S_{33}Ge_{45}$) or darkening (As_2S_3, AsSe, As_2Se_3). It was experimentally shown that the peculiarities of the laser pulse nonlinear absorption are the same in annealed or in previously darkened (by other light sources) ChG films.

The saturation of the interband light absorption (the Burstein-Moss dynamic shift of the fundamental absorption edge [13] in semiconductors, connected with the generation of a sufficient number of electron-hole pairs, and consequently, with the Fermi level shifts into the conduction band (the absorption edge shifts towards higher photon energies) cannot explain the recorded behavior of the excited amorphous semiconductor, which is connected with a considerable increase of the light absorption coefficient.

In the described experiments the measurements of the behavior of the reflected light from the surface of the NS films in a regime of laser pulse nonlinear transmission were done as well. It was ascertained that the oscillograms of the reflected pulses from the NS repeat the incident ones. This points to the real increase of the NS light absorption and to the impossibility of its explanation by "metallization" effects, connected with a carrier concentration increase in the exciting region and by melting of the material on the surface of this region.

A distinctive peculiarity of the amorphous semiconductors is the presence of a high (10^{16} - 10^{19} cm^{-3}) density of the localized states in the optical band gap [2]. As a rule, in the literature the light absorption peculiarities in the NS transparent region are connected with carrier occupation and redistribution on these states [14].

In a number of works this mechanism is also proposed for the explanation of the laser pulse nonlinear absorption at the NS interband excitation [15]. The hot carriers, created as a result of the NS light excitation with $h\nu \geq E_g$, relax (during a period shorter than 10^{-12} s, 300 K), giving as a rule the excess energy to the phonons and being captured on the localized states. Such localized carriers may be excited by light, giving their contribution to the light absorption. However, the laser pulse nonlinear absorption at NS interband excitation, described in this work, cannot be explained on the basis of the occupancy and redistribution of the carriers on the localized states with their next excitation in the conduction band. It is conditioned by the fact that the probability of the transitions "localized state-band" is some orders lower than the interband optical transitions probability [16]. In electrodynamics the coefficient of the interband light absorption can be written:

$$\alpha(\omega)=\frac{4\pi}{c\sqrt{\varepsilon_1}}\mathrm{Re}\,\sigma(\omega)\ , \tag{1}$$

where c - light velocity, σ - complex electrical conductivity, ε_1 - real part of the dielectric permittivity, ω - light frequency. In the NS microscopic theory the real part of the complex electrical conductivity is expressed [16]:

$$\mathrm{Re}\,\sigma(\omega)=\frac{\pi e^2}{3\omega\Omega}\sum_{\lambda'\lambda''}[n_F(E^v_{\lambda'})-n_F(E^c_{\lambda''})]\times$$
$$\times\delta(E^c_{\lambda''}-E^v_{\lambda'}-\hbar\omega)\left|\lambda_v'\middle|v\middle|\lambda_c''\right|^2\ , \tag{2}$$

where $n_F(E^v_{\lambda'})$ and $n_F(E^c_{\lambda''})$ - carrier density matrix on the $E^v_{\lambda'}$ and $E^c_{\lambda''}$ levels, respectively, λ' and λ'' - correspond to the states in the valence and in the conduction bands, $\delta(E^c_{\lambda''}-E^v_{\lambda'}-\hbar\omega)$ - express the energy conservation during optical transitions, $<\lambda_v'|v|\lambda_c''>$ - quantum mechanics transition rate matrix, which in NS have a monotonous dependence on λ' and λ''. Consequently, in NS interband transition processes is important not only the position of energetic levels ($n_F(E^v_{\lambda'})$, $n_F(E^c_{\lambda''})$), but the transition probability ($<\lambda_v'|v|\lambda_c''>$) also. As a rule, in experiments the $kT \ll E_g$, where T - sample temperature, which permits to write: $n_F(E^v_{\lambda'}) \cong 1$, and $n_F(E^c_{\lambda''}) = 0$. Consequently, the probability of the transitions "localized state-band" (from $10^{16} \div 10^{19}$ cm^{-3} to ~ 10^{23} cm^{-3}) is some orders lower than the interband optical transitions probability (from ~ 10^{23} cm^{-3} to ~ 10^{23} cm^{-3}) and, it is rather difficult to experimentally distinguish such a low contribution to the absorption (at the same light wavelength) from the interband absorption. This is even more difficult because the optical absorption spectrum of the investigated compounds and the energy dependence of the density of states in the optical band gap have a monotonous dependence in energy.

It is impossible to explain the laser pulse nonlinear transmission by the homogeneous heating of the NS films. The temperature coefficient of the optical absorption edge shift in these compounds

(dE_g/dT) is about 5 10^{-4} eV K^{-1} [2]. Therefore, in order to change the absorption light coefficient in the absorption edge region, by an amount equal to the one revealed in the experiment (~ two times), it is necessary to heat the NS sample to a temperature considerably exceeding the temperature of the material softening. More rigorous experimental demonstration was done, when an analogous laser pulse nonlinear absorption has also been registered at light quantum energies, considerably exceeding the NS optical band gap (E_g), where the contribution of the temperature shift of the absorption edge on the light transmission change is smaller.

The contribution of the sample temperature increase (as a result of the absorption of a fraction of laser radiation energy) on the light pulse nonlinear absorption may be revealed from the dependence of this absorption on the size of the excitation region, on the NS films thickness, and on the substrate material (with different heat conductivity). The investigations of the laser pulse nonlinear transmission dependence on the excitation region size was carried on. As a result of measurements no dependence of the optical hysteresis transmission curves from the diameters of the exciting beam (from 80 to 700 μm) was revealed.

The influence of the substrate material (with a different heat conductivity κ) and of the sample thickness on the characteristics of the laser pulse transmission has been studied. Substrates of glass (κ = 0.7 W m^{-1} K^{-1}), mica (κ = 0.24 W m^{-1} K^{-1}) and lavsan (κ = 0.17 W m^{-1} K^{-1}) were used. Within the experimental error (< 20%) no influence of the substrate heat conductivity, on the curves of the optical hysteresis transmission, has been revealed. Neither influence of the NS film thickness on the nonlinear absorption (while changing the thickness from 0.1 μm to 3 μm) has been revealed.

At the same time a significant influence of the NS film composition, its thickness, of the substrate material (heat conductivity), and of the excitation intensity has been revealed on the transmission restoration kinetics after the ending of the exciting pulse.

The impossibility to explain the laser pulse nonlinear transmission by the homogeneous heating of the NS films was demonstrated by femtosecond spectroscopy measurements also [11, 12]. Laser pulses with ~ 1,5 μJ energy were used, which can not provide a substantial heating during probe excitation with 100 fs pulses at 300 K.

The temperature behavior of the linear and nonlinear optical absorption in NS points to the important role which phonons play in the process of the interband light absorption. The temperature dependence of the light absorption coefficient is conditioned by strong electron-phonon interaction [16]. A new physical model of the nonlinear interaction of powerful light with NS, taking into account the light interaction with nonequilibrium phonons and localized vibrational excitations was proposed for the explanation of the experimental results.

Structure inhomogeneities in NS may lead to some spatial localization of the phonon vibrational modes, that is, they may be considered as quasiresonators for the vibration energy [5, 6]. The capture of nonequilibrium phonons by such quasiresonators leads to the increase of their effective lifetime [6, 7]. Some properties of the vibrational mode localization in quasiresonators can be similar to the electron localization in quantum well structures. Thermal phonons with a wavelength of the order of the distance between random inhomogeneities may additionally deform the random band potential in NS. This, in turn, leads to a change of the light absorption coefficient as a result of the appearance (or disappearance) of additional optical transitions channels between the earlier forbidden (or less probable) band states. Consequently, a considerable role in the amorphous semiconductor interband light absorption is played by the transitions with a simultaneous participation of light quanta and phonons [1, 2]. With such considerations in mind the coefficient of the interband light absorption may be thought to consist of two parts: $\alpha = \alpha_o + \beta N_a$, where α_o does not depend on the temperature, while the second part depends on the concentration of N_a phonons. The second part of the light absorption coefficient may change considerably by generating additional phonons as a result of the hot carrier (created by the light) energy dissipation with their consequent capture on localized states in the NS optical band gap. A part of the formed nonequilibrium phonons may "be captured" by the quasiresonators with a d_m size (where d_m is the average distance between the spatial inhomogeneities). Such quasiresonators may trap the vibrational modes with $\lambda_f \leq d_m$ (increasing their lifetime) and help to their further accumulation. Considerable deformation of the random (frozen) potential may take place at high concentration of nonequilibrium phonons (that is, a considerable deformation of the quasiresonator "walls"). On the one hand, it leads to the phonons nonlinear decay from the quasiresonator regions, and, on the other hand, to the appearance (disappearance) of additional channels of interband light absorption. The kinetic equations, describing the exciting and relaxation processes of electrons and phonons in NS have been written in [9]. Equations describe: the laser energy loss during the interaction with the medium, the medium polarization decay, the transition of the carriers (under the laser field) from the valence to the conduction band and vice versa, the carrier concentration changes in the conduction and in the valence band under the action of the external laser field and carrier relaxation with the emission of optical phonons, the generation of optical and acoustic phonons, the optical phonon decay process into acoustical phonons.

The nonlinear change of the light absorption coefficient may be reached in the NS at much lower irradiation intensity than in the crystalline ones. This is possible when transitions with the participation of local acoustic phonons play an important role. Here the system of kinetic equations presented in [9] simplifies. In the experiments the duration of the laser pulses was greater than the characteristic times of the optical and hot acoustic phonons decay. In this case it is possible to neglect the intermediate stages of the phonons decay, which arise during the optical and hot acoustic phonon decay, and take into consideration only the kinetic of the local acoustic phonons. Under such assumptions, corresponding to the experimental conditions, the equations may be written as follows:

$$\begin{aligned} &\frac{\partial I(t,x)}{c\,\partial t} - \frac{\partial I(t,x)}{\partial x} = (\alpha_o + \beta N_a) I(t,x); \\ &\frac{\partial N_c}{\partial t} = -\frac{N_c}{\tau_o} + (\alpha_o + \beta N_a)\frac{I(t,x)}{\hbar\omega}; \\ &\frac{\partial N_a}{\partial t} = \frac{\Theta N_c}{\tau_o} - \frac{N_a(N_a - N_a^o)}{\tau N_a^o}, \end{aligned} \qquad (3)$$

where $\theta = k\delta$ is the number of created acoustic phonons as the result of one hot carrier decay; $I = cE^2/2$ is the laser pulse intensity in the medium; $\beta = 2\omega T_2 p_{cv}^2 \gamma N_v^o/\hbar c$; N_a^o and N_a are the concentrations of the equilibrium and nonequilibrium acoustic phonons.

The analytical solution of equations (3), corresponding to the experimental conditions, is difficult to obtain. The kinetic equations were solved numerically for parameter values and initial conditions, corresponding to the experimental situation. The results of the numerical solution fit reasonable well the experimental dependencies. Fig. 5 gives the experimental and the calculated hysteresis dependencies of the pulse intensity, passed through an AsSe film (the thickness 0.6 μm), versus the corresponding intensity value at the input, for 7 ns laser pulse duration. Some differences between the calculated curves and the experimental points are determined by the discrepancy of the laser pulse shapes to the Gaussian shape as well as by the difficulty to take into account the intermediate stage processes of the hot carrier energy decay into acoustic phonons.

In support of the proposed model for the explanation of the laser pulse nonlinear absorption of ChG may also serve the fact that with the same parameter values the calculated curves coincide satisfactorily with the experimental ones in spite of the changing of the laser pulse duration or amplitude in the wide range.

An additional experiment was performed, in order to demonstrate the adequacy of the proposed model. The main point of the physical model of the nonlinear interaction of powerful light with NS, taking into account the light interaction with nonequilibrium phonons and localized vibrational excitations, is the phonon localization. So, we should investigate in the same conditions the peculiarities of pulse transmission in the medium with the localization possibilities (NS) and without localization (crystalline semiconductor). Such a medium is As_2S_3 : which exists in crystalline and amorphous state. The main peculiarities of the nonlinear light transmission in amorphous As_2S_3 thin films were presented.

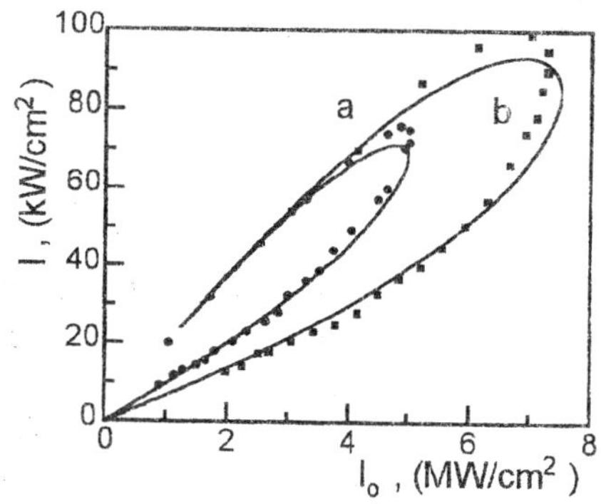

Fig. 5. Experimental (symbols) and calculated dependencies of transmitted light intensity versus input one (I_o). Pulse peak intensity: 5.0 (a) and 7.5 MW/cm^2. Parameter values: $N_a^o = 1.2 \cdot 10^{22}$ cm^{-3}; $\beta = 10^{-19}$ cm^2; $\alpha_l = 5.6 \cdot 10^4$ cm^{-1}; $\Theta = 50$; $\tau = 10^{-5}$ s; T=300K.

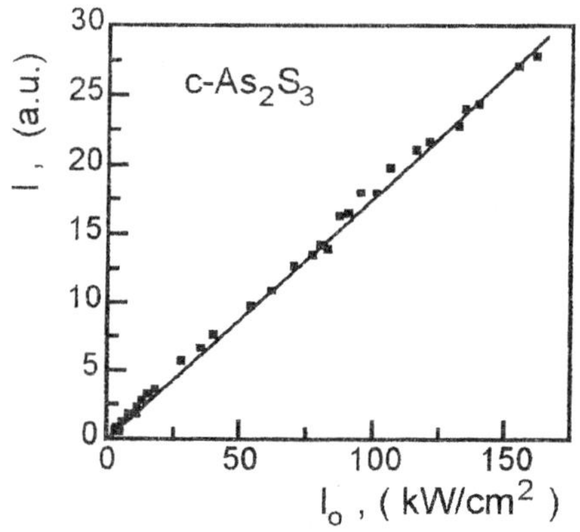

Fig. 6. The output laser pulses peak intensity versus input peak intensity in c-As_2S_3 film (thickness 5 μm), $\hbar\omega = 2.75$ eV.

The peculiarities of the pulse absorption in crystalline As_2S_3 films were measured in the similar excitation conditions (the wavelength of the laser generation was changed to λ = 450 nm). The orientation, corresponding to the minimum of the light transmission of c- As_2S_3 thin film (the light electric vector **E** is directed in the middle between a and c axes of the crystal) was chosen to get the absorption coefficient $\alpha = 7.3 \cdot 10^3$ cm^{-1}, comparable with amorphous material one. Fig. 6 shows the dependence of the output laser pulses peak intensity versus input one. As a result of measurement no dependence (in limits of experimental error < 20%) of light absorption in crystalline As_2S_3 from pulse intensity was revealed.

These measurements indicate that the clue to understand the physical mechanism lying at its base, is connected with the amorphous nature of the material. This fact additionally confirms the possibility of the application of physical model of the nonlinear interaction of powerful light with NS, taking into account the light interaction with nonequilibrium phonons and localized vibrational excitations.

The best experiment, which, from our point of view, confirm the possibility of application of the proposed physical model, is the measurements of nonlinear laser pulse absorption in a-Si:H and a-

SiC:H thin films [18-20]. It is well known, that the electronic spectrum in the band gap of a-Si:H is different from electronic spectrum of the chalcogenide glasses [2]. So, if light generation and redistribution of the carriers on the localized states with their next excitation in the conduction band play an important role, we should have different peculiarities of laser pulse absorption in a-Si:H and chalcogenide glasses.

The peculiarities of the pulse absorption in a-Si:H and a-SiC:H films were measured in the similar excitation conditions as in the case of chalcogenide glasses. The qualitative behavior of the characteristics of the nonlinear light absorption and, as well as, of the kinetics of the sample photoinduced darkening in the field of the laser pulses (at 300K) is the same as in the case of chalcogenide glasses. For example, in figure 7 the hysteresis-like dependencies of the output light intensity (passed through the a-Si:H sample) on the corresponding value of the input is shown and, figure 8 presents the characteristic times of the setting in of the new light absorption quasiequilibrium state (corresponding to the horizontal part) in the field of 7 ns laser pulses.

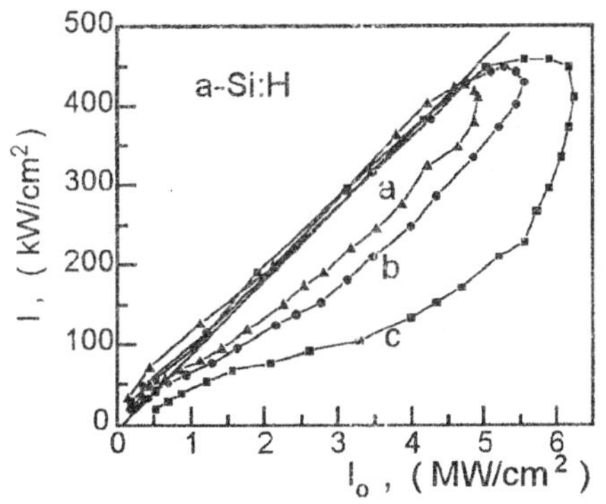

Fig. 7. Hysteresis dependencies of output light intensity versus input one. Incident peak pulse intensity: 6.3 (a); 5.7 (b) and 4.9 MW/cm^2 (c); pulse duration 7 ns, $\hbar\omega$ = 2.07 eV.

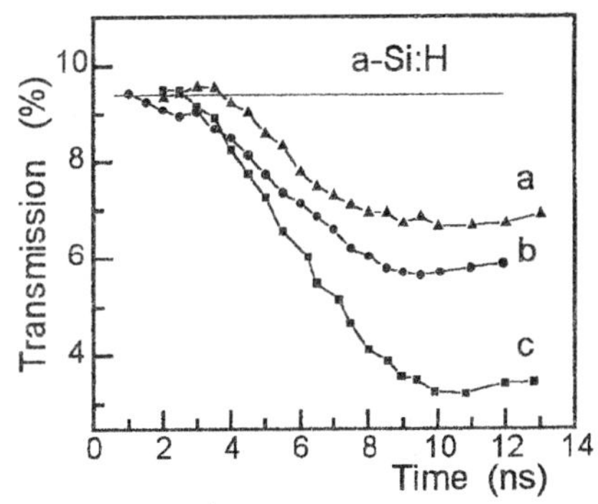

Fig. 8. Kinetics of film darkening during pulse irradiation. The parameter values - the same as in the Fig. 7.

The most characteristic peculiarities of the investigated nonlinear light transmissions in a-Si:H thin films are the following: a nonlinear increase of the light absorption coefficient; the threshold character of the nonlinear transmission; the change of the time profile of the laser pulses, leading to a hysteresis transmission dependence; the characteristic time constants of the sample photoinduced darkening, which does not exceed the laser pulse duration; the weak dependence on the laser pulse wavelength; the restoration of light transmission initial state in a microsecond time interval. These peculiarities of nonlinear light absorption can be understood in terms of a picture of generation of the nonequilibrium localized lattice vibrations with the emission of the extended phonons, created as a result of the a-Si:H light excitation with $\hbar\omega \geq E_g$.

4. Conclusions.

Nonlinear interaction of short and ultrashort laser pulses with NS have shortly been described. It was shown that these characteristics do not find a satisfactory explanation in the limits of the existing physical models proposed in the literature. A new mechanism of a nonlinear light absorption in NS, taking into account the interaction with nonequilibrium phonons and localized vibrational modes is briefly discussed. It is not really difficult to estimate that the number of created phonons and localized vibrational modes due to the energy dissipation of hot carriers is of the same order as the number of the equilibrium phonons using laser pulse energy density in the range 1-10 mJ cm^{-2}. A reasonably agreement of the calculated dependencies with the experimental ones is obtained.

Acknowledgment

We acknowledge the partial support of the NATO Programme for Priority Area on High Technology (Grant HTECH. EV 950772) for this research.

References

1. V. Chumash, I. Cojocaru, E. Fazio, F. Michelotti, M. Bertolotti. Progress in Optics, **36**, 1996.
2. M. Popescu, A. Andries, V. Ciumas, M. Iovu, S. Sutov D. Tiuleanu. Fizica Sticlelor Calcogenice, Ed. Stiintifica, Bucuresti & I. E. P. Stiinta, Chisinau, 1996, 486 pag.
3. V. K. Malinovskii, Iu. E. Nesterihin, V. N. Novikov, A. P. Sokolov, Avtometria, **2**, pp.3-9, 1986.
4. R. Orbach, J. Non-Crystall. Solids, **164-166**, pp. 917-922, 1993.
5. R. Baumgartner, M. Engelhardt, F. F. Renk, Phys. Lett., **94A**, pp. 55-58, 1983.
6. H. M. Gibbs, Optical Bistability: Controlling Light with Light, Academic Press, Inc., N-Y, 1985.
7. M. Bertolotti, V. Chumash, E. Fazio, A. Ferrari, C. Sibilia, Inst. Phys. Conf. Ser., **115**, pp. 169-172, 1991, *idem* J. Appl. Phys., **74**, pp. 3024-3027, 1993.
8. A. M. Andriesh, N. A. Enaki, I. A. Cojocaru, N. D. Ostafeichuk, P. G. Cherbari, V. N. Chumash, Sov. J.T.F. Lett., **14**, pp. 1985-1989, 1988.
9. A. M. Andriesh, O. I. Bogdan, N. A. Enaki, I. A. Cojocaru, V. N. Chumash, Buletin Ross.Acad.Sci., Phys. Ser, **56**, pp. 96-109, 1992.
10. M. Popescu, A. Andriesh, V. Chumash, N. Enachi, I. Cojocaru, A. Grozescu, J. Non-Cryst. Solids, **137-138**, pp. 973-976, 1991.
11. A. Andriesh, V. Chumash, I. Cojocaru, M. Bertolotti, E. Fazio, F. Michelotti and D. Hulin, in "Photonic Switching", SPIE Proceed., **1807**, pp. 126-135, 1992.
12. E. Fazio, D. Hulin, V. Chumash, F.Michelotti, A. M. Andriesh and M. Bertolotti, J. Non.-Cryst. Solids, **168**, pp. 213-222, 1993.
13. D. A. B. Miller, C. T. Seaton, M. E. Prise, S. D. Smith, Phys. Rev. Lett., **47**, pp. 197-200, 1981.
14. M. A. Kastner. Physical Properties of Amorphous Materials. Eds. D. Adler, B. B. Schwartz, M. C. Steele, Plenum Press, N-Y. and London, 1985, p. 381.
15. Z. Vardeny, J. Non-Cryst. Solids, **59-60**, pp. 317-324, 1983.
16. V. L. Bonch-Bruevichi, I. P. Zveaghin, R. Kaiber, A. G. Mironov, R. Anderlain, B. Esser, The Electronic Theory of Disordered Semiconductors, Moscow, Nauka, 1981, 384 p.
17. V. Chumash, I. Cojocaru, G Bostan, P. Cerbari, N. Mincu. "Optical Diagnostics of Materials and Devices for Opto-, Micro-, Quant. Electronics - OPTDIM'95", SPIE Proceed., Kiev, **2648**, p. 221-225, 1995.
18. G. N. Bostan, I. A. Cojocari, E. Carbunescu, V. N. Ciumas, Buletinul Academiei de Stiinte a Republicii Moldova, Fizica & Tehnica, **2**, p. 60-63, 1992.
19. A. Andriesh, V. Chumash, E. Carbunescu, I. Cojocaru, G. Bostan. Romanian Journal of Physics, **38**, No. 9 & 10, p. 897-902, 1993.
20. V. Chumash, I. Cojocaru, G. Bostan, G. De Cesare, S. La Monica, G. Maiello, A. Ferrari. "Optical Diagnostics of Materials and Devices for Opto-, Micro- and Quantum Electronics-OPTDIM'95",SPIE Proceedings, Kiev, **2648**, p. 23-28, 1995.

Session 17

DEVICE MODELING
(Poster)

THE SHORTED ANODE BRT: A MOS-THYRISTOR STRUCTURE WITH IMPROVED TURN-OFF PERFORMANCE

D. Flores, X. Jordà, M. Vellvehi, J. Fernández, S. Hidalgo, J. Rebollo and J. Millán.

Centro Nacional de Microelectrónica (CNM) - CSIC, Campus UAB, 08193 Bellaterra, Barcelona. Spain.

SUMMARY

This paper analyzes the static and dynamic electrical characteristics of BRT devices fabricated on homogeneous wafers, including a shorted anode structure. The use of homogeneous substrates allows the implementation of a direct path between the N^--drift region and the anode electrode. This allows a reduction of the transient losses and the turnoff time. The operation mode, the electrical characteristics and the switching behaviour of BRT devices with shorted have been analysed by means of 2D numerical simulations. The reduction of the transient losses has been corroborated experimentally by performing switching tests under resistive load.

1. INTRODUCTION

There have been reported several ways to control the current flowing through a thyristor structure via a MOS gate. These structures combine the benefits of both MOSFET and thyristor devices; i.e. a high input impedance and a higher current capability than that of the power MOSFET and IGBT devices, The MOS gate allows to turn on and off the thyristor by rasing the bipolar transistor base currents and breaking the latching condition, respectively. The first reported structure from this category was the MOS-Controlled Thyristor (MCT) /1/ with a process technology based on a triple diffusion. MOS-thyristor structures which fabrication process is compatible with those of power MOSFET and IGBT have been also reported. Among them, we can mention the Emitter Switched Thyristor (EST) /2/, the Insulated Base MOS-Controlled Thyristor /3/ and the Base Resistance MOS-Controlled Thyristor (BRT) /4/. These devices, fabricated on epitaxied substrates have been extensively analyzed in the literature. In order to obtain a high blocking voltage capability, a N-buffer layer is included to prevent the punch-through breakdown. When these MOS-thyristor devices are turned off, the excess charge stored in the N^--epitaxy can only be removed via a recombination process, which determines their turnoff time. The epitaxied MOS-thyristor devices have a high turnoff time, which limits their frequency operation range. It is necessary to reduce the transient losses when using these devices in high frequency applications.

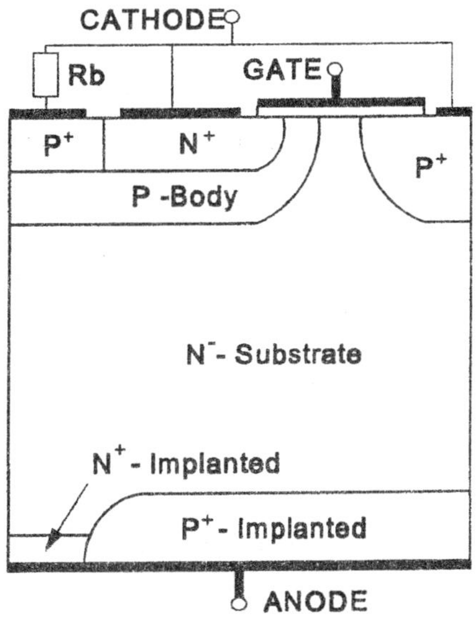

Fig. 1.- Cross section of the shorted anode BRT

These MOS-thyristor devices can also be fabricated on homogeneous N^--substrates, as in the case of the IGBT /5/ allowing the integration of a shorted anode structure in the back side of the die. The emitter of

0-7803-3223-7/96/$5,00 © 1996 IEEE

the lower bipolar transistor is locally implanted, and superficial N^+ short circuits are implemented into the N^--base in order to reduce the transient losses. This paper is addressed to the analysis of the electrical characteristics of homogeneous BRT devices with and without a shorted anode structure. The shorted anode BRT operation mode is analyzed by means of 2D numerical simulations and experimental results from multicell BRT structures are also presented. The shorted anode BRT electrical performance is compared with homogeneous BRT devices fabricated with the same process.

2. DEVICE CROSS SECTION AND OPERATION MODE

A cross-section of the shorted anode BRT structure basic cell is shown in Fig. 1, where R_b is the equivalent stripe P-body resistance. The lower bipolar transistor emitter is locally boron implanted and the N^--substrate is electrically connected to the anode terminal via the superficial N^+-implanted dots. The operation mode of the shorted anode BRT is similar to that of the BRT fabricated on epitaxied substrates /4/. The device is turned on by applying a positive gate bias which rises the lower bipolar transistor base current. It is turned off by applying a negative gate bias, which electrically connects the P-body diffusion to the P^+-diverter, creating a direct path for the hole excess extraction in the initial phase of the turn off process. Unlike the epitaxied BRT, there is an initial unipolar regime prior to the activation of the lower P^+N^- junction. When the electron current is high enough, the voltage drop of this junction reaches 0.7 V and the device enters the IGBT regime. Holes are injected from the P^+-anode diffusion into the drift region and are collected by the P-body, flowing under the N^+-cathode diffusion up to the cathode contact located at the edge of each cell stripe. When the voltage drop across the P-body/N^+-cathode junction reaches 0.7 V, the thyristor latches. Homogeneous BRT devices have a forward voltage drop higher than that of the epitaxied BRTs due to their wider N^--base region. Once the BRT is in the thyristor regime, holes injected from the P^+-anode diffusion and electrons injected from the N^+-cathode diffusion, through the P-body, create a charge excess into the base of the lower bipolar transistor.

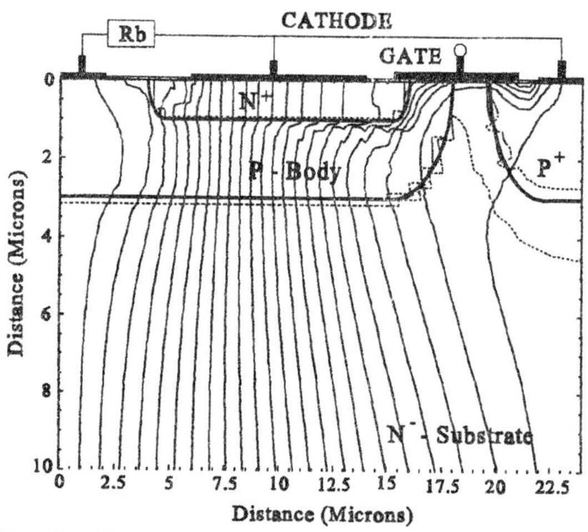

Fig. 2.- Simulated flowlines at the initial phase of the turnoff process. Top side of the basic cell

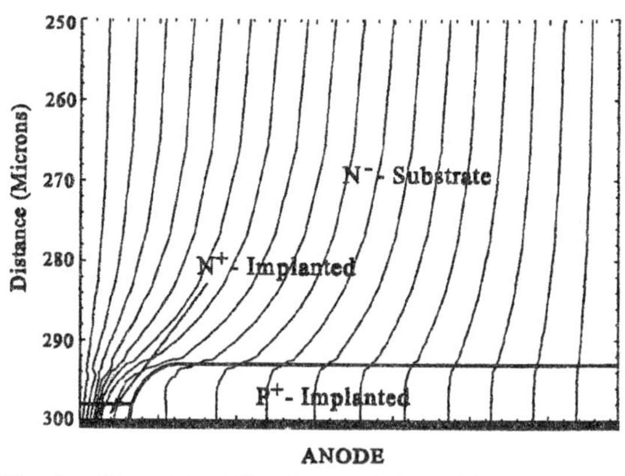

Fig. 3.- Simulated flowlines at the initial phase of the turnoff process. Bottom side of the basic cell.

When the BRT is turned off, the excess charge stored in the two bipolar transistor bases has to be removed. Hole charge excess is removed away from the P-body to the diverter, through the PMOS channel, breaking the latching condition. The turnoff time of the epitaxied BRT is mainly determined by the recombination process of the charge excess stored in the N^--drift region. This turnoff time can be reduced by killing the minority carrier lifetime using electron irradiation, etc., as it is usually done

IGBT. In the case of the homogeneous BRT devices, the inclusion of the shorted anode structure allows a significant reduction of the device transient losses since electrons are extracted directly from the N^--drift region to the anode electrode, at the initial phase of the turn off process.

The shorted anode BRT operation mode has been analyzed by means of 2D numerical simulations carried out with MEDICI /6/. A reduction of 70% of the transient losses in comparison with homogeneous BRTs without shorted anode has been obtained on simulated structures. The flowlines at the initial phase of the turnoff process at the top side of the basic cell are plotted on Fig. 2. From this figure one can observe the hole flow through the R_b resistance which still maintains the thyristor regime. Moreover, holes are extracted from the P-body, flowing through the PMOS transistor and the P^+-diverter up to the cathode. In Fig. 3 one can observe the flowlines during the turnoff process in the bottom side of the structure, showing the electron extraction ability of the shorted anode structure.

3. DEVICE FABRICATION AND EXPERIMENTAL RESULTS

Multicell BRT test structures have been fabricated on <111> oriented 45 Ωcm N^--type substrates 300 µm thick. The most important basic cell geometrical dimensions are: an N^+-cathode width of 20 µm, a cell stripe length of 250 µm, an NMOS channel length of 2 µm and a PMOS channel length of 1.5 µm. The structures, with an active area of 1.25×10^{-3} cm^2, have been implemented with a P-body surface concentration of 3×10^{17} cm^{-3} and a junction depth of 3 µm and a gate oxide thickness of 800 Å. The process technology is a conventional double diffused DMOS/IGBT process which requires a double side alignment step for the implementation of the shorted anode structure. The edge termination techniques integrated to obtain a high blocking voltage capability are either two floating field rings or a junction termination extension (JTE), consisting of a 100 µm length P^--diffusion with a surface concentration of 7×10^{15} cm^{-3} and a junction depth of 3.3 µm. Experimental forward blocking voltages higher than 1100 V have been measured on devices with JTE.

The experimental electrical characteristics, obtained with a gate bias of 12 V, of homogeneous BRT devices with and without shorted anode have been compared. The shorted anode BRT has a high resistive initial unipolar regime before entering the IGBT conduction mode. This high resistive regime is due to the resistance of the wide (300 µm) and low doped (10^{14} cm^{-3}) N^--region. When the electron current through the N^--drift region is high enough to forward bias the lower junction at the edge of the basic cell stripe, the shorted anode BRT enters the IGBT regime. It has to be pointed out that the bipolar base current is lower than that in the BRT structure without shorted anode since part of the electron current provided by the N-channel MOSFET maintains the lower junction forward biased. The homogeneous BRT devices with and without shorted anode have a forward voltage drop of 1.7 V at 50 A/cm^2, which is higher than that of the equivalent epitaxied BRTs (1.2 V) at the

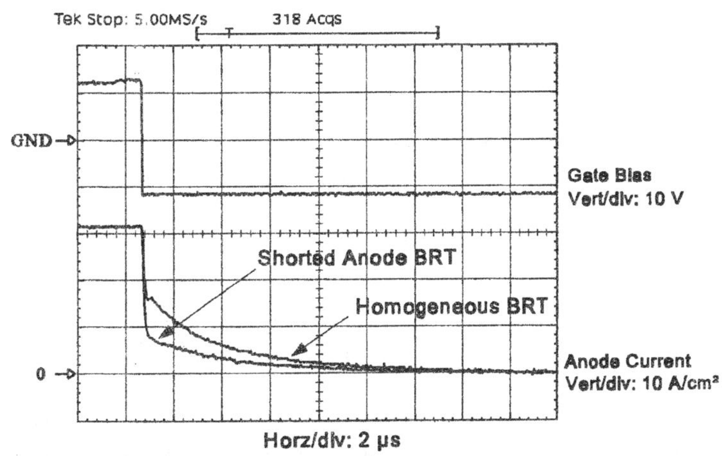

Fig. 4.- Tturnoff process of BRT devices with and without shorted anode under hard switching condition.

same current density, due to the wider N⁻-base. This current density value corresponds to the maximum controllable current density measured on the fabricated homogeneous BRT devices.

Transient measurements carried out on encapsulated BRT devices have confirmed their ability to be turned on and off by applying a gate bias pulse. Switching tests have been performed under resistive load (100 Ω), monitoring the anode current decay at the initial phase of the turnoff process when the gate bias is ramped down from 12 V to -12 V. The switching frequency was 1 kHz with a duty cycle of 50%. In Fig. 4 it is provided a comparison of the turnoff processes of BRT structures with and without shorted anode. The measured maximum controllable current density of homogeneous BRT devices with and withoutb shorted anode is in the range of 50 A/cm^2 at room temperature and in the range of 1000 A/cm^2 at 77 K. The lower waveforms of the figure show that the initial current decay in the shorted anode BRT is higher than that of the BRT without shorted anode due to the fast electron extraction from the N^--region through the superficial N^+-diffusion. The elimination of the minority carrier excess stored in the N^--drift region via a recombination process determines the current tail in both devices. As expected from simulation results, Fig. 4 shows that a significant reduction in transient losses can be achieved by including the shorted anode structure.

CONCLUSIONS

In this paper, the operation modes of standard and shorted anode BRT structures, both implemented on N^--wafers, are compared. These devices exhibit a higher forward voltage drop than those fabricated on epitaxied wafers due to the increase of the lower bipolar transistor base width. On the contrary, the use of homogneous wafers allows the inclusion of the shorted anode structure that connects the N^--base of the lower bipolar transistor to the anode electrode. As a consequence, electrons are extracted rapidly through this direct path at the initial phase of the turnoff process, reducing the transient losses in a 70%. Transient measurements have confirmed the ability of the shorted anode BRT turnoff process by using multicell devices with an active area of 1.25×10^{-3} cm^2.

Acknowledgments

This work was supported by the Comisión Interministerial de Ciencia y Tecnología (CICYT) under contract TIC 95-0078, and by the Comissió Interdepartamental de Recerca i Innovació Tecnològica (CIRIT) under contracts GRQ 94-8006 and SGR 95-00536.

REFERENCES

/1/ V. A. K. Temple, "The MOS Controlled Thyristor". in IEDM Tech. Dig., Abstr. 10.7, pp. 282-285, 1984.

/2/ B. J. Baliga, "The MOS-Gated Emitter Switched Thyristor". IEEE Electron Device Lett., vol. 11, pp. 75-77, 1990.

/3/ D. Flores, M. Vellvehi, P. Godignon, J. Fernández, S. Hidalgo, J. Rebollo and J. Millán "Analysis of the Insulated Base MOS-Controlled Thyristor Structure (IBMCT)" Proc. of the IEE PEVD'94, pp. 532-536, London, 1994.

/4/ M. Nandakumar, B. J. Baliga, M. S. Shekar, S. Tandon and A. Reisman, "A New MOS-Gated Power Thyristor Structure with Turn-Off Achieved by Controlling the Base Resistance". IEEE Electron Device Lett., vol. 12, pp. 227-229, 1991.

/5/ H. Akiyama, T. Minato, M. Harada, H. Pan, H. Kondoh and Y. Akasaka, "Effect of Shorted in Collector on Characteristics of IGBTs". Proceedings of ISPSD'90, pp. 131-136, 1990.

/6/ MEDICI was obtained from Technology Modeling Associates, Palo Alto, CA, USA.

The Trench Inversion Layer Emitter Thyristor (ILET)

Florin Udrea[a] and Gehan A.J. Amaratunga[b]

a. Department of Engineering, Cambridge University, Cambridge CB2 1PZ, England

b. Department of Electrical Engineering and Electronics, Liverpool University, Liverpool LG9 3BX, England

Summary

The Trench Inversion Layer Emitter Thyristor (ILET) is a novel power device structure [1,2] based on an entirely new physical concept that expresses the transition of an inversion layer from a majority carrier channel into a minority carrier injector. Numerical simulations of the ILET and the experimental evidence of the minority inversion layer injection mechanism are presented. It is found that the ILET has potential for achieving an optimum trade-off between the on-state resistance, off-state blocking voltage, turn-off time and safe operating control.

1. Introduction

The DMOS Insulated Gate Bipolar Transistor (IGBT) is presently the main MOS-controllable switching structure in the field of high voltage applications.

To improve the overall performance of the IGBT a trench variant of the IGBT, termed the Trench IGBT, in which the channel is formed on the side-walls of a vertical groove has been proposed [3,4]. Trench IGBT offers a substantially lower on-state resistance, improved Safe Operating Area (SOA) and a better trade-off between the on-state and short-circuit capability [4,5]. In this paper we analyse a combined thyristor-IGBT trench gated device, termed the Inversion Layer Emitter Thyristor (ILET) [1,2] which further improves the on-state performance of the Trench IGBT with no penalty in the breakdown capability or switching speed. The ILET is based on a new physical mechanism. This is the use of a MOS inversion layer as a minority carrier injector. Experimental evidence of the mechanism of inversion layer injection is shown for the first time.

2. The Inversion Layer Injection concept

The concept of Inversion Layer Injection refers to diffusion of carriers direct from the MOS inversion layer into the MOS substrate. The effect is possible when the junction formed between the substrate and inversion layer becomes forward-biassed.

To demonstrate the concept of Inversion Layer Injection, the structure shown in Fig. 1(a), termed Inversion Layer Diode (ILD) has been fabricated using a CMOS technological process [1]. This structure could be regarded as a MOS-controllable resistor. If the gate contact is shorted to the cathode contact there is no inversion layer and the device acts as a high resistance. The carrier transport is based in this case on drift of holes through the resistive p- epitaxial base (it is assumed that the parasitic cathode junction is not latched-up). When the anode voltage increases and a positive potential is applied to the MOS gate, the voltage drop between the p- epi substrate and inversion layer reaches the junction barrier potential. Consequently, injection of electrons from the inversion layer into the p- epi commences, as shown by the flow-lines of the simulation in Fig. 1(b). The injected electrons reach the inversion layer injector from the cathode via the channel formed in the p well. High electron injection from the inversion layer leads to conductivity modulation of the p- epi and an abrupt decrease in its resistivity.

The experimentally observed I-V characteristics for the Inversion Layer Diode are shown in Fig. 1(c). Prior to the forward-biassing of the inversion layer junction,

0-7803-3223-7/96/$5,00

the on-state resistance remains unchanged with increasing gate voltage. When the anode voltage reaches the threshold voltage associated with the forward-biassing of the inversion layer junction (the kink in the I-V characteristics in Fig. 1(c)) the on-state resistance decreases abruptly with increasing gate voltage.

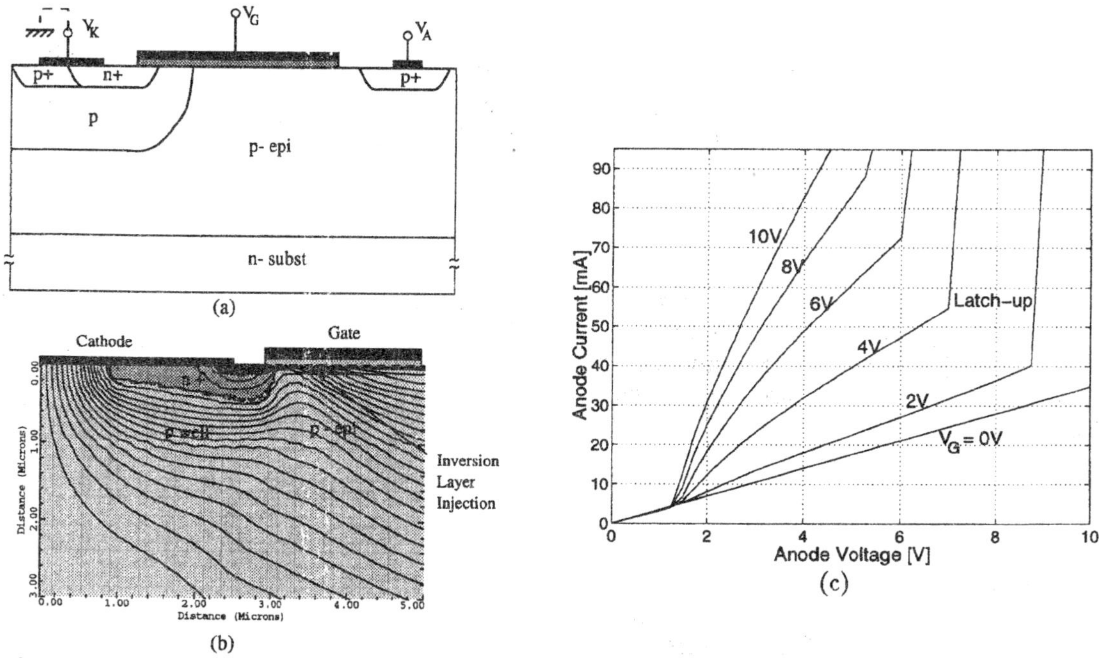

Figure 1: (a) Cross section of the Inversion Layer Diode (ILD), (b) Numerical simulations showing the current flow lines at the cathode side. The inversion layer injection can be clearly seen. (c) Measured output characteristics of the Inversion Layer Diode for different gate voltages. At low anode voltages, an initial resistive region independent on the gate voltage can be observed. Once the inversion layer junction is forward-biassed the resistance decreases function on the potential applied on the gate.

3. ILET Structure, Operation and Performance

An example of a 600 V ILET is shown in Fig. 2(a). The n+ buffer layer has a doping concentration of $8 \cdot 10^{16}$ cm^{-3} and a thickness of 6 μm while the wide n- base has a doping of 10^{14} cm^{-3} and a thickness of 50 μm. The p base is comprised of a sandwich of three layers, named p1, p2 and p3 with different conductivity levels.

The p1 base has a Gaussian doping profile with a peak concentration $N_{As} = 5 \cdot 10^{18}$ cm^{-3}. The effective channel length is given by the difference between the thickness of the p1 layer and the cathode junction depth which is typically 0.5 μm. The p2 base has an uniform doping profile in the range of 10^{14} cm^{-3} and a thickness of 4 μm. The purpose of the highly doped p3 base is to prevent punch-through of the device in the off-state and to push the potential lines away from the trench corners . This allows the substrate concentration of the p1 base and the channel length to be independent of the punch-through condition. The p3 base has a concentration of $10^{16} cm^{-3}$ and a thickness of 1.5μm (Fig. 2(b)).

The turn-on of the ILET is achieved by applying an appropriate positive voltage on the gate. The device remains in the on-state as long as the positive potential is kept on the gate. Initially, the channel formed in the p base allows electron injection from the cathode region into the drift region (n- base) followed by hole injection from the anode p+ region. This mode is identified as the IGBT mode. A numerical simulation of the two dimensional current distribution in this mode is shown in Fig. 3(a). As the anode current increases, the difference between the potential in the lowly conductive p2 base and the potential in the channel (expressed by the quasi-Fermi potential drop) becomes equal to the voltage barrier causing forward-biassing of the pn junction formed between the p2 substrate and the inversion layer. This is followed by heavy injection of electrons from the inversion layer into the p2 base bypassing the initial MOSFET channel, thereby initiating thyristor action (Fig. 3(b)). The thyristor structure is formed between the inversion layer in the p2 base (emitter), p2 and p3 bases (p base), drift region (n-base) and p+ region (anode).

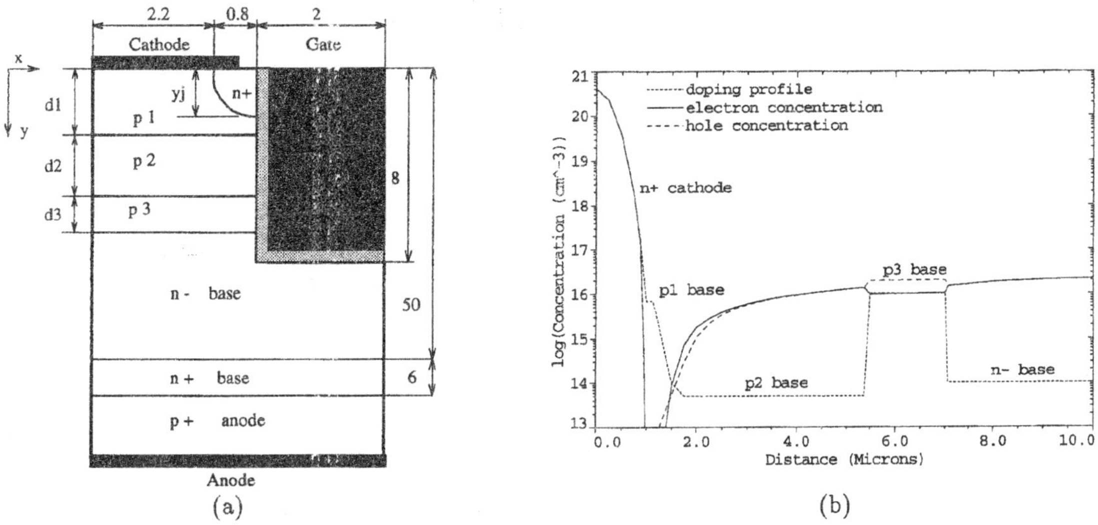

Figure 2: (a) Cross-section the ILET structure and (b) Doping profile, electron and hole concentrations in the cathode region and p base under high level injection (thyristor mode). It can be noticed that the n- base, p2 and p3 bases are highly modulated by thyristor action (MEDICI simulations for the anode current density = 200 A/cm^2)

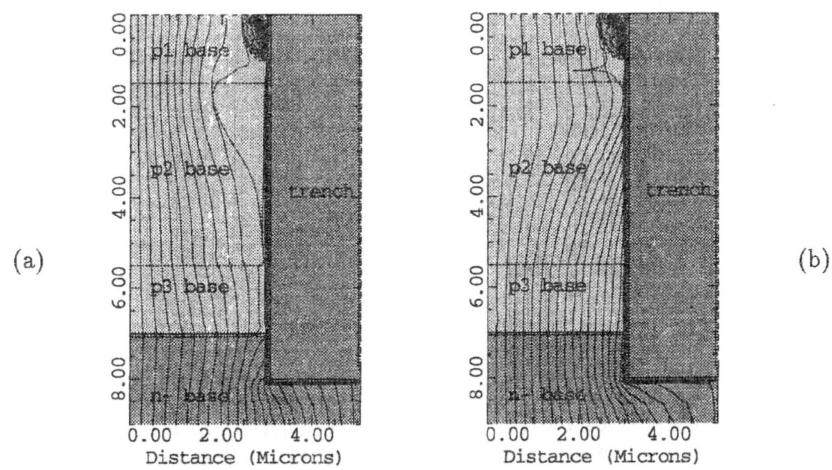

Figure 3: (a) Current distribution in the ILET in (a) the IGBT mode at 10 A/cm^2 and (b) in the thyristor mode at 200 A/cm^2.

Although both the base-emitter and base-collector junctions of the distributed NPN transistor are electrically connected by the inversion layer formed in the p3 base, due to the high resistivity of this path the fraction of the electron current flowing through it is relatively small. Consequently the p2, p3 and n- base layers are highly modulated by mobile carriers. At this stage the device operates in a thyristor mode having the emitter formed by an inversion layer. A two-dimensional representation of the action of Inversion Layer Injection is shown in Fig. 3(b). It can be seen that the Inversion Layer Emitter acts as a proper bipolar emitter modulating by at least two orders of magnitude the hole and electron concentration of p2 and n- bases (Fig. 2(b)).

The turning-off of the device is achieved by interrupting the positive gate voltage. This causes the removal of the inversion layer formed in the p base and hence the collapse of the thyristor emitter. The turn-off continues with the extraction of the hole charge through the p base to the cathode contact, and the slow phase of charge recombination and depletion advance at the anode side.

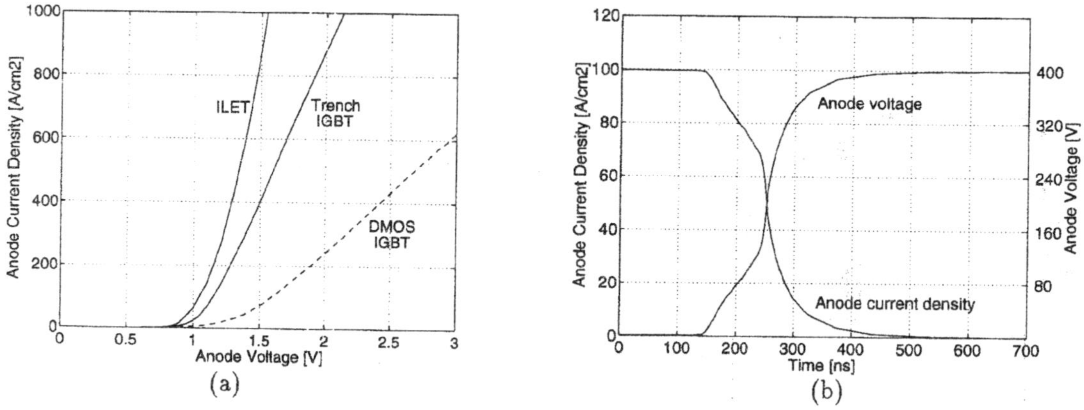

Figure 4: (a) A comparison between the I-V characteristics of 600 V ILET, Trench IGBT and DMOS IGBT for a lifetime of 0.1 μs; (b) ILET turn-off characteristics under resistive conditions (V_{out} = 400 V, initial current density I = 100 A/cm^2, V_G = 15 to 0 V in 150 ns)

The I-V characteristics for the 600 V ILET, Trench IGBT and DMOS IGBT are presented in Fig. 4(a). The ILET and Trench IGBT are considerably superior to the conventional DMOS IGBT due to increased channel density and enhanced carrier concentration at the cathode side as a result of a thyristor action (ILET) and respectively PIN diode action (Trench IGBT). At high current densities, the ILET has, however, a smaller on-state voltage drop than Trench IGBT due to a reduced effective channel length. Turn-off numerical simulations of the ILET have been carried out under resistive (Fig. 4(b)) conditions. The results indicate that the turn-off mechanism and performance of the ILET are very close to those of DMOS and Trench IGBTs.

4. Conclusions

A novel MOS-controlled power device concept in which the thyristor emitter is formed by an inversion layer is proposed and its utilisation in a trench gate device is discussed. The device exhibits on-state characteristics very close to those of a vertical thyristor, is fully controllable via a single unipolar MOS gate and capable of turning-off very high current densities. The Inversion Layer Injection mechanism upon which the ILET is based is demonstrated experimentally in a lateral MOS-controllable structure.

Acknowledgement

F. Udrea acknowledges the award of Tucker-Price postdoctoral, research fellowship from Girton College, Cambridge University. The authors acknowledge the help of the staff from Microelectronics Center, Southampton University in fabricating CMOS compatible lateral devices.

References

1. F. Udrea "Novel MOS-gated bipolar device concepts towards a new generation of power semiconductor devices", *Ph.D. thesis, Cambridge University,* 1995
2. F. Udrea and G.A.J. Amaratunga *International Patent with Philips Electronics* ,No. EP 633611 (in Europe), (1995) first filed in UK (1993)
3. D. Ueda *et al., Proc. Int. Conf. Solid State Device and Materials* p. 97, 1986.
4. H.-R. Chang and B.J. Baliga, *IEEE Trans. Electron Device* **ED-36**, p. 1824 (1989)
5. F. Udrea and GAJ. Amaratunga, *IEEE Trans. Electron Device,* vol. 42, p. 1356, July 1995.

STRESS INDUCED CHARGE EFFECTS IN AlGaAs/GaAs HETEROJUNCTION BIPOLAR TRANSISTORS

Ovidiu BAJDECHI , Faculty of Electronics, "POLITEHNICA" UNIVERSITY of BUCHAREST, Armata Poporului 1, Bucharest, ROMANIA
Patrick J. McNALLY , School of Electronic Engineering, DUBLIN CITY UNIVERSITY, Dublin 9, Dublin, IRELAND

Abstract - *A simulation of stress-induced charge effects in AlGaAs/GaAs heterojunction bipolar transistors has been performed. The results prove that this charge does not modify the transistor's behaviour even at high emitter stress. The charge was computed using an analytical model and the simulations have been done using a MEDICI 2-D device simulator, including the heterojunction advanced module.*

I. Introduction

During the fabrication of semiconductor devices there are mechanical stresses induced in the device layers. The stress is commonly due to the difference of volume thermal expansion coefficient between two overlaid materials (e.g. semiconductor and dielectric) [4]. For heterojunction bipolar devices, an important contribution could be added by the lattice mismatch of two different semiconductors which form one of the junctions. This mechanical stress has a bigger influence in GaAs-based devices then in Si-based ones, one of the reasons for this being the piezoelectric properties of the compound semiconductor [5]. For example, stress induced charge can modify the characteristics of MESFET's, introducing shifts (up to 300mV) of threshold voltage for non-uniformly distributed stresses in a given wafer.

The influence of stress induced charge in heterojunction bipolar transistors (HBT's) has not been investigated so rigorously. A recent paper [1] presents experimental results about the orientation effect on AlGaAs/GaAs HBT's. The authors of the paper made a link between the transistor's orientational dependence of current gain vs. collector current and the stress-induced charge at the emitter edges. The stress at the emitter-base interface can be considered as being due to the effect of thermal expansion or compression of dielectric layers over/near emitter material. In this work a full device simulation was carried out for the transistor structure used in [1], thereby allowing a thorough analysis of the effects of piezoelectric charge on the characteristics of the device. The results show no important modification in the current gain dependency vs. collector current due to stress induced piezoelectric charge. The electric field and the modification of base-collector junction depth due to supplementary charge has also been considered.

II. Experimental results

The device used in [1] was an HBT with $Al_xGa_{1-x}As$ with x between 0 and 0.3 as emitter material and GaAs as base and collector material. The AlGaAs emitter and GaAs collector were Silicon doped (n-type) with concentrations of $2\cdot10^{18}$ cm^{-3} and $3\cdot10^{16}$ cm^{-3} while the GaAs base was Carbon doped (p-type) with a concentration of $2\cdot10^{18}$ cm^{-3} . Three emitter orientations of $[011]$, $[010]$, $[01\bar{1}]$ were compared in respect of the current gain vs. collector current, ideality factor and base current density vs. the ratio of emitter perimeter to emitter area. The wafers used were (100) surface orientated GaAs wafers.

The most important result quoted in [1] was a significant difference in the maximum values and shape of the current gain characteristics for transistors with emitter orientations of either $[010]$ or $[011]$ when compared with $[01\bar{1}]$ orientated transistors on the other hand. The original measured data in this respect are reproduced in Fig.1. Obviously, the $[01\bar{1}]$ orientation is the worst one, showing the

0-7803-3223-7/96/$5,00

smallest maximum current gain and the highest emitter size effect. These effects are presumed to be due to the presence of charge induced by the dielectric stress at the emitter edges.

III. Device Simulation

While the structure used in [1] was a 3-dimensional one, involving two orthogonal current directions, a 2-dimensional simplified structure was used for the simulations outlined here by adding the collector contact to the bottom of the structure. This geometry modification is not very important because the most sensitive part of the device - the base - is working under the same conditions as in the real device. The simulated device is shown in Fig. 2.

In this geometry, the device simulation is stored in a TIF (Technology Interchange Format) file, and the stress-induced piezoelectric charge was introduced using an analytical model for the stress induced charge density described in [2]:

$$\rho_{pz} = \gamma_b \sigma_f d_f \left(\frac{x_1 y(x_1^2 - \beta y^2)}{r_1^6} - \frac{x_2 y(x_2^2 - \beta y^2)}{r_2^6} \right) \qquad (1)$$

with: ρ_{pz} - piezoelectric charge density

σ_f, d_f - stress in the dielectric film and the film thickness

x_1, x_2, y - the relative coordinates of the point to the edges of the stressed layer,as described in [2]

$r_1 = \sqrt{x_1^2 + y^2}$, $r_2 = \sqrt{x_2^2 + y^2}$

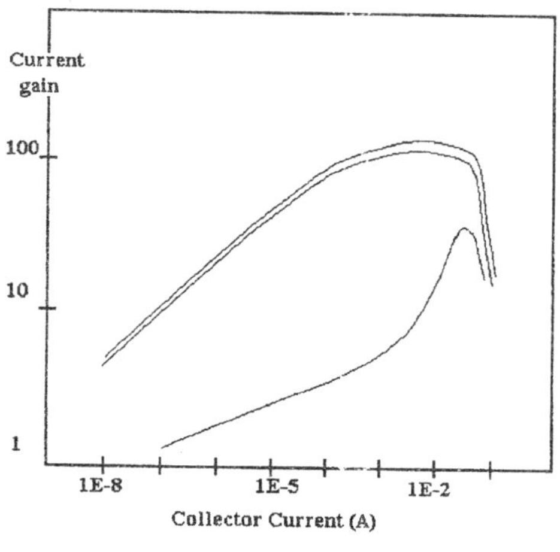

Fig. 1. Experimental curves

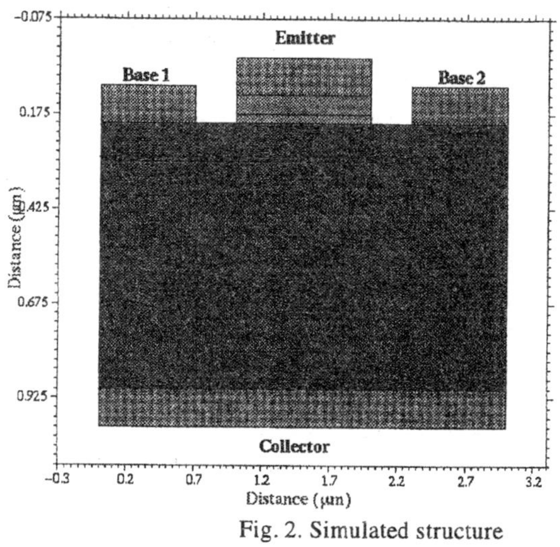

Fig. 2. Simulated structure

This formula was used for the FET transistors with $[011]$ drain-to-source direction and compressive stress in the gate. For the tensile stress in the gate, the sign of the charge changes. Using the coordinate transformation described in [3] and the relative position of $[011]$ and $[01\bar{1}]$ directions in GaAs, the formulae for the HBT with $[011]$ emitter direction is identical to (1), though of opposite sign.

For the $[010]$ transistor, the piezoelectric charge is zero due to the zero value of the piezoelectric coefficient for this direction in GaAs.

The emitter and collector contacts were voltage biased with fixed voltages of 0 and 10 Volts, while the base contacts were current biased in a range between 10^{-10} Amps/μm and 10^{-4} Amps/μm. This allowed the extraction of current gain for different collector current values while avoiding convergence problems.

IV. Results and Discussion

The structure containing the induced piezoelectric charge for the $[01\bar{1}]$ heterojunction bipolar transistor, considering compressive stress of 10^8 Pa at the edges of the emitter, is shown in Fig. 3. The charge peak is about 10^{16} cm^{-3} and is situated in the base region which is doped 10^{18} cm^{-3}. The position of the peak is under the emitter edge. Under the emitter, in the most sensitive section of the base – wherein almost all the emitter-collector current flows – the charge has a negative value. This implies that the charge is added to the doping acceptors, driving the collector - base junction deeper into the collector region. This effect could be observed starting with emitter stress values as high as 10^9 Pa. The modification of depth is non-uniformly distributed over the line of the junction, following the profiles of piezoelectric charge, so the maximum modification is reached under the emitter and in two lateral regions starting under the emitter edges.

Current gain vs. collector current curves were simulated for devices experiencing a compressive stress of 0, 10^8, 10^9 and 10^{10} Pa at the emitter edges due to dielectric layers near it, for the $[01\bar{1}]$ transistor, and are shown in Fig. 4. There is a very small difference between the curves for 0, 10^8 and 10^9 Pa emitter stress because the charge introduced by such stresses have little impact on the doping profile at

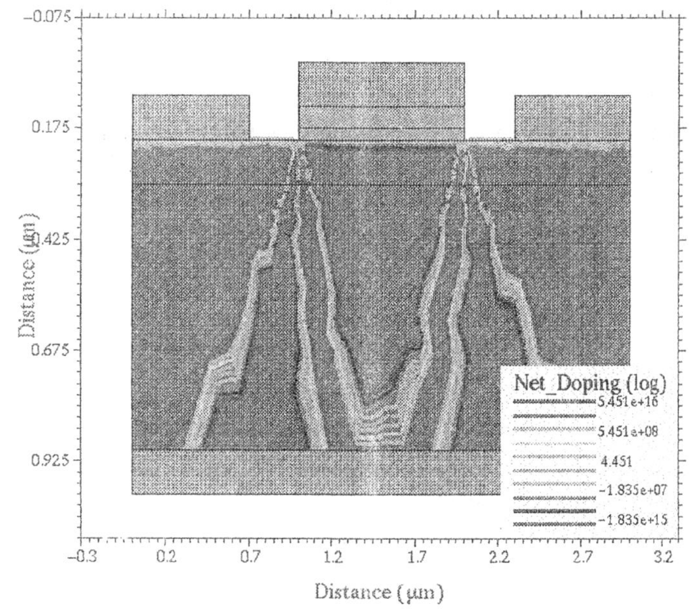

Fig. 3. Stress induced charge

the base - collector junction. For an imposed stress of 10^{10} Pa (which implies a linear force of 10N at the edges of the 0.1μm thick emitter) the modification of the position of base - collector junction is as high as 0.2μm (200%) downwards under the emitter, but still the offset introduced in the curve is smaller then 6%. It must be noted that this overstress was applied just to get a clear shift of the current gain curve, and such stress levels are not common in the real structure. This overstressed simulation also shows that even at very high imposed stress the shape of the curve is not changing, in accord with the experimental observation [1].

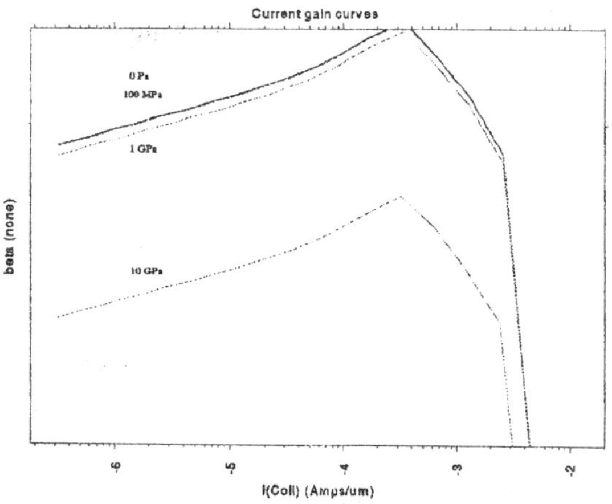

Fig. 4. Current gain curves for different stress values

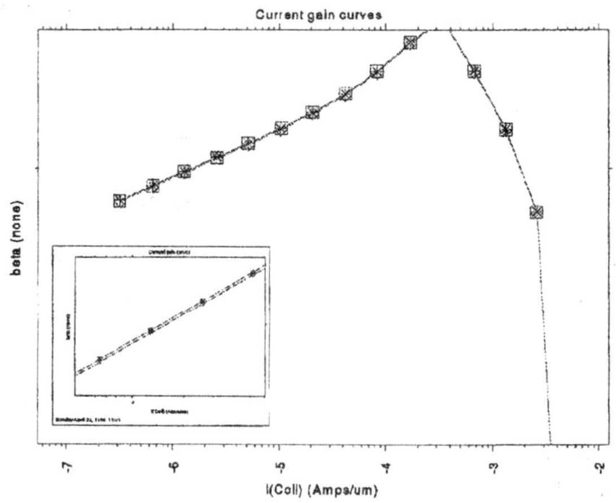

Fig 5. Current gain curves for different HBT orientation

Fig. 5 shows the results of the simulations for β vs. I_C for all three emitter directions considered in the experiments of [1]. All three curves were obtained for an applied stress of 10^8 Pa. One can see that the shift of the curve for the $[01\bar{1}]$ direction is larger then the one for [011] (see inset) because for the [011] direction the charge under the emitter is a positive one and is effectively subtracted from the base doping thus reducing the depth of base - collector junction. However this reduction is smaller in absolute value then the increase of the depth produced by the charge for the $[01\bar{1}]$ transistor because of the greater base doping. The shift is zero for [010] because the piezoelectric charge is zero for this emitter direction.

During the simulations the effect of doping on the carrier lifetime was considered, yet the piezoelectric charge in the base being so small it did not affect the base current in a visible manner. The model used for recombination was Shockley - Read - Hall with total impurity concentration dependent carriers lifetime.

V. Conclusion

A simulation has been carried out on the effects of stress-induced piezoelectric charge in the vicinity of the base of an AlGaAs/GaAs heterjunction bipolar transistor, for different emitter stress values and emitter orientations. The shifts introduced by this charge in the β vs. I_C curves are different for the [011] and $[01\bar{1}]$ directions, but even at large stress values do not explain the experimental curves for $[01\bar{1}]$ transistor. It must therefore be concluded that this mechanism is not responsible for altering the aforementioned device performance.

REFERENCES

1. H. Ishida and D. Ueda, " Orientation Effect on AlGaAs/GaAs Heterojunction Bipolar Transistors ", *IEEE Electron Device Letters*, vol. 16, p. 448, 1995

2. P. M. Asbeck , C. P. Lee and M. F. Chang, " Piezoelectric Effects in GaAs FET's and Their Role in Orientation-Dependent Device Characteristics ", *IEEE Transactions on Electron Devices*, vol. ED-31, p. 1377, 1984

3. J. C. Ramirez, P. J. McNally, L. S Cooper, J. J. Rosenberg, L. B. Freund, T. N. Jackson, " Development and Experimental Verification of a Two-Dimensional Numerical Model of Piezoelectrically Induced Threshold Voltage Shifts in GaAs MESFET's ", *IEEE Transactions on Electron Devices*, vol. ED-35, p. 1232, 1988

4. T. Ohnishi, T. Onodera, N. Yokoyama, H. Nishi, " Comparison of the Orientation Effect of SiO_2- and Si_3N_4-Encapsulated GaAs MESFET's ", *IEEE Electron Device Letters*, vol. EDL-6, 1985

5. V. Swaminathan, A. T. Macrander, " Materials Aspect of GaAs and InP Based Structures ", Prentice Hall, 1991

MEDICI SIMULATION OF 6H-SiC OXIDE RAMP PROFILE (ORP) SCHOTTKY STRUCTURE

G. Brezeanu, J. Fernandez[*)], J. Millan[*)], J. Rebollo[*)],
M. Badila[**)], G. Dilimot, P. Lungu

University POLITEHNICA Bucharest, Romania
[*)] Centro Nacional de Microelectronica Barcelona, Spain
[**)] Research Institute of Electronic Components, Bucharest, Romania

SUMMARY

A MEDICI simulation of 6H-SiC Schottky structure which uses the oxide ramp etching technique in order to attenuate edge effects is reported. The results of simulation show a uniform reverse current density and volume breakdown at Schottky structure with 300V blocking voltage can be obtained for a maximum 5 degs. ramp angle of 1μm oxide.

1. INTRODUCTION

Silicon carbide (SiC) based semiconductor electronic devices and circuits are currently being developed for advantageous use in high-temperature, high power and/or high radiation conditions under which Si or GaAs cannot adequately perform /1/.

6H-SiC power MOSFET's and diode rectifiers would operate over higher voltage and temperature ranges, have superior switching characteristics and yet have die sizes nearly 20 times smaller than correspondingly rated silicon-based devices which would offset the higher silicon carbide material cost. These improvements arise from the inherent material property advantages that 6H-SiC enjoys over silicon, namely a higher breakdown electric field (≥ 8 times that of Si) that permits much smaller drift regions (i.e., much lower drift region resistance), a higher thermal conductivity (≥ 3 times that of Si) that allows a better heat dissipation and a wide bandgap energy (3eV for 6H-SiC) that enables higher junction operating temperatures /1/.

Because silicon carbide semiconductor technology is relatively young and unoptimized, there are several crucial device fabrication issues that must be solved before truly advantageous SiC power devices can be realised experimentally. Through the fabrication of the first 2000V 6H-SiC rectifier diodes has recently made progresses on some of these issues /1/.

One simple technology in the manufacturing of Si power Schottky diodes uses the oxide ramp etching technique for the oxide surrounding the Schottky contact window in order to attenuate edge effects. This technology requires a comparatively smaller number of high temperature process steps and the temperatures involved are lower (≤ 900 °C) /2/

The results of simulation 6H-SiC oxide ramp profile Schottky structure using MEDICI program are reported in this paper.

0-7803-3223-7/96/$5,00

2. SCHOTTKY STRUCTURE

The simulated structure (Fig. 1) has a 10μm thick and $8 \cdot 10^{16}$ cm^{-3} doping n-type epilayer on n^+ ($2 \cdot 10^{18}$ cm^{-3}) substrate. Ti is used as Schottky metal. The reduction of the electric field at the contact edge, almost to the value of Schottky field is achieved by ramp etching of the oxide at the contact circular window (10μm diameter) under small angles.

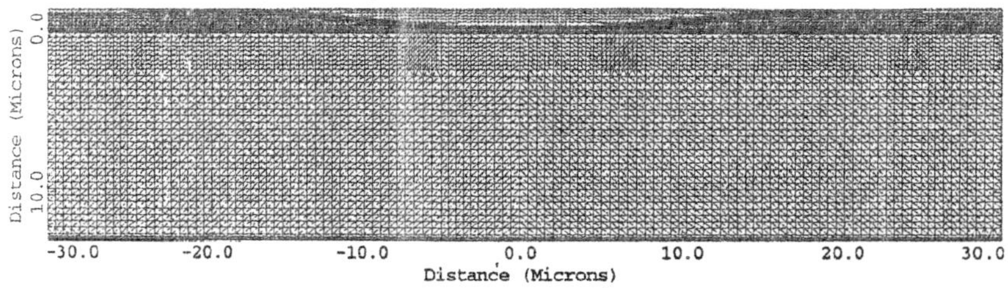

Fig. 1 - Nonuniformity grid used at Medici simulation of the 6H-SiC oxide ramp profile Schottky structure (Ramp = 5 degs.)

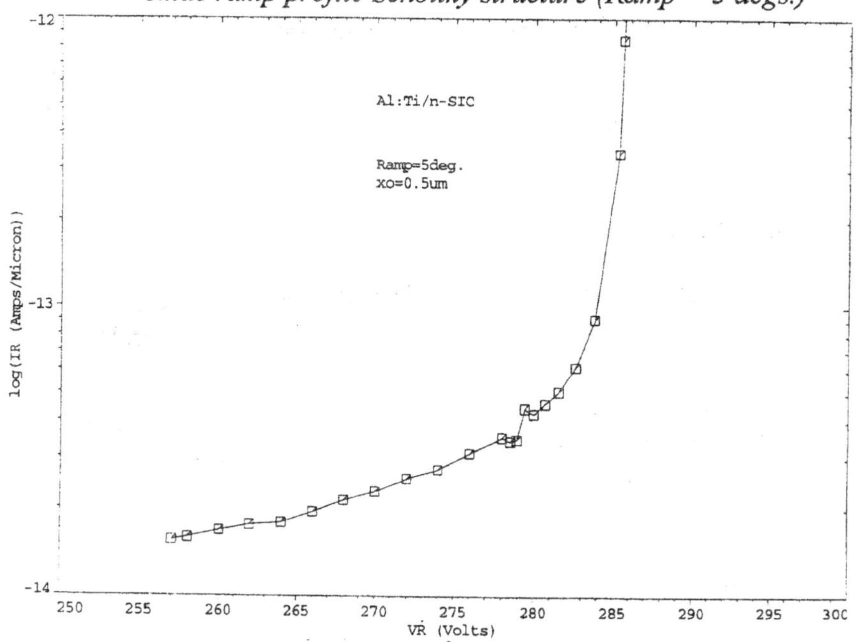

Fig. 2 - Simulated reverse characteristics of the 6H-SiC oxide ramp profile Schottky structure

The simulated reverse I-V characteristics (Fig. 2) show a very low leakage current and a breakdown voltage around 300V.

3. RESULTS OF SIMULATION. DISCUSSION

Fig. 3-5 show the electric field and current density distributions of 5 degs. oxide ramp angle and two values of the oxide thickness (x_o).

In the contact windows the electric field monotonously increases from Schottky field value to a maximum field at the oxide edge (Fig. 3). When the ramp increases the maximum field value increases too. Under oxide layer the electric field in semiconductor has values smaller than Schottky field (Fig. 3).

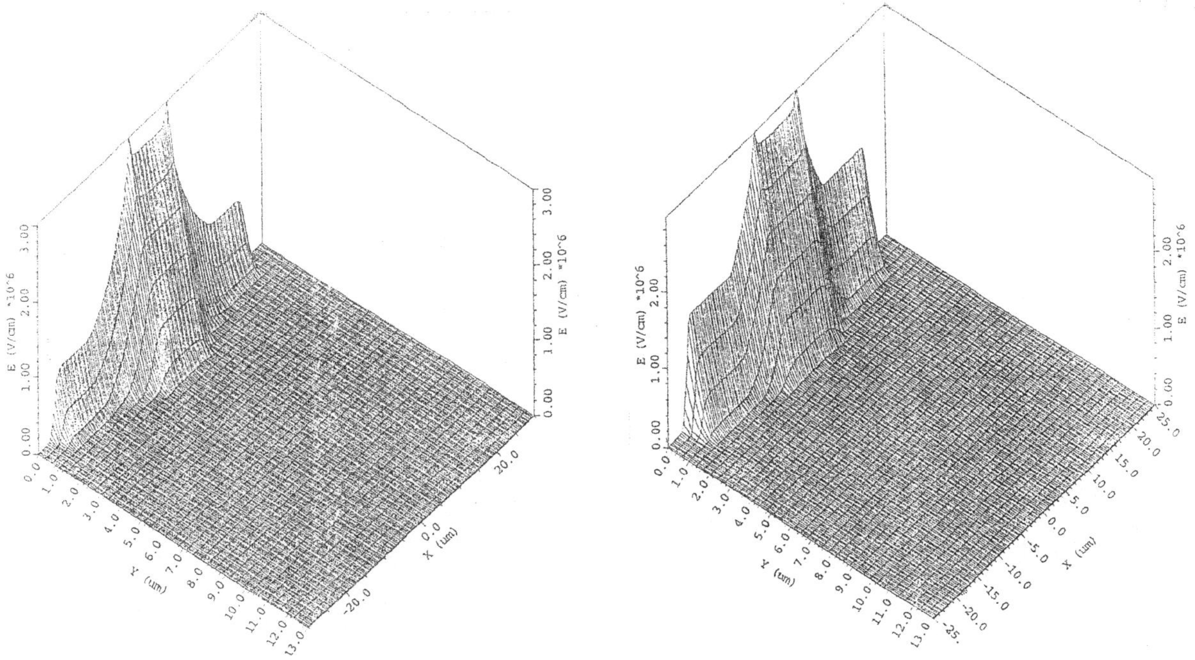

(a) (b)

Fig. 3 - The bidimensional electric field distribution at 250 V for the 6H-SiC oxide ramp profile Schottky structure (Ramp = 5 degs.):(a) x_o = 1μm, (b) x_o = 0.5μm.

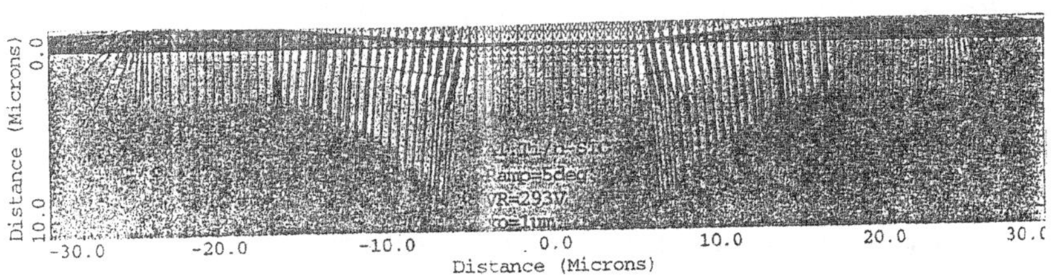

(a)

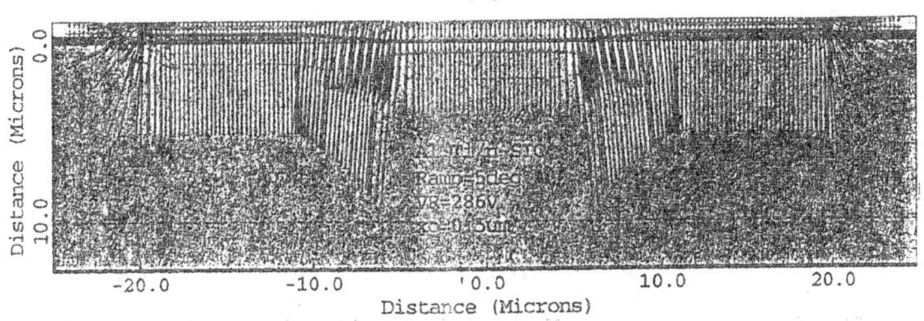

(b)

Fig.4 - The electric field vectors at 290 V for the 6H-SiC oxide ramp profile Schottky structure (Ramp = 5 degs.):
(a) x_o = 1μm, (b) x_o = 0.5μm.

If the ramp angle is reduced (≤ 5 degs.) and for x_o = 1μm the structure has parallel plane vectors of electric field (Fig. 4(a)) and current in the contact region. As a result one obtains a uniform current density (Fig. 5(a)) and a volume breakdown.

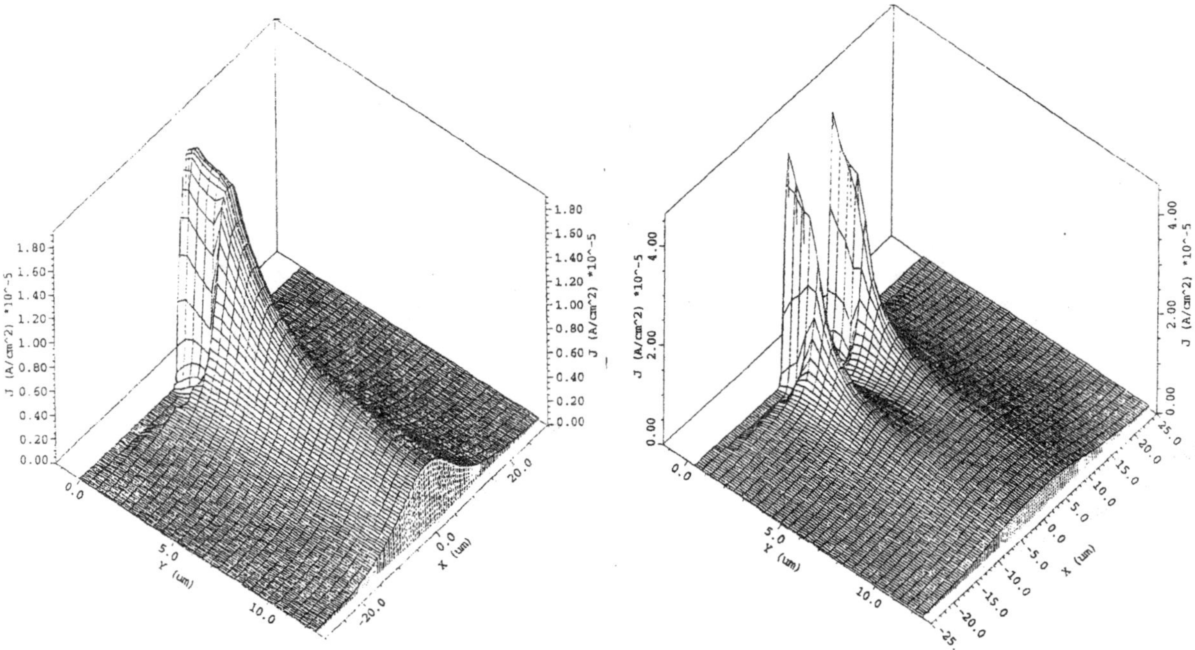

(a) *(b)*

Fig.5 - The reverse current density at 290 V for the 6H-SiC oxide ramp profile Schottky structure (Ramp = 5 degs.): (a) x_o = 1μm, (b) x_o = 0.5μm.

If the oxide thickness is reduced (x_o = 0.5μm), electric field vectors are getting crowded in ramp region (Fig. 4(b)) and a high current density at the edge (Fig. 5(b)) and a corner breakdown results. For ramp ≥ 10 degs., the agglomeration of electric field and current vectors occur and so a corner breakdown is obtained even for x_o = 1μm.

The MEDICI simulation of the same Si structure with 90 V blocking voltage lead to Ramp ≤ 10 degs. and x_o ≥ 0.5μm for a uniform current density and a volume breakdown. /3/

4. CONCLUSIONS

A MEDICI simulation has been presented to determine the oxide ramp angle and oxide thickness in a given 6H-SiC ORP Schottky structure. A ramp ≤ 5 degs. and x_o ≥ 1μm one obtain for 6H-SiC Schottky diode with 300V blocking capability.

REFERENCES

/1/ M. Ruff, H. Mitlehner; IEEE Trans. El. Dev. vol. 41, p. 1040-1054, 1994.

/2/ M. Badila, G. Brezeanu; Microelectron. Journal vol. 27, p. 67-72, 1996.

/3/ G. Brezeanu; Activity Report at CNM Barcelona, Spain, 1995.

A Critical Review of Series Resistances Extraction Methods in Advanced Bipolar Transistors

A.E. Popescu[1,2], A.M. Ionescu[1,2], A. Rusu[2], A. Chovet[1], D. Steriu[2] and B. Tudor[2]

[1]LPCS/ENSERG, Grenoble, France

[2]'Politehnica' University of Bucharest, Electronics Dpt., Romania

Abstract

This paper gives a critical review of some of the most widely used series resistance extraction methods based essentially on static curves revealed by state-of-the-art bipolar transistors: polysilicon-emitter Bipolar Junction Transistors (BJT's) with 'normal' (Al) and $TiSi_2$ contacts. An unusual current-dependence of series resistances in polysilicon emitter transistors is also reported.

1. Introduction

Due to the reduction of geometrical dimensions in advanced integrated circuits, the contribution of parasitic effects (such as series resistances) becomes critical for the microelectronic device behaviour. As a consequence, an increased interest was paid last years to the development and accuracy improvement of conventional methods associated to the evaluations of base, emitter and collector resistances (r_B, r_E and r_C, respectively). The interest in methods based on static curves comes from the simplicity of experimental set-up and modelling. The need in the progress of such methods is directly related to new technological solutions for bipolar transistors such as: polysilicon emitters (eventually with an interfacial insulating layer between polysilicon contact and single crystal) [1] and/or vertical BJT's/SIMOX [2]. Systematic tests show that conventional extractions are not always able to offer accurate results on the contribution of parasitic resistances in the mentioned devices.

2. Methods and experiment

Various extraction methods [3] are used in order to evaluate r_B, r_E and r_C in bipolar transistors (NPN polysilicon-emitter transistors for BiCMOS technology with 'normal' or $TiSi_2$ contacts, Figs. 1 and 2):

(i) ***Open-collector method*** [4]- the linear plot of V_{CE} vs. I_B, measured with the set-up of Fig. 3a (and using a HP4155A parameter analyser), has an intercept of $(kT/q)ln(1/\alpha_R)$ with the V_{CE} axis and a slope of r_E. Non-linearity of the extraction plot comes from the current-dependence of α_R (usually) or r_E (rarely, but possible in our case: polysilicon-emitter transistors). An original point is the use of this method in order to calculate r_C, by reversing the emitter and collector contacts (Fig. 3 b).

(ii) ***'Depolarisation' method*** [5]- this procedure takes into account the shift of the experimental base current from the 'ideal' extrapolated straight line $I_B(V_{BE})$ in log scale, due to the depolarisation induced by r_B and r_E (in the forward active mode - FAM). The plot of $\Delta V_{BE}/I_B$ vs. current gain, β_F, has r_E as slope and intercepts the vertical axis at r_E+r_B (Fig. 4). The weakness of this method comes from the hypothesis that r_B and r_E are constants.

(iii) *Ning-Tang method* [6] - it allows the separation of the intrinsic and extrinsic base resistance contributions (r_{Bi} and r_{Bx}, respectively) using a specific current modelling and

0-7803-3223-7/96/$5,00

supposing that r_{Bi}/β_F is constant. The plot of $(kT/q\ I_C)\ ln(I_{B0}/I_B)$ vs. $1/\beta_F$ has r_E+r_{Bx} as slope and r_E+r_{bi}/β_F as vertical intercept. This method appears unsuitable for the investigated polysilicon emitter transistor because no accurate linear region of the mentioned plot is found.

(iv) ***'Slope' method*** [7] - two slopes can be defined in the output characteristics (I_C-V_{CE} at V_{BE}=constant) providing the collector resistance values in saturation, r_{Csat}, and in normal, r_{Cnorm}, operation modes, Fig. 5. For all the investigated devices it is verified that the collector resistance value is strongly dependent on the device operation point.

(v) ***Impact ionization method*** [8,9] - this recent method is based upon the fact that, due to impact ionization within the base-collector space-charge-region, at the collector-base voltage, V_{CBZ}, the base current and therefore the voltage drop on r_B are reduced to zero, Fig. 6. If high-level injection is avoided, base and emitter series resistances are calculated as:

$$r_B = \frac{V_{BE} - V_{BEZ} + (kT/q)ln[(1+V_{CB}/V_A)/(1+V_{CBZ}/V_A)]}{I_B} \tag{1}$$

$$r_E = \frac{V_{BEZ} + (kT/q)ln[(1+V_{CBZ}/V_A)/(I_E/I_{ES})]}{I_E} \tag{2}$$

The method has the advantage of a simple measurement set-up and the possibility to extract series resistances as a function of current level. Typical r_B and r_E plots are reported in Fig. 7. It is worth noting that the method can be adapted in order to evaluate the collector resistance. This method is found extremely useful and accurate in polysilicon-emitter BJT's.

(vi) ***Optimal extraction*** [9] - the aim of this method is to minimize a square error function using data and the modelling associated (Gummel-Poon, in our case) for a specified mode of operation of the bipolar transistor. All the parameters (for example: I_{SS}, β_{Fmax}, I_{KF}, C_2, N_{EL}, N_F, V_{AF}, r_B, r_E, in FAM) are simultaneously extracted as a vector solution of the function minimum. The method is accurate and fast as long as the analytical model is fully exact and there is no redundancy in the parameters. In our case this procedure is not found very successful because of the unusual current dependence of r_B and r_E resistances (the thin oxide interfacial layer between polysilicon and Si being probably responsible for this).

(vii) ***Noise measurements*** [10] - the main source of base current (1/f) fluctuations can be attributed to dynamic exchanges of carriers with the Si/SiO_2 interface at the periphery of E-B junction and less to the interfacial oxide layer. At low current level it appears possible to evaluate the contribution of series resistances; an interesting point is to conclude about the relevance of the white noise. It was verified that $TiSi_2$ contacts give no significant change in 1/f noise level.

3. Discussion

The above presented methods allow us to reveal the following results concerning the investigated polysilicon emitter BJT's:

-series resistance extraction in such devices must take into account a current-dependence for all series resistances: r_B, r_E and r_C ; it appears that r_E resistance *decreases with the current level* (typically by 50-150% for the whole current range) and this can be related to the current conduction by tunnelling through the thin oxide layer at the polysilicon/silicon interface (see Fig. 1);

-$TiSi_2$ contacts reduce (compared to 'normal' contacts) r_C and r_E by a factor of 4-6 without contributing essentially in the 1/f noise;

-optimal extraction and Ning-Tang method are not able to offer accurate results; this re-confirms that non-conventional current dependences for all series resistances are very probable;

-other extraction methods ('depolarisation', 'slope') can offer partially significant results if the extraction is conducted without changing strongly the device bias point;

-the most accurate and suitable series resistance extraction method is found to be the impact-ionization method since it offers the extracted value at a given bias an permits to tune easily all dependences. Fig. 7 shows typical dependences of r_B and r_E revealed by this method.

The range of the estimated series resistances values are (for a typical device with emitter width and length, $W_E = 12$ μm and $L_E = 1.2$ μm, respectively): (i) for $TiSi_2$ contacts: r_E: 50-100 Ω, r_C: 20-100 Ω, r_B: 400-1kΩ and (ii) for 'normal' contacts: r_E: 250-400 Ω, r_C: 150-1000 Ω, r_B: 600-1.5 kΩ.

4. Conclusion

A systematic critical experimental verification of the most used series resistance methods for bipolar transistors was conducted on polysilicon emitter BJT's with a thin interfacial oxide layer. It was found that for this type of transistor all series resistances are current-dependent. The impact ionization method appears as the most accurate for series resistance evaluation. We verified that $TiSi_2$ contacts strongly reduce emitter and collector resistances.

References:

1. A. Mounib et al., *IEEE Transaction on Electron Devices*, vol. 42 (1995), 1647-1652.
2. B. Tudor et al., *Microelectronics Journal*, 25 (1994), 443-450.
3. D. Schroder, *Semiconductor material and device characterisation*, John Wiley&Sons (1991).
4. W. Filensky and H. Beneking, *Electron. Letts.*, vol. 17 (1981), 503-504.
5. J. Logan, *Bell Syst. Tech. J.*, vol. 50 (1971), 1105-1147.
6. T. H. Ning, D. D. Tang, *IEEE Trans. on Electron Devices*, ED-31 (1984), 409-412.
7. I. Getreu, *Modelling the bipolar transistor*, Tektronix, Beaverton, OR, 1976.
8. G. Verzelessi et al., *Proceedings of ESSDERC '93*, Grenoble (1993), 433-436.
9. G. Verzelessi et al., *IEEE Electron Device Letters*, vol. 14 (1993), 431-434.

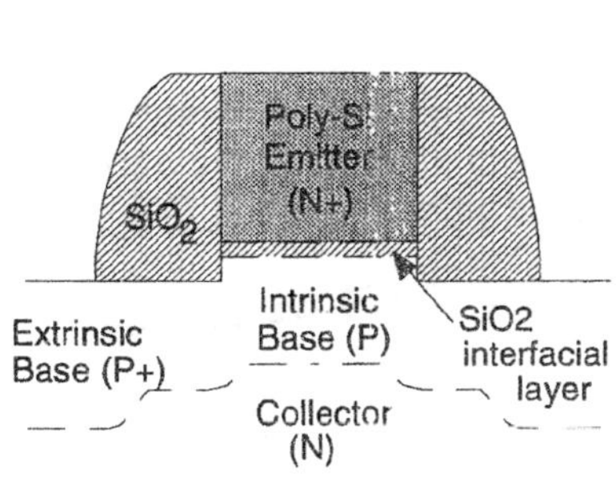

Fig. 1 Cross section of the polysilicon emitter bipolar transistor.

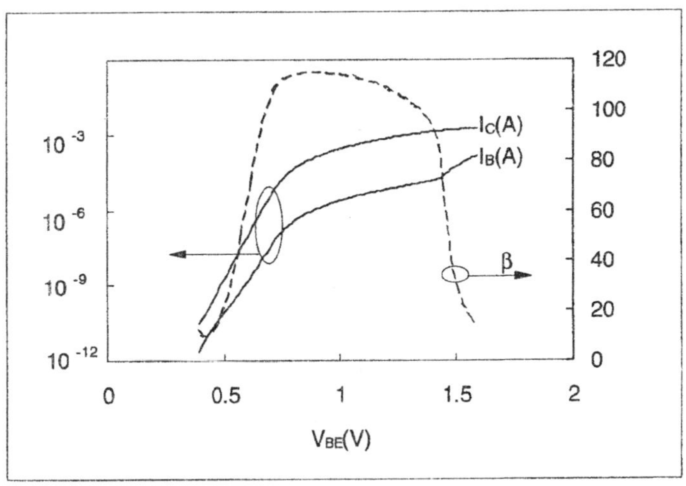

Fig. 2 Typical static characteristics for a polysilicon emitter BJT.

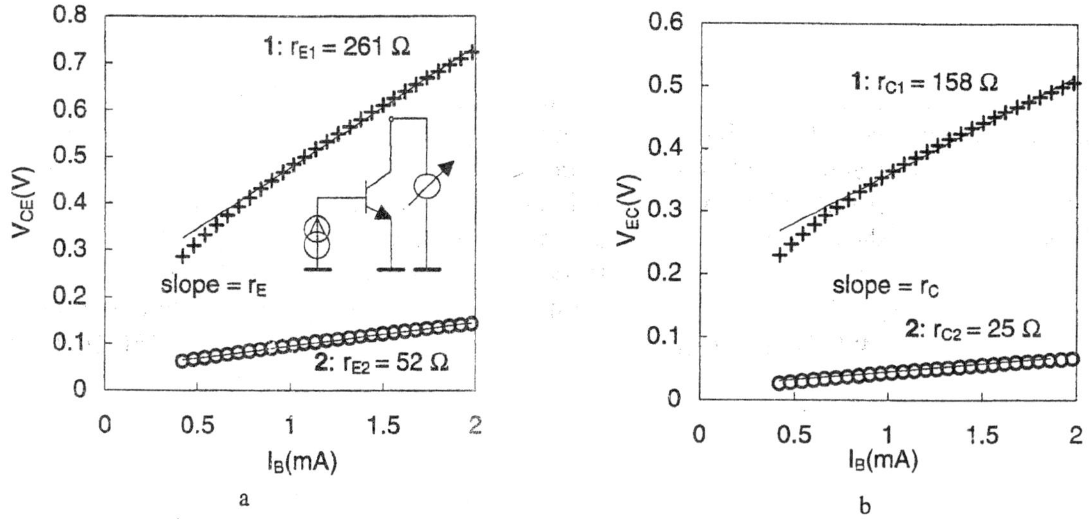

Fig. 3 a. Typical plots associated to the open-collector method for r_E evaluation (1: 'normal' contacts, 2: $TiSi_2$ contacts); b. r_C evaluation by reversing the emitter and the collector contacts.

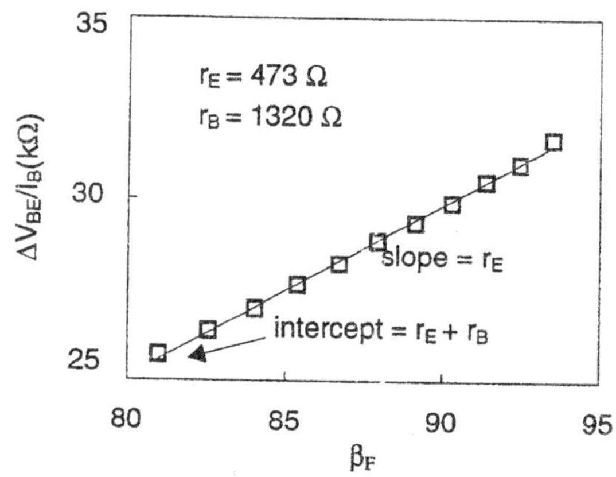

Fig. 4 Typical plot for the 'depolarisation' method.

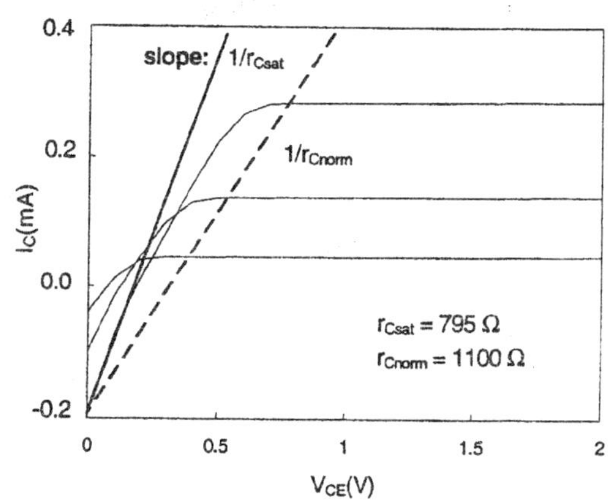

Fig. 5 Typical slopes in output static characteristics, $I_C(V_{CE})$, for collector resistance extraction.

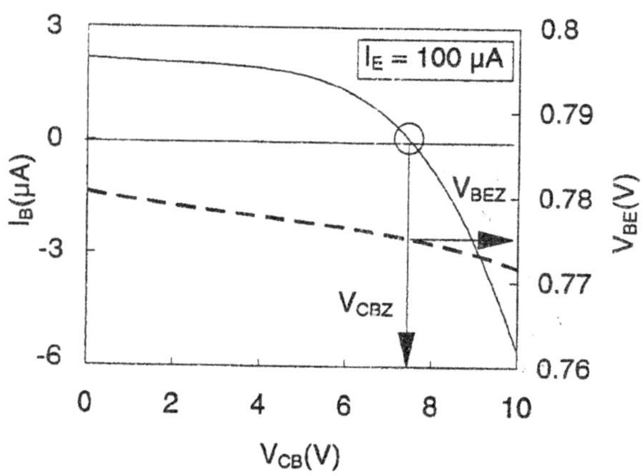

Fig. 6 V_{CBZ} and V_{BEZ} voltages evaluation in impact ionization method.

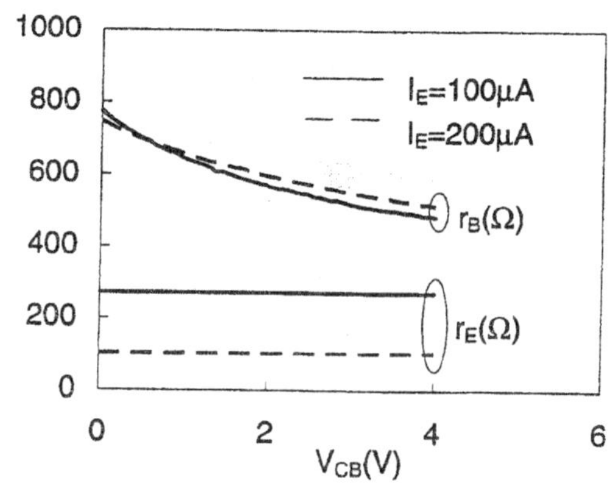

Fig. 7 Series resistances, r_B and r_E, vs. V_{CB} using data from Fig. 6.

Experimental investigation of noise sources in silicon carbide Schottky barriers

L.Anghel, T.Ouisse, T.Billon*, P.Lassagne* and C.Jaussaud*

Laboratoire de Physique des Composants à Semiconducteurs (LPCS)
UMR-CNRS 5531, ENSERG, 23, rue des Martyrs, 38016, Grenoble, France
**LETI (CEA-Technologies Avancées), DMITEC-CEN/G, 85X F38041, Grenoble, France*

Summary

The excess low frequency noise of silicon carbide Schottky diodes has been systematically measured on n-type silicon carbide devices with Ti gate. The noise results have been related to general properties of these devices such as barrier height and doping level, extracted from varying temperature measurements. The 1/f low frequency noise closely follows a model proposed by T.G.M.Kleinpenning and is thus most probably due to mobility fluctuations in the depletion region of the Schottky barrier.

I Introduction

Due to its wide bandgap, high breakdown field and high thermal conductivity, silicon carbide (SiC) has become a serious candidate for high power/high temperature applications, and material quality has been greatly enhanced for the last five years [1]. In particular, the fabrication of high quality Schottky barriers is a key aspect for developing various SiC components [2]. Devices such as SiC MESFET's and power Schottky diodes would represent serious competitors with respect to silicon and gallium arsenide, in the field of fast power switches and high frequency/high power transistors, respectively. Owing to the relatively recent improvement of material quality, a detailed investigation of their properties and performance is therefore required. Noise analysis can be useful not only for improving device performance but also for gaining substantial information on the physical properties of the SiC material. The aim of this paper is to present a systematic evaluation of the low frequency excess noise in SiC 6H Ti Schottky diodes fabricated at LETI, and then to give a quantitative interpretation of the obtained data.

II Devices

The Schottky diodes have been fabricated at LETI, by depositing Ti on 6H SiC epitaxial layers grown on 6H SiC substrates bought from Cree Research Inc. Device area was ranging from 100µm×1µm to 800µm×100µm. I(V) and C(V) characteristics were monitored from room temperature up to 300°C. The low frequency (LF) noise measurements (100mHz-100kHz) were performed at room temperature using an ONO-SOKKI spectrum analyser loaded by a low noise NF 5305 voltage amplifier and an EG&G model 181 current-voltage converter.

0-7803-3223-7/96/$5,00

III Current-voltage characteristics

Typical I(V) curves are given in Fig.1a. For all devices the ideality factor always remained very close to 1, as can be seen in fig.1b. This indicates that the thermionic model can be safely used for modelling the data in the forward conduction regime. To extract the barrier height, we used either plots of I_F/T^2 vs 1/T and I_F vs V_F. Both methods gave similar values, close to 0.96eV, in agreement with previous results published by other groups [2].

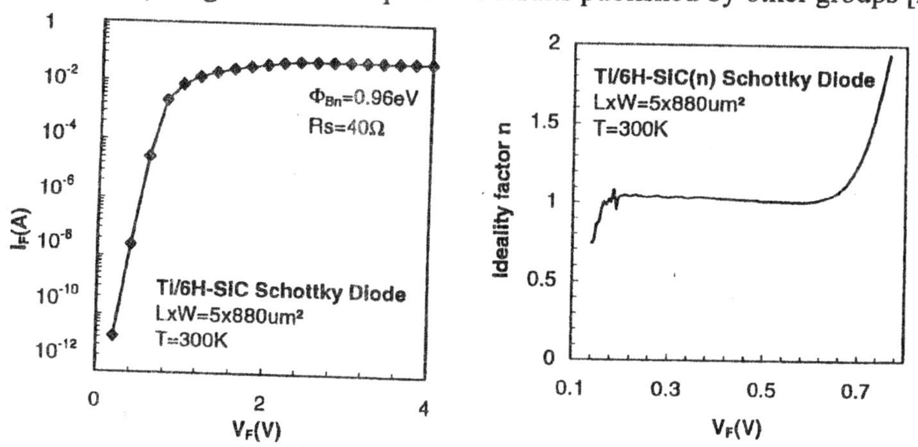

Fig.1: (a) typical current-voltage characteristics and (b) assessment of the ideality factor by plotting $(q/kT)(I_F/(dI_F/dV_F))$ vs V_F.

In contrast with the forward bias regime, the activation energy corresponding to the reverse current does not correspond to the barrier height, and varies from to, depending on V_R. This probably means that part of the reverse current is induced by leakage through the sample edges.

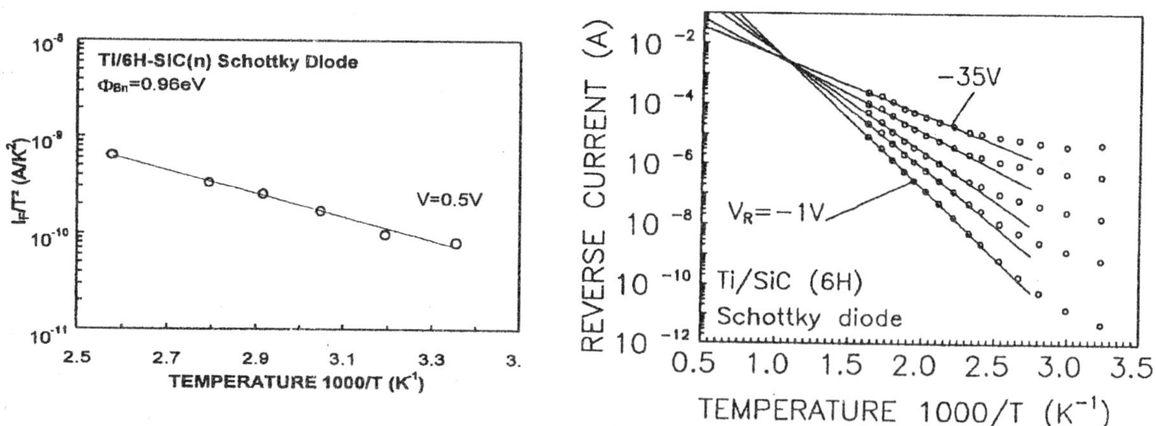

Fig.2: (a) Richardson's plot and (b) reverse current vs temperature for various V_R values.

IV Noise results and interpretation

Typical noise spectra and noise level are given in fig.3

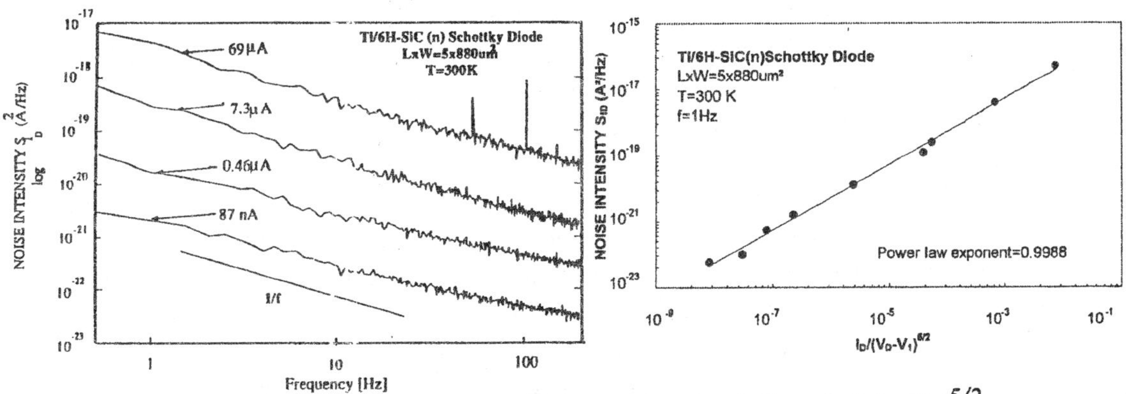

Fig.3: (a) noise spectra and (b) noise level at 1 Hz vs $I_F/(V_F-V_1)^{5/2}$

As it can be seen, at low frequency the diodes exhibit 1/f noise, and in this regime the power spectral density of the current fluctuations is described by the following empirical relation:

$$S_I(f) \approx \frac{I_F{}^{\beta}}{f^{\gamma}},$$

where I_F is the forward current, f the frequency , $1<\beta<1.4$ and γ remains very close to 1, in agreement with experimental results obtained with schottky diodes fabricated with other semiconductors [3].

Several theories have been proposed in the literature for accounting for the 1/f noise appearing in schottky barriers, most of them invoking carriers' number fluctuations: Hsu [4] has interpreted this noise in terms of trapping and multistep tunnelling processes in the depletion region. these processes introduce a random fluctuation in the occupancy of the trap centres, which finally lead to local fluctuations of the barrier height, and eventually to current noise; besides, other experimental findings showed that 1/f noise in Schottky barriers could also be generated at the periphery of the diode, through the device edges. Still another model invokes a modulation of the barrier height [3], through trapping or generation-recombination (GR) noise in the depletion region, the charge density fluctuations in the space charge region also inducing local changes in the barrier height. Then, if the GR centers are located at the mid-gap, it was shown that the current noise should be of the form $S_I \approx I_F^{3/2}/f^2$ [3]. Alternatively, if the GR centers are located above the mid-gap and only transitions between the conduction band and g-r centers are taken into account (trap states uniformly distributed in space as well as in energy and with a constant capture cross section), the total current noise may exhibit a 1/f behaviour and then must vary as [3]:

$$S_I \approx \frac{I_F{}^2}{f}\left[\frac{kT}{q(V_F - V_1)}\right]^{1.5}$$

where the difference V_F-V_1 can be calculated from Richardson's equation:

$$V_F - V_1 = \frac{kT}{q}\ln\left(SA^*T^2\frac{N_D}{N_C I_F}\right)$$

In our case, $S=44x10^{-6}cm^2$, $A^*=154A^2/cm^2K^2$, $N_D=10^{17}cm^{-3}$ and $\mu=200cm^2/Vs$. In all our experimental results, we never obtained $S_{IF}(I_F)$ dependencies compatible either with that of GR noise in the depletion region, or with trapping and multi- step tunnelling. Besides, mobility noise in the neutral part of the semiconductor should follow a law of the form $S_I \approx I_F^2/f$ [3], which is also not supported by the experimental data.

Kleinpenning has proposed an additional mechanism for explaining 1/f noise in Schottky diodes [3]. Normally, an ideal diode with an I-V characteristic independent of the mobility of free carriers, does not generate mobility 1/f noise. However, it is now well established that the simplest form of Richardson's equation, or thermionic current model, must be corrected to take into account carrier drift and diffusion in the depletion region of the diode (see, e.g., ref.[5]). As a consequence, mobility fluctuations manifesting themselves in the depletion region could also be a source of 1/f noise. The only model to agree with our experiment is that of Kleinpenning [3], which gives as a result of the mobility fluctuations in the depletion region:

$$S_{I_F} = \frac{\alpha I_F}{16\pi f}\left(\frac{kT}{q(V_F-V_1)}\right)^{5/2}\left(\frac{q}{\mu m^*}\right)^2\left(\frac{m^*\varepsilon}{\pi N_D}\right)^{1/2}$$

α is the Hooge factor, which in our case was found to lie around $3x10^{-3}$, a usual order of magnitude for covalent semiconductors, owing to the doping level and the impurity scattering of our devices [6].

As can be seen in fig 3(b), a log-log plot of S_{IF} as a function of $I_F/(V_F-V_1)^{5/2}$ gives straight lines, and this was verified for all devices. Although a plot of S_{IF} vs I_F gives a power law exponent wich depends on the device area (between 1 and 1.4), in the somewhat more complicated plot of fig. 4 we found an exponent very close to unity for all samples. Hence our results are in very good agreement with Kleinpenning's model.

V Conclusion

The low frequency 1/f noise appearing in SiC 6H Schottky diodes has been investigated for devices with an ideality factor close to unity. None of the existing models based on carriers' number fluctuations agreed with the experiment. In contrast, our data nicely fit the model proposed by Kleinpenning a few years ago, who attributed the current fluctuations as the consequence of mobility fluctuations in the depletion region of the diode. The overall results therefore indicate that the density of traps close to the metal/semiconductor interface must remain within a reasonable range, and that at room temperature, generation-recombination processes are greatly reduced by the wide band gap (3eV) of this SiC polytype.

References

[1] R.F.Davis, G.Kelner, M.Shur, J.W.Palmour and J.A.Edmond, *Proc. IEEE* 79, p.677 (1991)
[2] A.Itoh, O.Takemura, T.Kimoto and H.Matsunami, *Inst. Phys. Conf. Ser.*, in press
[3] T.G.M. Kleinpenning, *Solid-St. Electron.* **22**, pp.121-128 (1979)
[4] S.T.Hsu, *IEEE Trans vol.* **ED-17** (7), pp. 496-506 (1972)
[5] S.M.Sze, *Physics of semiconductor devices* (Wiley, 1981)
[6] F.N.Hooge, *IEEE Trans. Electron Dev.* **ED-41** (11), pp.1926-1935 (1994)

Session 18

CIRCUIT MODELING & CAD (Poster)

MSIM2: Mixed-signal Short-channel IGFET Model for Circuit Simulations

H. J. Luo, C. S. Yeh, K. Hwang and R. Mendel

National Semiconductor Corporation
MS: D3-677
2900 Semiconductor Drive
Santa Clara, CA 95052
U. S. A.

Phone: (408) 721-5134
Fax: (408) 721-1472
Email: jluo@berlioz.nsc.com

SUMMARY

MSIM2 is a physical compact model specifically developed for mixed-signal and analog-intensive circuit simulations in deep sub-micron CMOS technologies. MSIM2 separates the drift and diffusion current components as continuous functions in all operating regions, and shows superior accuracy and continuity in the subthreshold and moderate inversion regions even for second-order derivatives. It has full temperature scalability (-55°C to 150°C) and geometry scalability (down to 0.25μm in W and L), and uses less parameters than other advanced models. MSIM2 also includes scalable charge and noise models, and has been verified against various processes from around the world.

I. Introduction

There has been a quest for a good compact MOSFET model for analog and mixed-signal circuit simulations due to the deficiencies revealed in traditional, digital-oriented models [1]. For precise and predictive analog modeling, full continuity and good accuracy of the current (I_{ds}) and small signal properties such as output resistance (r_{ds}) and transconductance (g_m) are required over a wide range of device geometry and temperature. Existing models compromise between the continuity and accuracy, especially in the moderate inversion region. MSIM2, on the other hand, meets both the continuity and accuracy requirements but uses less parameters than BSIM3 [2] and MOS9 [3]. In addition, it satisfies all the analog benchmark tests proposed by Tsividis *et al* [1].

II. MSIM2 DC Model

MSIM2 includes all the known physical mechanisms necessary for sub-half-micron devices, such as short/narrow channel effects (including two-dimensional phenomena), drain induced barrier lowering, nonuniform doping, bulk charge variation, channel length modulation, impact ionization, subthreshold conduction, mobility reduction and carrier velocity saturation. MSIM2 considers the drift and diffusion currents as continuous functions in all operating regions (any combination of V_{ds}, V_{gs} & V_{sb}),

$$I_{ds} = I_{drift} + I_{diffusion} + I_{sub}$$

where I_{drift}, $I_{diffusion}$ and I_{sub} are physically derived from long channel devices with extensive modifications for sub-half-micron devices. Figure 1 illustrates the comparison between measurement data (circles) and simulation results (lines). By measuring the total current and its derivatives around threshold to directly extract model parameters, MSIM2 produces superior accuracy and continuity in the moderate inversion region (Figures 2, 3). To accommodate the physical effects mentioned above, key parameters are then scaled with temperature, W and L from the reference device (normally wide and long), e.g.,

$$V_{th} = V_{tho}\,(1+TVT\Delta T) + VTL1(1/L-1/L_{ref}) + VTL2(1/L^2-1/L_{ref}^2) + VTW(1/W-1/W_{ref}) + VTWL(1/WL - 1/W_{ref}L_{ref})$$

0-7803-3223-7/96/$5,00

Figures 4 to 8 are the results for analog benchmarks, Figures 9 to 12 show its geometry scalability and Figures 13 and 14 show the full temperature scalability. Excellent agreement is observed between the modeled (lines) and measured (symbols) data. As seen in Figure 3, MSIM2 is accurate and continuous even for the 2nd order derivative. In Figure 5, MSIM2 generates the correct "Gummel Treetop" curve in the subthreshold [4], as opposed to the flat curve in [2] or the sharp roll-over in [3].

III. MSIM2 Parameter Extraction

An in-house parameter extraction and simulation tool, PRESTO, is used to directly extract all the model parameters with minimal local optimization and no global optimization at all:

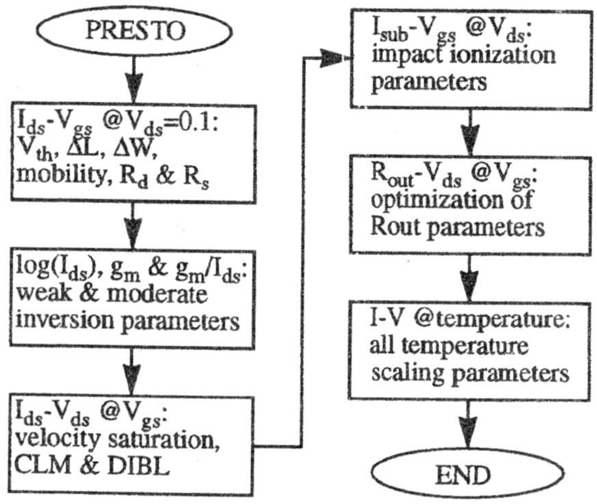

IV. MSIM2 AC and Noise Model

As a complete MOSFET model, MSIM2 includes a physical, charge based capacitance model and a semi-empirical scalable noise model.

V. Conclusions

It is shown that MSIM2 is an excellent compact model for mixed signal circuit simulations. MSIM2 produces superior results in the subthreshold and moderate inversion regions due to the physical approaches taken to model the drift and diffusion currents. Detailed model formulation will be presented at the meeting along with sample circuit verifications and comparisons among different models.

References

1. Y. P. Tsividis and K. Suyama, "MOSFET Modeling for Analog Circuit CAD: Problems and Prospects," *IEEE J. Solid State Circuits*, vol. 29, pp. 210-216, March 1994.
2. P. K. Ko and C. Hu, *BSIM3v3 Manual*, University of California at Berkeley, 1995.
3. R. M. D. A. Velghe, D. B. M. Klaassen and F. M. Klaassen, *MOS Model 9*, Unclassified Report NL-UR 003/94, Philips Electronics, 1995.
4. *Compact Models Workshop*, Sematech, August 1995.

Figure 1

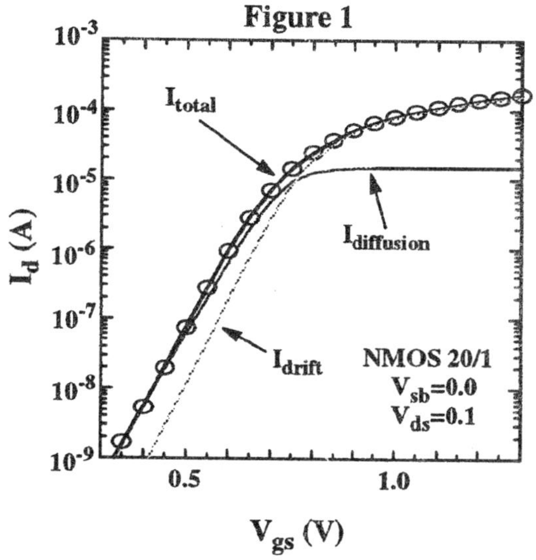

Figure 2

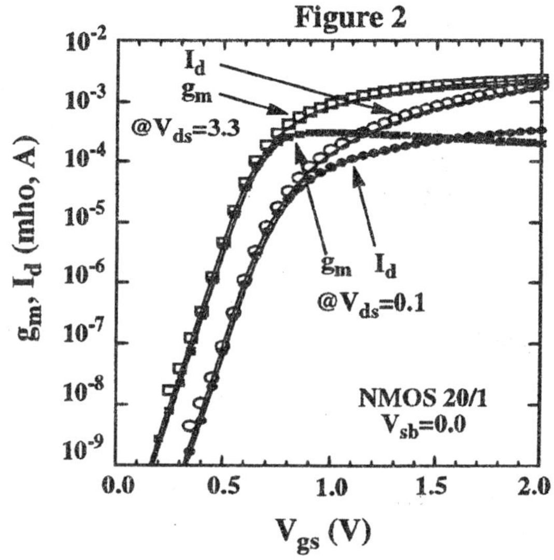

Figure 3

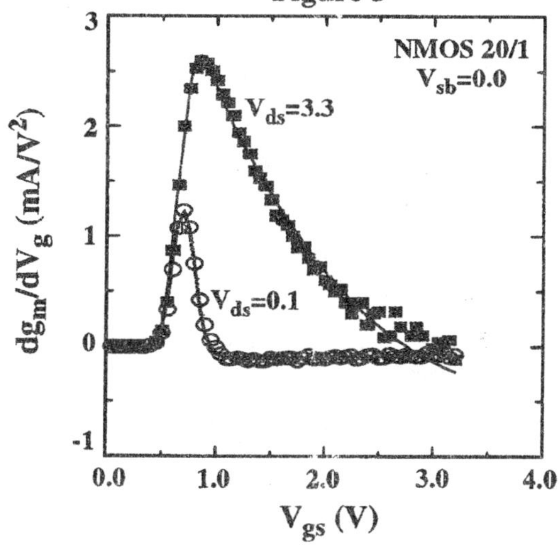

Figure 4

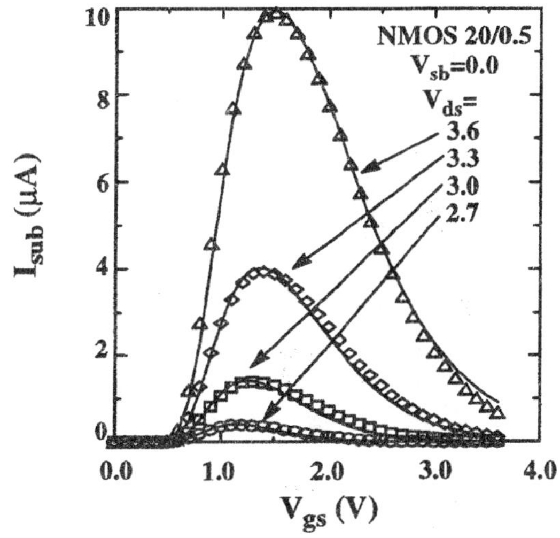

Figure 5

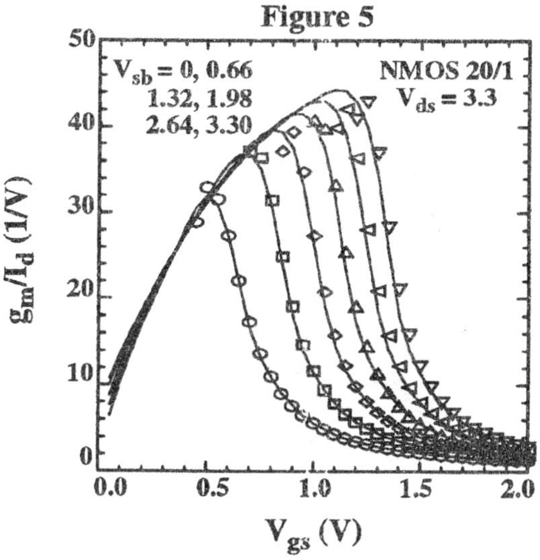

Figure 6

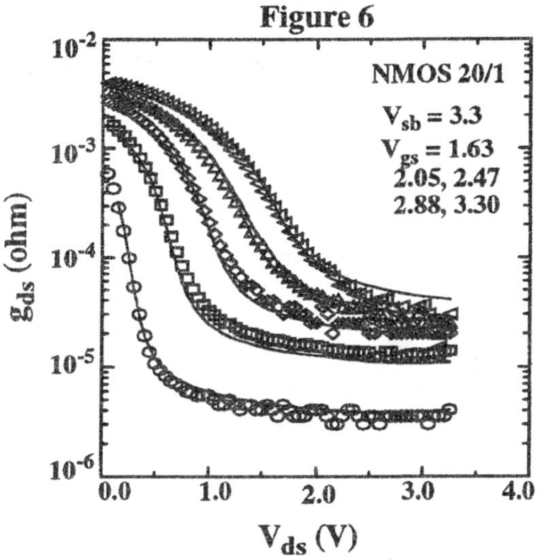

Figure 7

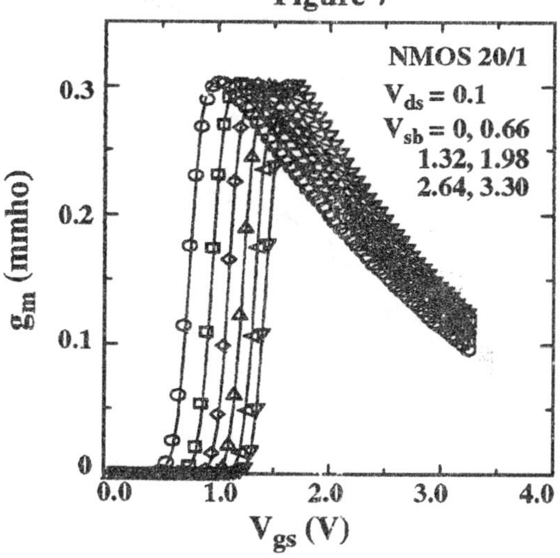

Figure 8

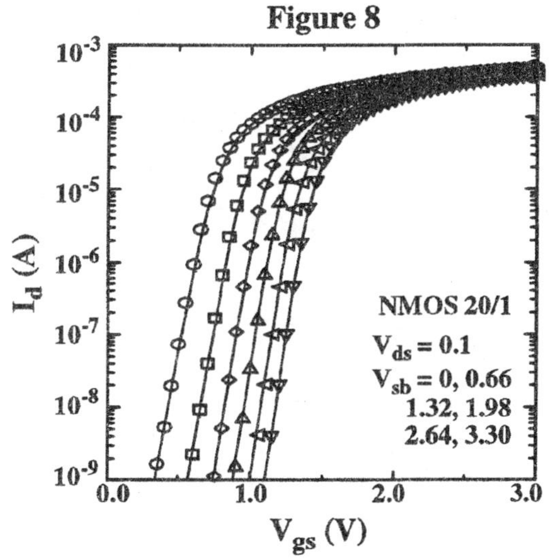

Figure 9

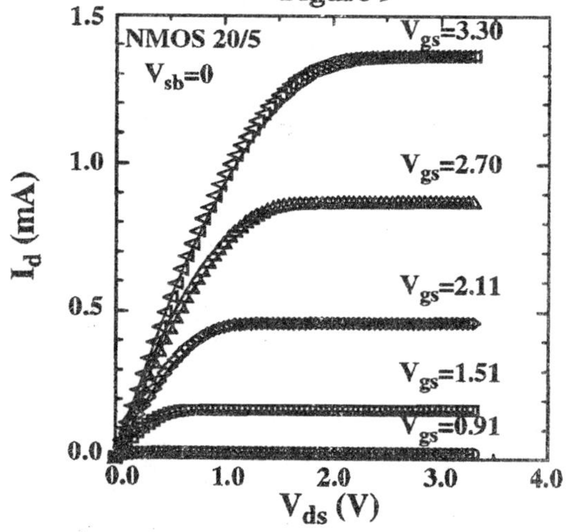

Figure 10

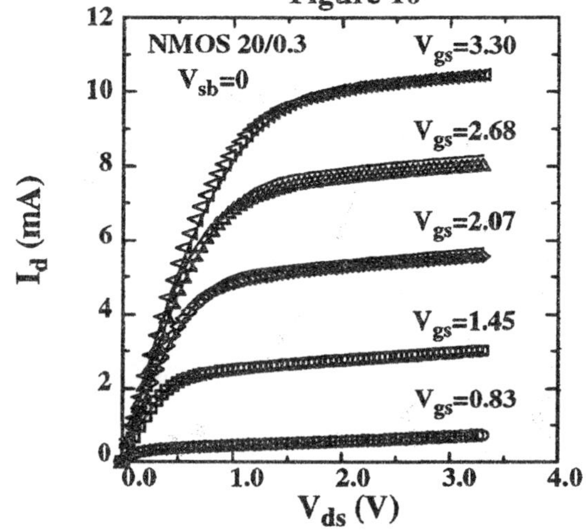

Figure 11

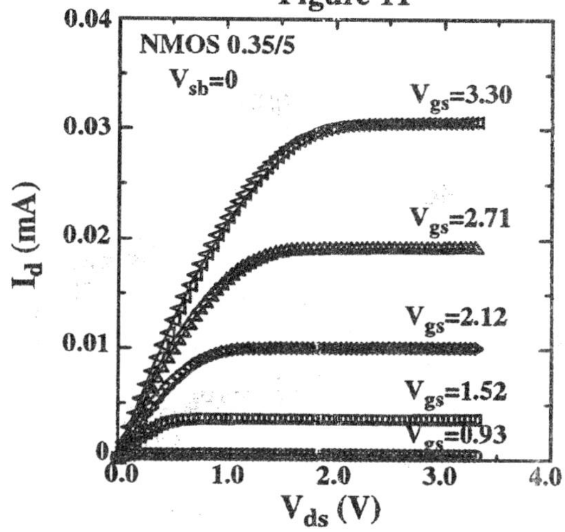

Figure 12

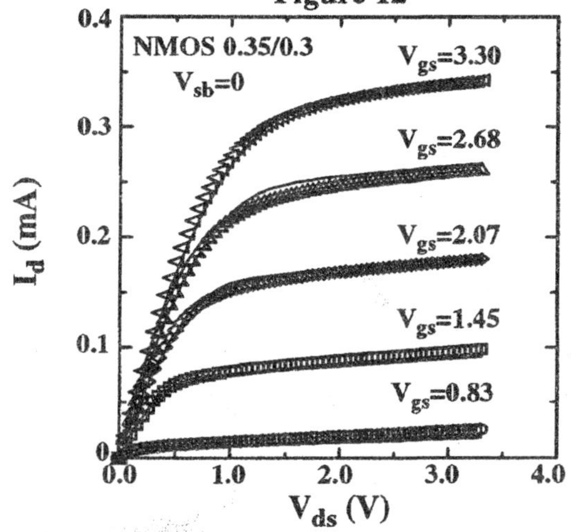

Figure 13

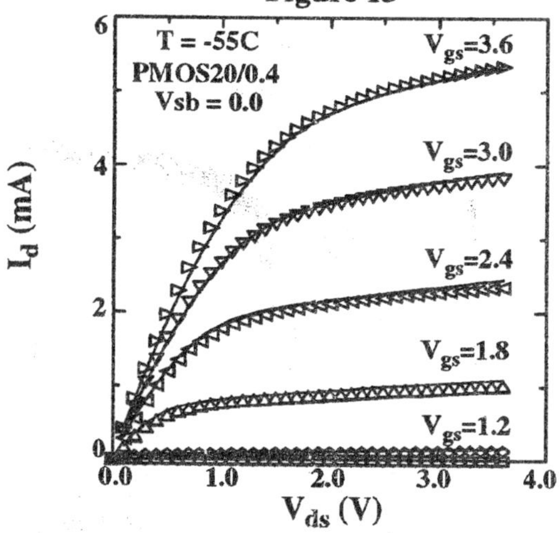

Figure 14

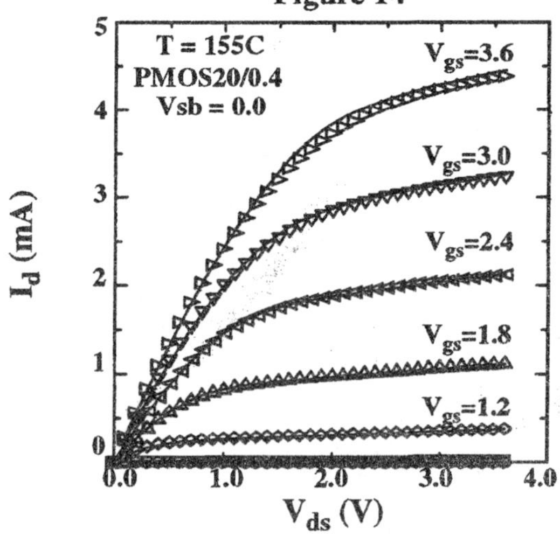

The Utilization of the Electrical Model Synthesis Technique for the MOS Interface Trap Parameters Measurement

S. Sh. BALTIANSKI

Penza State Technical University
Krasnaya Str., 40, Penza, 440017, RUSSIA
Tel.: (8412) 698429, (8412) 667292 (Home)

SUMMARY

A method using the well-known conductance technique is developed. The basis of the approach is the representation of the approximate function as a rational form. This form allows the decomposition of the approximate function into elementary fractions. The extension of the energy range is provided and the error estimation carried out for the extracted interface trap parameters.

1. INTRODUCTION

The high performance MOS technology requires greater precision information about interface trap (IT) properties. Figure 1 shows the model for MOS structure taking into consideration an interface trap [1]. In the diagram, C_i is the capacitance of a dielectric, C_D is the capacitance of a space charge layer and R_S is the resistance of a quasineutral region. The circuits C_{it} R_{it} reflect interface trap properties.

A date acquisition process requires representation of the IT component as an equivalent circuit with the frequency dependent variables $C_p(\omega)$ and $G_p(\omega)$. The low signal equivalent circuit is given in Figure 2.

The measuring process is as follows: The device receives measuring information as two scalar components, $G_e(\omega)$ and $C_e(\omega)$, of the general MOS admittance $Y_e(\omega)$. In the next step we extract the experimental information for $G_p^{(ex)}$. Finally we provide the interpretation of the experimental data to obtain IT parameters D_{it} and τ_s using the working equation for $G_p^{(th)}$ given in [1]. This is the conventional method for more sensitive measurements in real manufacturing.

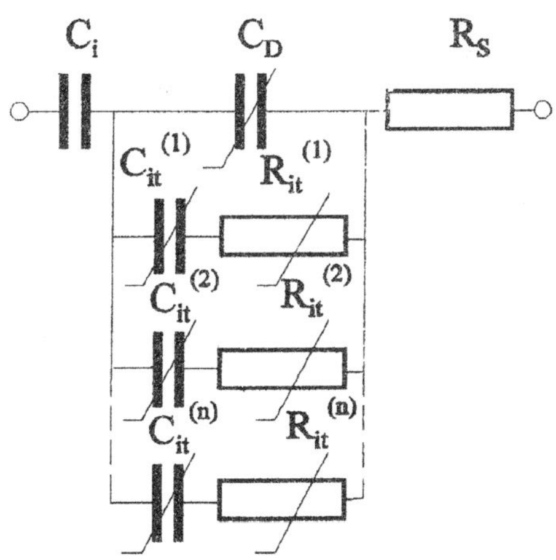

Fig.1 The basic electrical model of the MOS structure for Interface Trap parameters measurement.

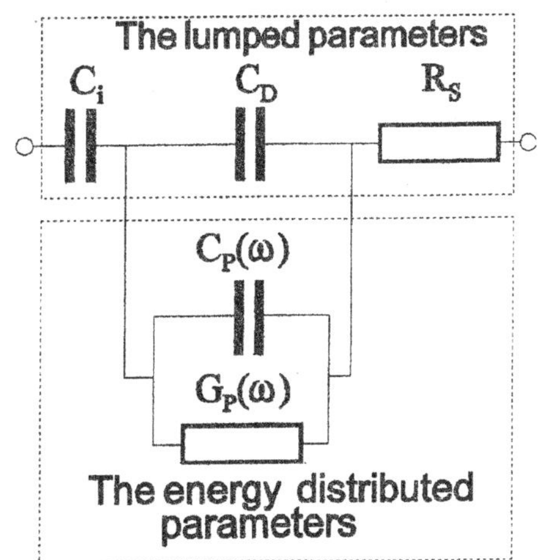

Fig. 2 The equivalent circuit in low signal approach.

0-7803-3223-7/96/$5,00

2. PROBLEMS OF THE CONVENTIONAL METHOD

In practice the last step described above has some difficulties. A large error or impossibility in extracting information about IT properties occurs when we have a small bandwidth for the experimental data. This situation arises near the accumulation or inversion region of MOS structures. In this case the algorithm converges badly or does not converge at all. There are different versions of the present technique but they only work for long frequency bandwidth.

It is important to note that in most cases the present technique does not provide the metrology test for the extracted IT parameters.

3. NEW PROCEDURE DESCRIPTION

The approach for the improvement of this technique has been suggested. We will describe the distributed parameters connected with IT (see equivalent circuit in Figure 2) by the equation:

$$Y_r(\omega,\overline{v}_s) = \int_{-\infty}^{\infty} \frac{j\omega C_r}{1+j\omega\tau_r \exp(\overline{v}_s)\exp(v_s)} P(\sigma_r, v_s) dv_s, \tag{1}$$

where $\overline{v}_s$ is the average dimensionless surface potential, P is the gaussian density of a probability of dispersion σ_r^2 arising due to surface potential fluctuation v_s. We call the C_r, τ_r and σ_r the initial parameters of (1). Using assumptions as described in [1], the informative scalar component is:

$$\frac{G_p^{(th)}}{\omega} = \frac{C_r\omega\tau_r \exp(\overline{v}_s)}{\sigma_r\sqrt{2\pi}} \int_{-\infty}^{\infty} \frac{\exp(\eta)}{1+\omega^2\tau_r^2 \exp(2\overline{v}_s)\exp(2\eta)} \exp\left(-\frac{\eta^2}{2\sigma_r^2}\right) d\eta, \tag{2}$$

where η is the deviation of the surface potential fluctuation.

In other words, we approximate the working expression for extracted IT parameters given in [1] by equation (2). In this case the main extracted parameter D_{it} is error free. The error of the τ_s is regular and is taken into account.

It has been suggested in [2] to express the function (2) as a rational function in the following form in the S-domain:

$$\frac{G_p^{(ex)}}{S} = \frac{-B_1S + B_3S^3 - B_5S^5 + \ldots - B_{2n-1}S^{2n-1}}{1 - A_2S^2 + A_4S^4 - A_6S^6 + \ldots - A_{2n}S^{2n}}, \tag{3}$$

where $S=j\omega$ and B_k, A_k are the coefficients extracted from experimental data by solving the corresponding system of equations. This form is a discrete approach to the function (2).

Given representation makes it possible to decompose the admittance (1) using (3) into elementary fractions in the S-domain [3]. As corollary we obtain the parameters of the distributed part of the equivalent circuit in Figure 2 in discrete form: $\tau_k = -1/S_k$, $C_k = -2K_k/S_k$. Here S_k are the poles and K_k are the residuals of (3).

We determine the initial parameters of the equation (1) by the equations:

$$C_r = \sum_{k=1}^{n} C_k, \tag{4}$$

$$\tau_r = \exp\left[\left(\sum_{i=1}^{n} K_{bi}\right)^{-1} \sum_{i=1}^{n} (K_{bi}\ln\tau_i)\right], \tag{5}$$

$$\sigma_r = \sqrt{\left(\sum_{i=1}^{n} K_{bi}\right)^{-1} \sum_{i=1}^{n} \left[K_{bi}(\ln\tau_i - \ln\tau_r)\right]}, \tag{6}$$

where K_{bi} is defined as $K_{bi} = C_{k(i)} / C_{k(min)}$.

Finally, we define the IT parameters:

$$D_{it} = C_r/q, \tag{7}$$

$$\tau_{it} = K_\sigma \tau_r , \tag{8}$$

where K_σ is the coefficient for a more accurate definition of the time constant. It is determined as a function of σ_r by the preliminary simulation and comparison of the functions (2) with the working equation given in [1].

The coefficients of (3) often do not allow us to use the conventional decomposition technique [3] in real measurements. In this case we approximate equation (2) by the composition of an elementary admittance using the least squares method. The equal of the partial capacitance can be taken into account. This limitation corresponds to the slow change of D_{it} around a determined surface potential.

It is important to note that the discrete representation of (2) allows provision for metrology test of the extracted IT parameters. The experimental model parameters are determined similarly by the extracted IT and lumped parameters of the MOS structure. This model corresponds to the linear image in Figure 1. The experimental model measurements are carried out in the same conditions. The error estimate is defined by comparing the value of the theoretical initial parameters with the experimental ones used in equations (4)-(6).

4. EXPERIMENTAL RESULTS AND DISCUSSION

The automatic data acquisition device provides experimental data in the frequency range from 100 Hz to 100 kHz with 90 logarithmic uniform steps and the separate frequency 1MHz. The test-signal amplitude is 25 mV. We estimated errors less than 0.1% of the acquired admittance component data C_e and G_e.

An example of the experiment is presented. Figure 3 shows the 1 MHz C-V and G-V curves with sections of gate voltage V_G in which we acquire the frequency depending data: $C_e(\omega)$ and $G_e(\omega)$. Figure 4 presents the extracted parameter $G_p(\omega)$. Figure 5 shows this function after interpolation and extrapolation by equation (3). One can see the effective bandwidth increase of about 3 logs of frequency in Figure 5 as compared to Figure 4. The curves 1, 2 and 7 cannot be used by conventional technique.

The final parameters are determined using the described procedure and presented in Table 1. Also the metrology estimate has been done. Figure 6 shows the spectrum IT with the error band for the new technique. We did not provide error estimate of the conventional technique [1]. The energy range of the extracted D_{it} has been increased in comparison with the conventional technique. The error band shows the potential sensitivity of the applied measuring devices.

Table 1

Number of the curve	The initial parameters			The Interface Trap Parameters				
	C_r [pF/mm^2]	τ_r [sec]	σ_r	$D_{it}\cdot 10^{11}$ [cm^{-2}eV^{-1}]	δD_{it} [±%]	τ_{it} [sec]	$\delta\tau_{it}$ [±%]	E [eV]
1	169.9	$7.8\cdot 10^{-7}$	0.7	1.05	7	$2.0\cdot 10^{-6}$	10	-0.29
2	189.9	$1.6\cdot 10^{-6}$	1.3	1.19	3	$4.0\cdot 10^{-6}$	7	-0.25
3	240.0	$3.2\cdot 10^{-6}$	1.6	1.50	1.5	$8.0\cdot 10^{-6}$	5	-0.22
4	260.6	$6.3\cdot 10^{-6}$	1.6	1.63	1.5	$1.6\cdot 10^{-5}$	3	-0.19
5	259.4	$4.3\cdot 10^{-5}$	1.8	1.62	1.5	$1.1\cdot 10^{-4}$	5	-0.16
6	237.0	$1.6\cdot 10^{-4}$	1.6	1.48	3	$4.0\cdot 10^{-4}$	7	-0.12
7	187.2	$2.5\cdot 10^{-3}$	1.5	1.17	5	$6.3\cdot 10^{-3}$	10	-0.10

Utilization of the described technique increases the possibility of greater precision and a larger bandwidth for measurement devices like HP4284 and HP4285 in these applications. The error estimate of the technique allows accurate comparison of extracted IT parameters under various conditions.

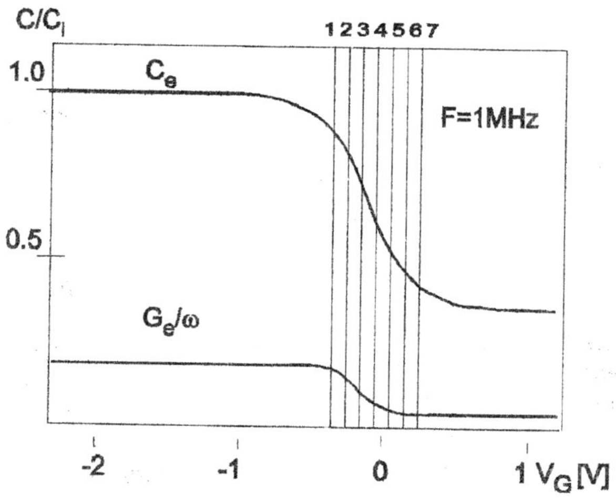

Fig.3 The C-V and G-V curves with the sections of the gate voltage where the frequency depending data have been acquired.

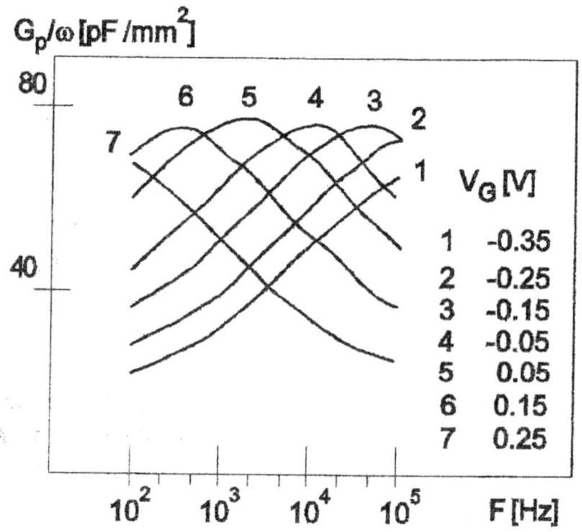

Fig.4 The extracted parameter $G_p(\omega)$.

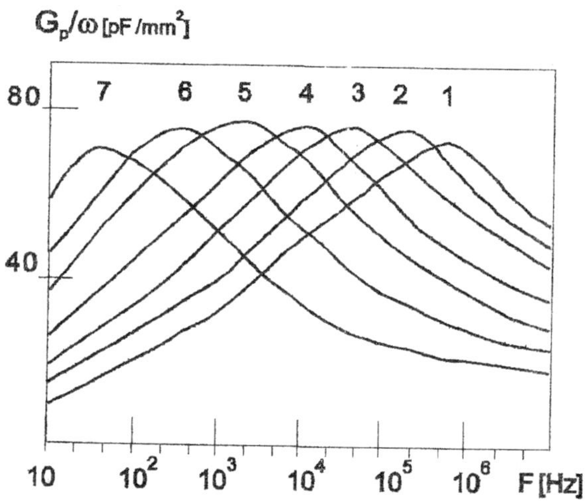

Fig.5 The data from the Figure 4 after interpolation and extrapolation by the equation (3).

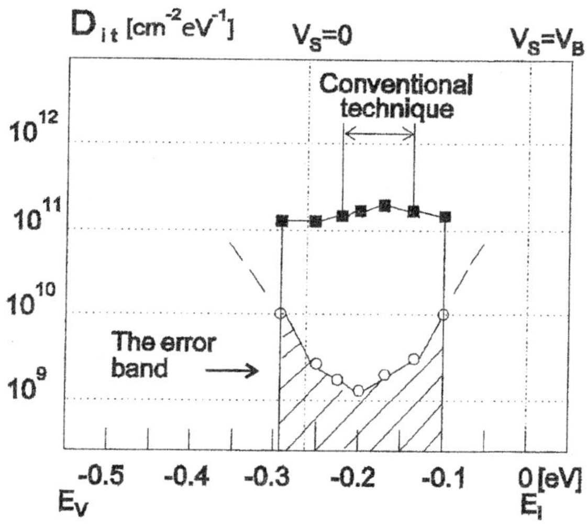

Fig. 6 The extracted D_{it} with the error band.

5. CONCLUSIONS

The described technique is based on the synthesis theory of passive circuits and formalizes the measured procedure. The main advantages of the suggested approach are as follows: we can restore information about IT lengthening the bandwidth in the frequency domain, therefore extending the effective measuring devices bandwidth. In addition we can provide a metrology test by the experimental model increasing the reliability of the information extracted about IT parameters.

REFERENCES

1. E.N. Nicollian, J.R. Brews, MOS Physics and Technology.- N.-Y., Willy, 1982.
2. S.Sh. Baltianski, "The methods and apparatus of the MOS structure admittance distributed parameters measurement", Ph.D. Thesis (in Russia), Politechnique Institute of Penza, March, 1993.
3. F. David, Jr. Tuttle, Network Synthesis. -N.-Y., John Willey & Sons, Inc., 1958.

A NOISE MODELLING APPROACH FOR ACCURATE TIME DOMAIN ANALYSIS

Şerban-Mihai POPESCU Sorin-Andrei SPANOCHE

POLITEHNICA University of Bucharest
Bvd. I.Maniu, 1-3, RO-77202, tel. +40-1-3122452

Abstract - This paper describes some theoretical aspects of the time domain simulation in the presence of noise sources. It is demonstrated that the commonly used techniques are giving distorted results when using adaptive timestep. A correct method of noise modelling will be explained. Based on this method, a practical implementation of a simulator capable to do accurate transient analysis with noise is announced.

1. Noise Modelling in the Time Domain

Including noise models in the transient circuit simulation was disregarded in most standard simulators up to now. Only few simulators at the system level (the Simulink™ extension of the MATLAB™ mathematical modelling package) and custom simulators for specific circuit topologies [1, 2] offer a noise source based on a floating point random number generator with normal distribution, zero average and variable dispersion. These random number generators use standard algorithms [3] and are called for each new transient analysis time point.

However, none of these simulators use a fixed time step. As a consequence, this introduces noise that is modulated by completely unrelated signals in the analyzed circuit.

At the same σ, the noise spectral density is lower when the time step is smaller. If noise is taken into account using this method, the simulation results can be artifacts.

A possible solution [4] to generate stationary noise is based on calling the random number generator with a *fixed* sampling rate. This corresponds to the definition of the discrete Gaussian white noise source [5].

For a discrete time simulator this representation suffices. However, the circuit simulators use adaptive time step for simulation efficiency. All signals (including noise) are treated as continuous time waveforms. Let's consider the case where each noise sample is kept constant for the whole noise sampling period T. For this signal the average power is σ_q^2, and the selfcorrelation function is:

$$\left\langle w_q(t_1)\, w_q^*(t_2)\right\rangle = \begin{cases} \sigma_q^2 (1 - |t_1 - t_2|/T) & \text{for } |t_1 - t_2| \le T \\ 0 & \text{for } |t_1 - t_2| > T \end{cases} \qquad (1)$$

The power spectral density can be calculated by applying the Fourier transform to the selfcorrelation function (Wiener Theorem):

0-7803-3223-7/96/$5,00 © 1996 IEEE

$$N' = |\mathcal{F}(\langle w_q w_q^* \rangle)| = 2\sigma_q^2 T\left(\frac{\sin(\pi f T)}{(\pi f T)}\right)^2 \qquad \mathbf{(2)}$$

To conclude, N Gaussian noise sources can be obtained by calling the qth (q=1,N) random number generator of zero average and σ_q dispersion, using a fixed T sampling rate and injecting into the simulator the resulted stair signal. The power spectral density is $N'_q=2T\sigma_q^2$. This approximation holds up to a frequency $f_{1\%}=1/(7T)$ for a voltage error of less than 1%.

2. Practical Aspects of a Monte Carlo Time Domain Analysis Method.

Designers usually need to have a simulator that offers reproductible noise sources with negligible crosscorrelation, different statistical distributions (at least Gauss), zero average and power spectral density that is user definable.

In order to develop such a simulator, we shifted the noise generator away from the simulation tool. This solution has several advantages. First, it allows the designer to perform numerical tests on the noise generator's output for statistical correctness. Second, the designer can reuse the same noise trace for design iterations (i.e. comparing circuit characteristics for the same input stimuli). Third, the same noise signal can be fed to different simulators for calibration or real life simulation. Moreover, during the simulation phase in presence of noise, the computation overhead induced by the noise source is minimized. To ease manipulation and storing, the noise traces are saved in files in a form of time-voltage or time-current pairs and fed at the simulator inputs as file-based voltage/current sources.

Based on the random number generators from [6], we developed a noise generator program, *gnoise*. It creates high-quality noise traces; the used algorithm, following the theory guidelines, involves five steps:

1) by applying a reproductible procedure, the random number generators are initialised;

2) the sampling rate and the number of samples are determinated, based on the simulation requirements (circuit bandwidth, noise spectral density, simulation length, etc.);

3) the primary white noise waveforms are generated;

4) the white noise waveforms are numerically filtered to get the noise spectral density in the specified bandwidth;

5) finally, statistical computations are performed to validate the most significant statistical figures (mean and dispersion).

The noise model interface to the circuit simulator is another important implementation decision. A new circuit element, **ASRC**, similar to the SPICE **PWL** source, was implemented. The difference is that **ASRC** reads, as the simulation flows, the time-voltage (or time-current) pairs from a separate file. This approach is an efficient one because, unlike **PWL**, the **ASRC** device does not read and store its waveform in memory so the simulator's memory requirements are low even if the noise traces tend to be large.

We decided to implant the **ASRC** source into XSPICE, a special Spice3 clone that provides "hooks" for user-defined compiled devices [7, 4]. The **ASRC** source has to control

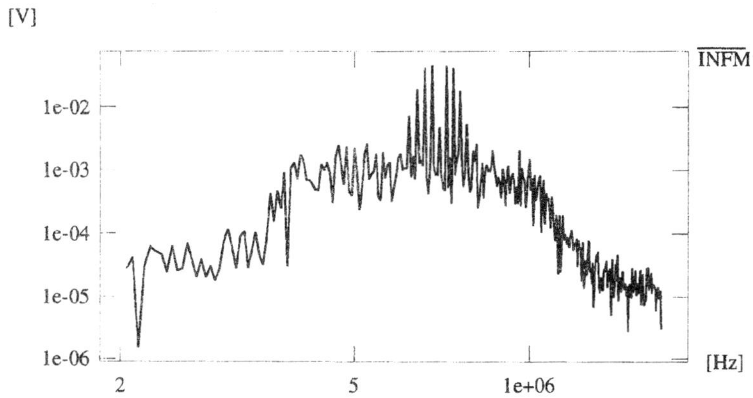

Fig. 1 FM signal and noise spectrum.

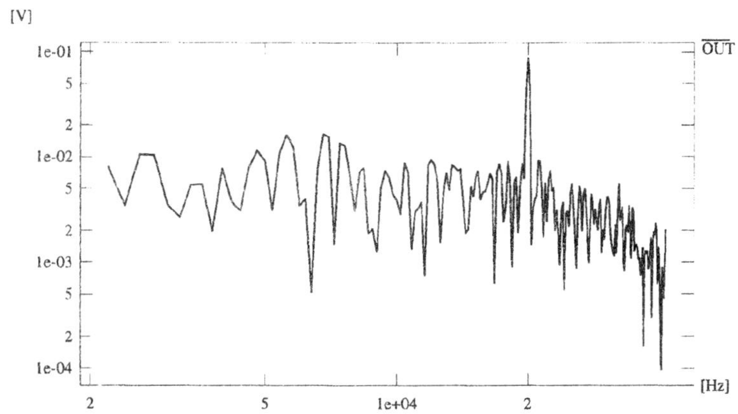

Fig. 2 PLL output signal spectrum.

on the fly the analysis time step acceptance mechanism and to insert breakpoints for an accurate simulation. Parameters like output voltage gain, output voltage offset, interpolation degree (currently, orders of 0 and 1 are supported) have been added, transforming the **ASRC** into a versatile device.

The noise source - simulator interaction was studied in the case of a realistic circuit (the performances evaluation of a PLL based demodulator). Figure 1 presents the spectrum of a noisy FM signal that is applied to the PLL input. As figure 2 shows, by using the developed simulator it was possible to determine, with a high degree of confidence, fundamental parameters (pull-in time, output SNR, etc.) extremely difficult to be evaluated with the traditional simulation tools.

3. Conclusions

Some aspects related to the time domain simulation in the presence of noise have been studied. It was clearly proven that, in order to preserve the spectrum density distribution of the noise, a fixed sampling rate method has to be used. Practical details concerning the implementation of a simulation tool for time domain analysis in the presence of noise are given, as well as several exemples illustrating the usefulness of the developed simulator.

References

[1] Liu, E.W.Y., A.L. Sangiovanni-Vincentelli, "Behavioral Representations for VCO and Detectors in Phase-Lock Systems", *Proc. CICC*, p. 1231–1234, mai 1992

[2] Demir, A., E.W.Y. Liu, A.L. Sangiovanni-Vincentelli, "Time-Domain non-Monte Carlo Noise Simulation for Nonlinear Dynamic Circuits with Arbitrary Excitations", *Proc. Int. Conf. ICCAD*, p. 598–603, nov. 1994

[3] Knuth, D.E., *The Art of Computer Programming*, vol.2, Addison-Wesley Publishing Company Inc., 1981

[4] Spânoche, S.A., *Analog Integrated Circuit Design Automation: Methods and Tools* PhD thesis at POLITEHNICA Univ. of Bucharest, pp.161-177, apr. 1996

[5] Engberg, J., T. Larsen, *Noise Theory of Linear and Nonlinear Circuits*, John Willey & Sons, 1995

[6] L'Ecuyer, P., S. Cote, "Implementing a Random Number Package with Splitting Facilities" , *ACM Trans. on Math. Software*, p. 98–111, 1991

[7] Popescu, Ş.M., unpublished works, 1996

Monte Carlo Hardware Simulator for Electron Dynamics in Semiconductors

Andy Negoi and Jacques Zimmermann

LPCS - ENSERG, UMR CNRS 840, 23 Rue des Martyrs, BP 257, 38016 Grenoble France

Abstract

An integrated circuit dedicated to the simulation of a charged particle in a semiconductor using the Monte Carlo method is presented. This circuit should be the basis brick of a semiconductor device hardware simulator. We detail the various phases of the physical processes involved and how to transfer them in terms of circuit architectures. The CAD tool used was Alliance 3.0. We present the very first practical realization on FPGA.

1 Introduction

In 1990, Milne and Cockshott [MILN] described a machine simulating the movement of a spatially extended collection of identical parameterized bodies so as to simulate, for instance, the behaviour of hard spheres in interaction with themselves as an approach to the problem of collective effects in molecular dynamics. The main advantage of this machine is in the speed of computation, with respect to a general purpose computer. Milne and Cockshott only describe the principles of the architecture of their machine. They say nothing about the parameterized bodies but that they are made up of digital integrated circuits. Here, for the first time, as far as we know, we present an ASIC making an elemental Monte Carlo simulation of one charged particle subject to an external electric field and in interaction with a crystal lattice potential. This ASIC could be one of the parameterized bodies for such a machine, or the basis brick of such a dedicated hardware realizing the simulation of a semiconductor device. A first model has been realized on FPGA.

2 Monte Carlo model of charge transport in semiconductors

We summarize the principles of Monte Carlo method very briefly. We used a simplified micro-dynamic model of free electrons in the conduction band of silicon. The physical principles of Monte Carlo simulation are described in detail in [JACO] for instance. This approach of charge transport has the advantage to allow an easy study of fluctuations in one- or many-particle systems as shown in [ZIMM].

A particle, as specified by six degrees of freedom, three wave-vector $\vec{k}$ (or velocity $\vec{v}$) components and three space components, is simulated, subject to two kinds of events. First the particle is accelerated (or decelerated) for some time δt in an electric field according to Newton's law as

$$\vec{k}(t+\delta t) = \vec{k}(t) + \frac{q}{\hbar} \cdot \vec{F} \cdot \delta t \tag{1}$$

then eventually followed by a collision with some imperfection of the medium in which it moves (lattice vibrations or phonons). At every time, the particle kinetic energy is known from the conduction band shape. In our case, we used the simple model described by

$$E(\vec{k}) = \frac{\hbar^2}{2m^*} \cdot \vec{k}^2 \tag{2}$$

where m^* is the effective mass.

0-7803-3223-7/96/$5,00

The energy is the only parameter needed to specify the collision probabilities. In our simplified model, we consider three types of collisions: one elastic P_1, one inelastic with emission P_2 and one inelastic with absorption of energy P_3. The corresponding probabilities are:

$$P_1(E) = A_1 \cdot \sqrt{E} \cdot \delta t \tag{3}$$

$$P_2(E) = \begin{cases} A_2 \cdot \sqrt{E - \hbar\omega_0} \cdot \delta t & \text{for } E > \hbar\omega_0 \\ 0 & \text{for } E \leq \hbar\omega_0 \end{cases} \tag{4}$$

$$P_3(E) = A_3 \cdot \sqrt{E + \hbar\omega_0} \cdot \delta t \tag{5}$$

where $\hbar\omega_0$ is the energy of the phonon and A_1, A_2, A_3 are constants. δt is the elemental step of the times where the collisions are tested. It must be chosen in such a way that the total collision probability is correctly normalized (this normalization depends, as we will see later, on the size of the registers used for the pseudo-random number generator). The algorithm operates in the following way. Once $\vec{k}$ is known from (1), the energy is computer with (2) and the collision probabilities with (3). Then a PRN[1] is generated and compared with $Q_3 = P_3 + P_2 + P_1$, $Q_2 = P_2 + P_1$, $Q_1 = P_1$ Four events may occur. If PRN $> Q_3$, no collision occurs, we go back to (1) for the next δt. Or one of the Q's is smaller than PRN, in which case a collision occurs, 1 or 2 or 3. Then we perform two successive rotations of $\vec{k}$ after loss (or gain) of energy has been taken into account. The two angles β and θ are specified by their probabilities: $P(\beta) = a \cdot \sin(\beta) \cdot \delta\beta$ and $P(\theta) = b \cdot \delta\theta$. We generate two fresh PRN and take $\cos(\beta) = 1 - 2 \cdot PRN1$ and $\theta = 2\pi \cdot PRN2$. Once the components of $\vec{k'}$ after collision has been obtained around the former $\vec{k}$ before collision, we transform $\vec{k'}$ again in the original coordinate system linked to the external electric field. We then go back to equation (1) for the next time step.

For a given field, the time average of the velocity along the field direction gives us the drift velocity. The instantaneous velocity of the particle appears in a register as a random variable giving access to the fluctuations and thus to the hot carrier diffusion noise if the field is high. In the next section, we present how to implement this algorithm into an integrated circuit.

3 From physical model to integrated circuit

We designed the circuit so as to take advantage of the parallelism at the operator level.

To achieve our goal, we designed the architecture of the circuit to contain one block for each elemental operation type in the algorithm. The blocks work in parallel, synchronized by a finite state machine. The arithmetic operators are based on serial algorithms based on shifts and additions or subtractions, proven that this approach generates a smaller layout surface and they are portable between different technologies. In figure 1 we show the structure of the datapath of the processor.

The operation scheduling is performed using the ASAP[2] method. The algorithm uses only integers and the length of the operators is chosen depending on the necessary accuracy imposed by a good description of the physical parameter. All the memories for stocking the variables are grouped in a register file of 32 bit length.

Once we have introduced the simulation parameters (14 numbers of 16 bit length each), the processor can work independently, because all the variables are stocked in the register file. This particularity is important, because another advantage of this approach results: the computation time is no longer limited by the access to external memories. After each iteration, the circuit can provide the new position and the new wave vector to an external control unit which can perform, for example, a time integration to computing the drift velocity of the simulated particle in the field's direction.

In figure 1 we describe a bus-oriented architecture. As elemental arithmetic operators we use a multiplier, a square root and division block, an adder/subtractor, a sine block and a random number generator. The multiplier is based on the shift and add method. The sine is computed for a 7 bits

[1] Pseudo Random Number
[2] As Soon As Possible

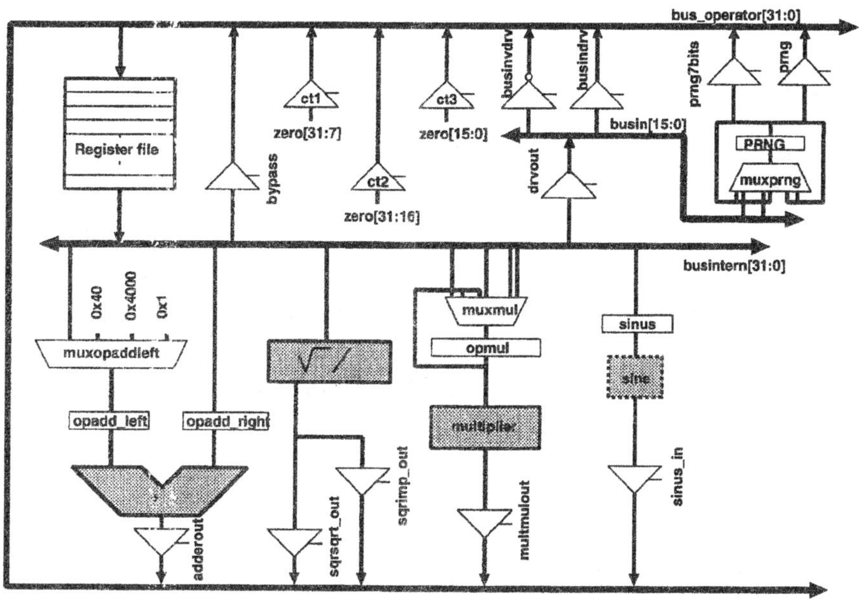

Figure 1: The architecture of the processor.

argument and outputs 7 bits result. We perform also the cosine with this block, known that $\cos(x) = \sin(x + \frac{\pi}{2})$. Due to the seven bits length, we could implement this operator with standard cells from a behavioural description. Both the square root and the division use a subtractor, so we designed these operations in a single block. The random number generator is based on a shift register with reaction and to ensure the statistical independence the next number is taken after 32 shifts from the previous (32 is the length of the register). We have chosen such a generator because we need to restart the same sequence to obtain the same results.

The complexity of the processor made the interpretation of the results of logical simulation on separated blocks very difficult. We thus performed this simulation on the entire procesor. The results are presented in the next section.

4 Simulation results

We have performed the logical simulation of the circuit at the gate level. Because we described them in different representations, we could not simulate the finite state machine and the datapath separately. The interaction between the two parts being so complex, it was necessary to simulate the behaviour of the entire circuit.

We used a procedural language to describe the test patterns. The outputs are shown in figure 2. Here, $\vec{k}$ is represented with integers, showing the results of logical simulation of the circuit. Now we look at how to convert the contents of the registers into real physical units. The physical phonon energy was $\hbar\omega_0 = 40meV$, represented by 400 in binary. The energy unit thus is 0.1 meV. Since the register for the energy E and $|\vec{k}|^2$ is the same, we have $\frac{\hbar^2}{2qm^*} \cdot (2^{32} - 1) = \frac{2^{32}-1}{10000}$. Within $\frac{m^*}{m_0} = 0.28$ and $q = 1.6 \cdot 10^{-19}C$, we find $\hbar = 2.857 \cdot 10^{-27}J \cdot s$. For the time scale now. What physical δt is represented by one iteration of the circuit? At some energy E, the collision probability is $\mathcal{P}(E) = \lambda(E) \cdot \delta t$. $\lambda(E)$ represents the probability per unit time that a particle with energy E undergo a collision. All references on how this quantity can be computed can be found in full details in [JACO]. At the same energy in the circuit we have for the same probability[3] $Q(E)/(2^{20} - 1)$. Equating, we find $\delta t = \frac{Q(E)}{(2^{20}-1)\cdot\lambda(E)}$. For electrons in undoped silicon, we find $\delta t = 1.36 \cdot 10^{-15}s$ taking $A_1 = A_3 = 125$ and $A_2 = 625$.

[3] Because we use only 20 bits of the random number generator

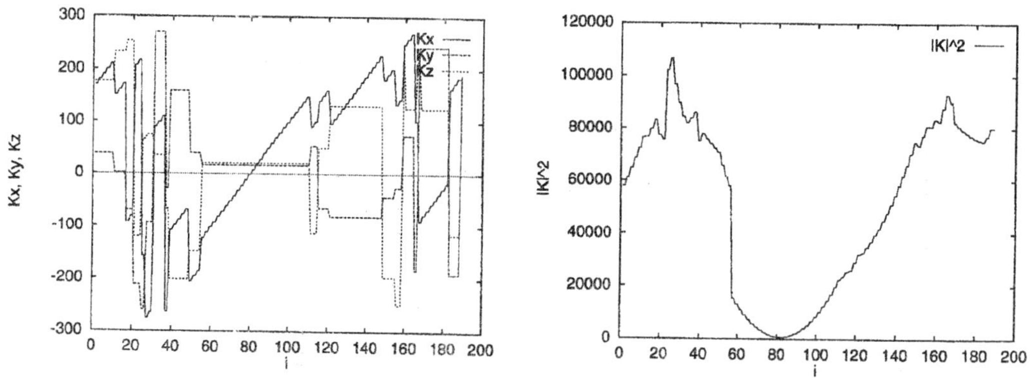

Figure 2: k_x, k_y, k_z and $|\vec{k}|^2$ vs. iteration number.

5 Conclusions and future work

The hard-coded implementation of a Monte Carlo simulation for one charge particle subjected to an electric field in a semiconductor was presented. The architecture the ASIC processor was described and synthesized entirely with the Alliance CAD kit (that is a free software developed in an university initiative). This architecture presents several dedicated operators working in parallel. The goal was to obtain a circuit optimized for surface rather than speed. The total estimated silicon surface is about $60mm^2$, in a technology $1\mu m$ DLM. Simulations were performed at logical gate level, revealing that a complete simulation loop takes an average of 800 clock cycles. Our estimated clock frequency is about 25 MHz.

The processor will be the basis brick to make a hard-coded simulator, implemented as a a parallel architecture for the simulation of multiple particle systems, for example a MOS transistor.

References

[JACO] C. Jacoboni, P. Lugli, "The Monte Carlo Method for Semiconductor Device Simulation", Spriger-Verlag, Wien - New York, 1989.

[ZIMM] J. Zimmermann, E. Constant, Solid-State Electronics, vol. 23, pp. 915-925, 1980.

[MILN] G.J. Milne, W.P. Cockshott, "Method and Apparatus for Simulation of a Physical Process" (patent), PCT WO 91/08560.

Combination of Loosely Coupled and Compactly Coupled Integration of CAD Tools and Their Applications in SAW Devices Design and Fabrication

Peigang Li, Jiaqin Zhang and Yuquan Hao
Institute 23 of the Second Academy of the China Aerospace Corporation

Abstract

This paper discusses the loosely and compactly coupled integration of CAD tools and their applications in the design and fabrication of surface acoustic wave (SAW) devices. Three core modules are developed as design and mask pattern verification module, database module, and device characteristics simulation module. All the operations are controlled under Microsoft Windows GUI interface. This leads to a reduced design/fabrication cycle, workload and cost. With the addition of knowledge-based library for intelligent design and Lotus Notes database for distributed and networked engineering database management (EDM) and effective control of documents, a new enterprise-leveled, computer-integrated design and manufacturing system (CIDM) can be established for monolithic and hybrid device designs (not confined to SAW device only).

Key words: SAW device simulation, CAD tools integration, database, mask pattern, MS-Windows, knowledge base.

1. INTRODUCTION

Nowadays, SAW devices for pulse compression are widely used in radar systems. These devices have different types and performances. Open-looped design and fabrication method is commonly used, that is, a lot of cycles of design-fabrication-measurement are needed for a successful device. This leads to increased manpower and cost and bad effect on systems development. In order to solve this problem, we have developed a close-looped design technology known as the integration of CAD tools for SAW device design and fabrication. Our objective is to make the design program standardized and fabrication process normalized and

0-7803-3223-7/96/$5,00

therefore increase the device performances. If combined with knowledge-based library and field-bus control of fabrication equipment, an intelligent, computer-integrated design and manufacturing (CIDM) system can be created for every kind of devices (e.g. MMIC, Thick and Thin Film, MCM).

2. Function Requirements and Technical Development

2.1. Function requirements of the CAD system

- Input information (take SAW delay line as an example)
- Data types
- Output information
- Integration of all modules and data sharing

2.2. Technical Development

According to the current conditions of SAW device simulation and data processing, we choose the combination of loosely coupled and compactly coupled integration of CAD tools as the SAW device design method. In a loosely coupled integration, the effectiveness and completeness of these tools are maintained and every program exchanges data via a centered database; while in a compactly coupled one, two programs connect with each other by writing data files into one another. The key to the integration of applications is to develop a common database and a pre-processor for all analysis programs. Technically we develop the CAD integration system as follows:

- designing engineering database to manage the data and other information;
- coexistence of loosely and compactly coupled integration;
- unified interactive operation under MS-Windows graphical user interface (GUI);
- utilization of current good design and verification tools;
- combination of design and fabrication.

3. System Configuration

The block diagram of the system configuration is shown in Figure 1.

Figure 1 Block diagram of the CAD integration system

The core system consists of three modules:

- Mask pattern design and verification
- Database
- Characteristics simulation

4. Conclusion

The CAD integration system can be further extended by the addition of knowledge-based intelligent design method and effective specification of the design parameters, process parameters and process equipment. In that way, the workload can be reduced, close-looped design cycle be attained and a new enterprise-leveled computer-integrated design and manufacturing (CIDM) system be built for monolithic and hybrid devices (not merely for SAW devices).

5. References

[1] D.Durbeck, J.H.Chern, and D.S.Boning, "A System for Semiconductor Process Specification", IEEE Trans. on Semiconductor Manufacturing, p.297, Vol. 6, No. 4, November, 1993.

[2)] Texas Instruments Technical Journal, September-October 1992.

[3] M. J. McCollister and S. M. Riche, "The Development of A First Pass Verification Module for a SAW Filter Design Automation System", IEEE Ultrasonics Symposium Proc., p.133, 1991.

[4] D. Karagiannis and L. Marinos, "Integrating Engineering Applications via Loosely Coupled Techniques: A Knowledge-Based Approach", Expert Systems With Applications, p.303, Vol. 8, No. 2, 1995.

[5] U. Roy, B. Bharadwaj, S. Sarathy, and P. Graham, "Development of An Intelligent Product Design System: Integration Strategies", Applied Artificial Intelligence, p.563, September, 1995.

The authors can be reached at P. O. Box 3923, No. 37, Beijing, 100854 P. R. China or the email address at zengxy9@hns.cjfh.ac.cn or caiwt@bepc2.ihep.ac.cn.

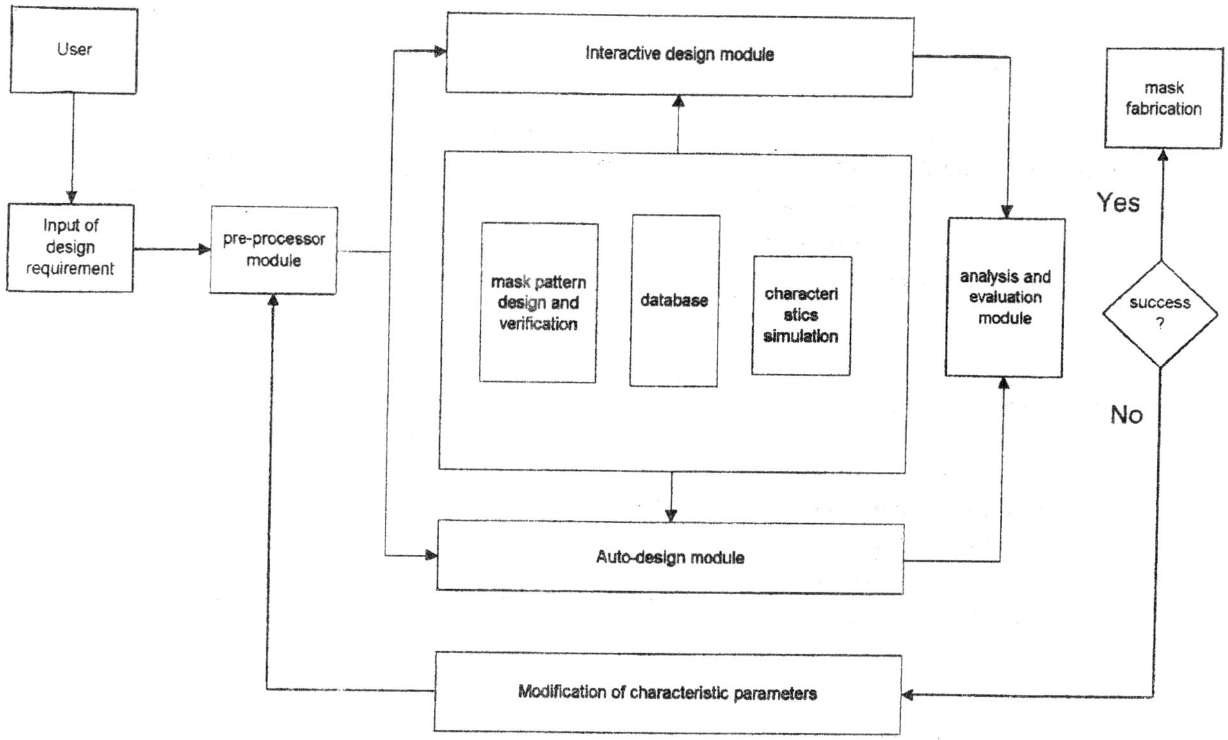

Figure 1 Block Diagram of the CAD Integration System

APA: A DESIGN SYSTEM FOR PLL BASED DEMODULATOR CIRCUITS

Sorin-Andrei SPANOCHE Şerban-Mihai POPESCU

POLITEHNICA University of Bucharest
Bvd. I.Maniu, 1-3, RO-77202, tel. +40-1-3122452

Abstract - This paper describes several implementation issues of a competitive design system for PLL based circuits. The APA system offers automated circuit characteristics estimation that is done by transient simulation. New and efficient solutions for noise and distortion circuit characteristics estimation are demonstrated. The efficiency/flexibility compromise is addressed by using different simulators.

1. The Circuit Characteristics Estimation Problem

An important part of the design methodology is the estimation of circuit characteristics of the object under design. For analog ICs, this task is a very complex one. The large variety of useful circuits did not allow systematic methods and standard for a universal definition for all characteristics. Also, each kind of circuit has many electrical characteristics. Once these characteristics are identified for a circuit, estimation implies: test circuit specification, circuit analysis and qualitative or quantitative characteristic extraction. Automated analysis is done by electrical or behavioral circuit simulation. Standard circuit simulators allow most of the necessary types of analysis.

Special cases are the distortion and noise simulations. The theoretical foundation for such simulations is limiting the use of standard simulators to a narrow class of circuits. For example, modulators, demodulators, frequency converters cannot be easily simulated for noise and distortion using small signal based algorithms from SPICE [1].

2. Distortion Characteristics Estimation using Transient Simulation

When one wants to use transient simulation results for distortion characteristics estimation in the frequency domain, there is a need for a good spectral analysis program. Spectral analysis can be done either internally, using SPICE's **.FOUR** card or by means of a post-processor [2]. The number of time points taken into account and the usually linear interpolation error lead to a noise floor, named hereafter *method's noise*.

Both the fast Fourier transform (FFT) routines from the usual SPICE post-processors and the internal discrete Fourier transform (DFT) routine cannot be easily used when the input frequency is unknown or is not fixed. The same holds for signals that are not directly related or for a noise signal with small selfcorrelation [3]. The solution in such a case is to use a window function that attenuates the signal discontinuities at the ends of the simulated period. This technique leads to a lower method's noise.

A program (*pdft2*) for automated estimation of distortion characteristics from transient simulation traces was adapted [4] from the spectral analysis algorithms used by [3, 5]. Special computing steps for different distortion characteristics and signal to noise ratio (SNR) were added. This program can be directly applied to a signal with irregular time step.

The Fourier integral can be rewritten (see eq. 1) as a sum for each sampling time slot. Each element of this sum estimates by linearly interpolating the input signal f (i.e., $f=a_k t+b_k$ for the time slot $[t_k,t_{k+1}]$).

The distortion characteristics that can be directly calculated by *pdft2* include THD, HD_k,

0-7803-3223-7/96/$5,00

$$F(\omega_i) \approx \sum_{k=0}^{N-1} \left(\left[\frac{a_k \cos(2\pi\, \omega_i t)}{(2\pi\, \omega_i)^2} + \frac{f(t) \sin(2\pi\, \omega_i t)}{2\pi\, \omega_i} \right]_{t_k}^{t_{k+1}} + \right.$$
$$\left. + j \times \left[\frac{a_k \sin(2\pi\, \omega_i t)}{(2\pi\, \omega_i)^2} - \frac{f(t) \cos(2\pi\, \omega_i t)}{2\pi\, \omega_i} \right]_{t_k}^{t_{k+1}} \right) \tag{1}$$

IMD, IM_k and TDFD.

3. Signal to Noise Ratio Estimation using Transient Simulation

One needs both the signal and the noise power to calculate the signal to noise ratio: $SNR = P_{signal}/P_{noise}$. These values can be calculated either in the time domain or by spectral analysis.

A solution to the SNR computing using a mixed approach can be the following step sequence:

1 spectral analysis for the fundamental and harmonics of the output signal that lie in the bandwidth of interest
2 signal power estimation by summing the components at step 1
3 calculation in the time domain of the RMS and average values of the output signal
4 SNR computing using equation (2)

$$SNR = \frac{\sum_{k=1}^{n} V_k^2}{M_{rms}^2 - M_{avg}^2 - \sum_{k=1}^{n} V_k^2} \tag{2}$$

Large errors can occur for the denominator if an inappropriate integrating method is used. The same kind of linear approximation for the waveform has to be used as in the DFT algorithm. M_{avg} can be calculated using the same DFT algorithm for $\omega_i=0$. This is equivalent to the trapezoidal integration method. The window function has to be taken also here into account. Keeping the same notations as in Section 2 M_{rms} can be calculated using the following equation:

$$M_{rms}^2 \approx \frac{2}{wspan} \sum_{k=0}^{N-1} \left[\frac{y^3(t)}{3a_k} \right]_{t_k}^{t_{k+1}} \tag{3}$$

were *wspan* is the raised cosine window span.

This integration method was included in the *pdft2* program for SNR estimation [4]. This solution avoids computing many spectral components and is better in terms of run time than a SNR estimation done only in frequency domain. A numerical experiment pictured in fig. 1 shows that *pdft2* results are valid over a 6 decade range of voltage with no special trimming.

4. Bit Error Rate Estimation using Transient Simulation

Another characteristic of interest for a digital output system is the bit error rate (BER). This is defined as the probability to get a wrong output bit/word. When BER cannot be analytically calculated using appropriate statistical models, it can be estimated by statistical experiments.

These experiments use transient simulation with noise sources. When the target BER and/or estimating error are low, the number of bits/words that have to be simulated is very high. For example, for BER lower that several percents, the bit number is proportional to 1/BER. For

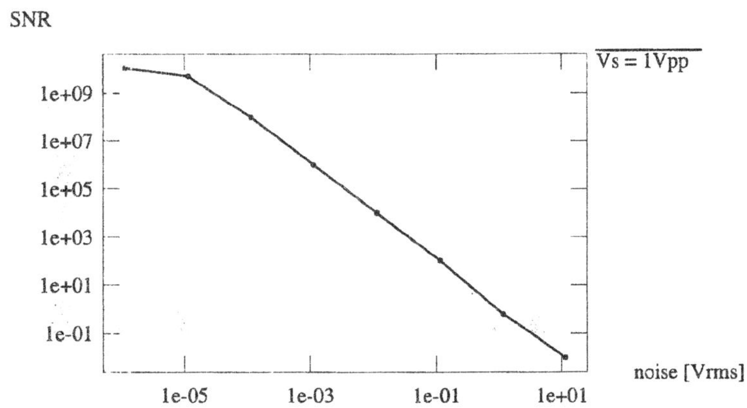

Fig. 1 SNR calculated by *pdft2* as a function of input noise.

large systems the simulation cannot be done in an acceptable time but at higher levels of abstraction. The exact size of the statistical experiment is done using mathematical statistics theory.

Two prototypes [6, 4] of high level simulators with noise were created for PLL based demodulator circuits simulation. The first one is based on XSPICE. Several high level models for PLL components were created in the C language. Even if the flexibility of this simulator is very high (new models/component types can be easily included) it has the drawback of very long simulation run time. This is mainly because of the complicated data structures and large code and data pools.

The second simulator (*ipll*) is more efficient due to it's custom nature. The integration method used by *ipll* is very simple (trapezoidal with fixed time step), corresponding to a discrete time model of the circuit. The simulation time for a typical 5000 demodulated bits decreased from 10 days to 50 minutes. This run time is acceptable even for design iterations.

The analysis of such large data streams has to be analyzed in terms of good/bad value. Manual analysis is tiresome. The *errok* program [4] was created for direct BER estimation from a reference signal (*ref*) trace and the demodulated output signal (*out*).

Direct comparison of *ref* and *out* is impossible because *out* is an asynchronous signal, with variable delays induced by noise. The starting point of a BER estimation algorithm is the error definition. Fig. 2 presents the BER estimation results for a PLL based FSK demodulator as a function of frequency deviation (Δf) at constant input SNR.

5. A Simulation and Automated Circuit Characteristics Estimation System for FM and FSK PLL Based Demodulators

The programs and algorithms described in this paper are building a simulation and circuit characteristics estimation system for PLL based circuits that are used in FM or FSK demodulation [4, 6]. The structure if this system is depicted in fig. 3. The designer can describe and simulate the circuit in a friendly environment that accomodates fast and accurate estimations of distortion characteristics, SNR and BER. Efficient simulation at higher levels of abstraction is carried out by custom simulators.

The described system is fit for use in the design of modern PLL based FM and FSK demodulators. It offers both flexible system and circuit level simulation and system level efficient

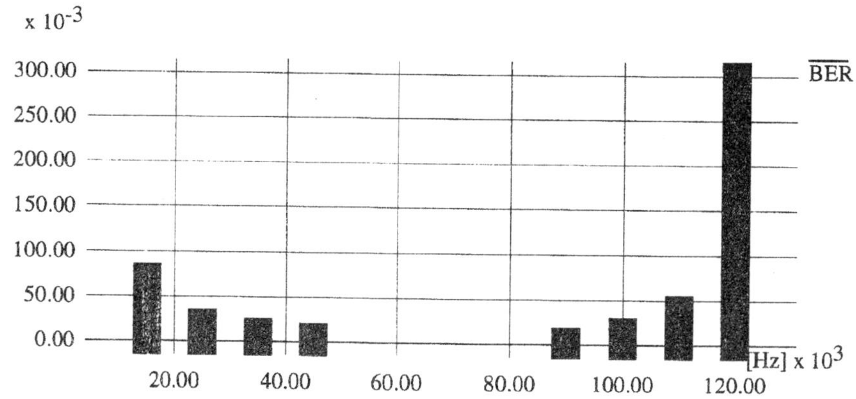

Fig. 2 BER for a FSK demodulator as a function of Δf.

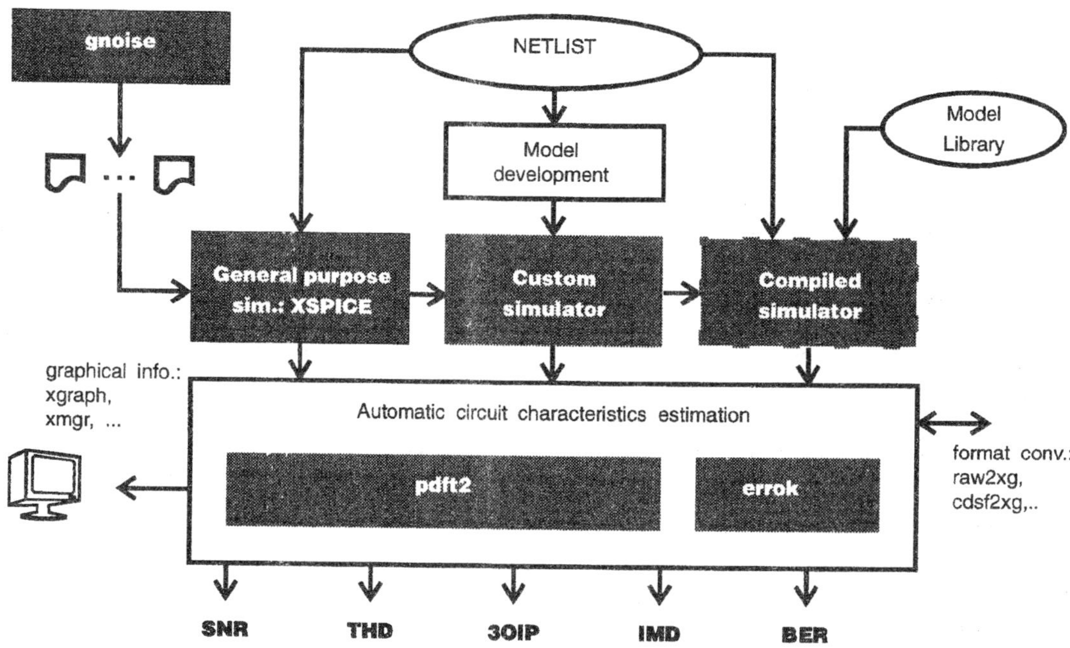

Fig. 3 The PLL based circuits design environment.

simulation. Circuit characteristics are estimated automatically, relieving the designer in the difficult and error prone design process.

References

[1] Demir, A., E.W.Y. Liu, A.L. Sangiovanni–Vincentelli, I. Vassiliou, "Behavioral Simulation Techniques for Phase/Delay–Locked Systems", *Proc. CICC*, pp. 453–456, may 1994

[2] Vladimirescu, A. *The SPICE Book*, John Wiley & Sons Inc., 1994

[3] Najm, F., K. Mayaram, "Modern Guide to Spectral Analysis with SPICE", *IEEE Circuits and Devices Magazine*, Vol.11, No.5, pp. 10–16, sep. 1995.

[4] Spânoche, S.A., *Analog Integrated Circuit Design Automation: Methods and Tools* PhD thesis at POLITEHNICA Univ. of Bucharest, pp.161-177, apr. 1996

[5] Press, W.H., B.P. Flannery, S.A. Teukolsky, W.T. Vetteling, *Numerical Recipes in C – The Art of Scientific Computing*, Cambridge University Press, 1988

[6] Popescu, Ş.M., unpublished works, 1996

Session 19

OPTOELECTRONIC DEVICES (Poster)

LOW-DIMENSIONAL INHOMOGENEITY OF LATTICE OF $Cd_xHg_{1-x}Te$ NARROW-GAP SEMICONDUCTORS

A.I.Belogorokhov, L.I.Belogorokhova, A.B.Danilin, and L.M.Liberant

Centre for Analysis of Substances, 1 Elektrodnaya str., 111524 Moscow, Russia. Fax: +7 (095) 176-3958. E-mail: Alexei@ansubs.msk.ru

1. Summary

The infrared optical transmission and photosensitivity of doped, compensated, narrow-gap semiconductor material $Cd_xHg_{1-x}Te$ have been studied as functions of temperature under the same experimental conditions in the range from 5K to 400K. The strong anomalous changes in the position of the fundamental absorption edge and the maximum and relative photosensitivity of the material in the range 50K to 65K are accounted for by collective effects in disordered semiconductors. Potential-contrast scanning electron microscopy has been used to estimate the size of the *p*-type regions formed in the crystal at liquid nitrogen temperature.

2. Introduction

Narrow-gap semiconductor solid solutions are conventionally used for the production of optoelectronic devices intended for operation in the middle and far infrared (FIR) range. The advantage of the materials is the possibility of varying their band gap E_g by properly choosing composition and doping, applying an external force, etc. Nonetheless, the parameters of the IR devices based on such materials as $Cd_xHg_{1-x}Te$ are limited by the high residual free-carrier concentration, which may be as high as 10^{18} cm^{-3} due to the presence of a large number of vacancies and interstitials that are generated in the material during growth as a result of deviations from the stoichiometric composition. Also, this may be caused by uncontrolled residual impurities. The free-carrier concentration can be reduced to $\sim10^{16}$ cm^{-3} by additional annealing or introduction of compensating dopants. This additional treatment of the material produces, however, a random field, which may affect optical absorption at photon energies comparable to E_g, charge transport phenomena, and the redistribution of internal energy in the crystal.

This work reports experimental data which, on the one hand, defy explanation in terms of the modern theory of collective effects in solids, and, on the other hand, suggest factors which may noticeably influence kinetic processes in crystals.

3. Experimental

We used *n*- and *p*- type single-crystal $Cd_xHg_{1-x}Te$ samples with $0.19 < x < 0.30$. The carrier concentration n, p (cm^{-3}) and mobility μ, (cm^{2}V^{-1}s^{-1}) at liquid nitrogen temperature estimated from Hall measurements were $n = (0.8\div6.3)\times10^{14}$ cm^{-3}, $\mu = (0.8\div5)\times10^5$ cm^{2}V^{-1}s^{-1} for the *n*-type material, and $p = (1.4\div0.9)\times10^{15}$ cm^{-3}, $\mu = (500\div700)$ cm^{2}V^{-1}s^{-1} for the *p*-type material. The material was converted to *n*-type by annealing it in saturated mercury vapor. The sample surface was prepared in several stages, including mechanical grinding, polishing with ASM1/10 diamond

0-7803-3223-7/96/$5,00

abrasive powder, and chemical etching in a Br_2 methanol solution to remove the $60\mu m$ damaged layer. The composition of the material and the concentration of impurities were characterized using laser mass - spectrometric analysis. The composition variation across the sample thickness $d = 1mm$ was within $\Delta x \sim 0.02$. The surface roughness was estimated using scanning electron microscopy. After the above treatment, the height of the surface irregularities did not exceed $1\mu m$. The optical spectra were obtained using a Bruker IFS-113v fast-Fourier spectrometer, within the wave number range 10 to 5000 cm^{-1} at temperatures from 5 to 400K. The optical absorption was studied along with the spectral dependence of the photosensitivity of the material over the same temperature range and under the same experimental conditions. The current through the sample was minimized to exclude sample heating.

4. Results and discussion

The aim of this work was to study the behavior of the electromagnetic radiation absorption coefficient at $E < E_g$. The free-carrier concentration was derived from the reflection coefficient in the region of plasma and plasmon-phonon oscillations.

The samples can be divided into two groups, "normal" (N) and "anomalous" (A). Indeed, if the temperature is lowered from 300 to 60K, the fundamental absorption edge $\lambda(\omega)$ shifts monotonically to lower energies in accordance with the well-known behavior of $dE_g/dT > 0$ for $x < 0.45$. At T ~ 55K $\lambda(\omega)$ shifts abruptly toward higher energies, formally suggesting an abrupt increase of the band gap for the carriers by approximately $E_g/2$. At the same time, electrical measurements, i.e., the Hall effect, did not reveal in these samples any deviations of the mobility and resistivity from the values typical of CdHgTe with an appropriate composition and free-carrier concentration. This behavior is natural because the samples were heavily doped, compensated crystals, with mercury vacancies acting as acceptor centers representing intrinsic defects, and residual uncontrollable impurities showing themselves as donors. Analyzing the kinetic processes and conduction phenomena of these systems, one should take into account the role of large-scale potential fluctuations and the fact that the free-carrier concentration in such materials is usually not high, which makes their mean interaction energy far lower than the kinetic energy. If the Fermi level E_F lies close to the conduction band minimum or falls in the gap, the conduction then involves localized states, while at low temperatures, heavily doped, compensated semiconductors should exhibit variable-range Mott-type hopping conductivity. At high temperatures ($T \sim 55$ K, which is considered here), conduction over localized states close to the localization threshold dominates over hopping between strongly localized states with localization radii much smaller than the hopping range. Here the localization radius may substantially exceed the average distance between centers. Therefore, a semiconductor with a random field and smooth band bending, though being microscopically homogeneous, will be inhomogeneous on the intermediate hopping range scale. Depending on sample preparation technology, these inhomogeneities may play different roles. Assuming that the probability of electron tunneling through potential hills is small, and that potential wells contain many electron levels, we may consider the unbounded motion of an electron with a given energy in a bent band using the classical curved-band approach [1]:

$$\mathbf{E}_{c,v}(\mathbf{x}) = \mathbf{E}_{c,v}^{(0)} + \mathbf{U}_{c,v}(\mathbf{x}),$$

where $E_{c,v}^{(0)}$ are the semiconductor band edges in the absence of random field fluctuations, and $U_{c,v}(\mathbf{x})$ is the electron potential energy in a random field. We should now introduce E_c^{cl} and E_v^{cl}, i.e., the lassical percolation levels of the electrons and holes, respectively. Electrons with energies $E > E_c^{cl}$ may move around or surmount potential hills. If, however, $E < E_c^{cl}$, electron motion over classically accessible regions $E > E_c(\mathbf{x})$ is bounded because these regions are isolated from one anotherThen we should evaluate the fraction of the total semiconductor volume that is accessible to an electron of energy E in the classical approximation. Here the conductivity of the system does not include quantum corrections if the fraction of the classically accessible volume is greater than some critical value U_c which depends on the statistical properties of the random field. For Gaussian potential [2], $U_c = 0.17$. Thus, this difference is connected with the scale of the random field fluctuations. Assuming that the superposition of a random potential leads to smooth band bending, we can describe some features of the kinetic coefficients that are related to nonequilibrium carrier recombination on the plasma oscillations induced by the resonance behavior of the free carrier system. If nonequilibrium electrons and holes in the bands thermalize faster than they recombine, these carriers tend to concentrate in the valleys and hills of the potential relief, respectively, and, hence, they are spatially separated, which, in turn, reduces the recombination rate. This is easy to understand because electrons have to overcome the potential barrier, to do which they must have additional energy, far higher than the percolation level. This requires the condition of strong electron–phonon interaction be satisfied. If, in addition, the process can be made to be resonant, conduction over strongly localized states may occur in semiconductors with an insulating phase at $T \sim 60K$. At lower temperatures, the free-carrier concentration may exhibit the so-called frozen-out photoconductivity of nonequilibrium carriers persisting for a long time.

A number of publications offer different opinions concerning the nature of defects in CdHgTe. It has been assumed, for example, that the predominant intrinsic defects in these crystals are mercury vacancies, which can be either neutral, or single-, or double-ionized. Attempts to attribute the abrupt shift in the temperature dependence of the fundamental absorption edge to effects produced by tails of the density of states in the band gap failed. It should also be noted that the impurity band models do not work here either.

4. Conclusions

Thus, a basis for a selfconsistent description of the anomalities in the behavior of the fundamental absorption edge of heavily doped, compensated, narrow-gap semiconductor materials was found. We believe this to be the first observation of such a clear manifestation of the internal fluctuation field or its effect on the kinetic properties of a semiconductor. The present approach, however, is not completely adequate because it leaves many problems unresolved, such as the nonmonotonic behavior of random field fluctuations magnitude with a decrease in temperature. As was noted, one should develop new combined mechanisms that should account for the abrupt shift of the absorption coefficient caused by temperature variations of only 9 – 15K, without involving phase transition models.

To broaden the scope of experimental material, we studied the temperature dependence of photoconductivity in these compounds. It should be pointed out that, with decreasing temperature, data derived from the absorption spectra of these materials agree with photoconductivity data. We believe these results to be of special interest. Moreover, the

experimental techniques used by us allowed the behavior of the phonon plasmon systems to be studied under practically the same conditions. This approach clarified the nature of an abrupt change in the magnitude of the internal fluctuation-induced random field at a certain temperature. In our opinion, temperatures $T < 65K$ are high enough for the formation of a stable cluster from a pair of interstitial tellurium atoms in the CdTe sublattice and a hole localized in their vicinity, as a result of a local change in the lattice polarizability. Such a cluster can be formally considered as a counterpart of the H-center in alkali halide crystals or as a modified small-radius polaron. In order for the free energy of a crystal to reach a minimum, the crystal should contain, at the same temperatures, stable cadmium atom vacancies with electrons localized thereon. The diffusion mobility of electrons at $T \sim 65K$ is high enough for the formation ofg complexes of the above types. These can be considered as counterparts of the Frenkel exciton in a polarized medium. Correct description of the processes occurring in this point of the crystal should probably include vibronic interaction, because a change in the phonon density caused by a change in masses and force constants notably affects the strength of vibronic coupling [3]. At the same time, the quadratic terms of vibronic interaction control the variation of the force constants and thus affect the local lattice dynamics.
The samples of series N and A were studied on a scanning electron microscope in reflection mode using the potential contrast technique. A detector of secondary electrons with energies lower than 25 eV was used and could resolve surface potential variations of $\delta U < 0.5$ V. For the mass contrast technique, we used a detector of reflected electrons with energies close to that of the incident beam, i.e., 10 to 25 keV. Note that small δU do not affect the signal of scanning electron microscope. At the same time, the electron reflection coefficient increases with increasing atomic number Z of elements located at distances < 1000 A from the sample surface. While the sensitivity of this method to composition variations of materials is quite low, the predominant factor of contrast formation in both signal detection techniques is the topographic relief of the sample surface. We measured the variations of surface potential across the sample. The A-series samples were found to exhibit random alternations of p- and n-type regions at $T <$ 78 K. Thin elongated dark regions with n-type of conductivity could be resolved against the light background of the p-conducting material, the size of the p-fluctuations being less than 10 μm. Further decrease in temperature to 55–60K would probably reveal a clearly pronounced mosaic pattern indicating a nonuniform distribution of internal potentials in the crystal, which, in their turn, reflect a strong fluctuation field in the material. It should be stressed that the samples of the N series exhibited a uniform distribution of the potential over the sample, which did not differ from that obtained electron reflection.
Thus, the above experiments are completely consistent with the optical measurements and shed additional light on the results obtained in our work.

5. References

1. V.L. Bonch-Bruevich, I.P. Zvyagin, R. Keuper, A.G. Mironov, R. Enderline, and B. Esser, *Electronic Theory of Disordered Semiconductors,* Nauka, Moscow (1981).
2. B.I. Shklovskii and A.L. Efros, *Electronic Properties of Doped Semiconductors,* Nauka, Moscow (1979).
3. A.I. Belogorokhov, M.P. Kulakov, V.A. Kremerman, A.L. Natadze, Yu.B. Rozenfeld, and A.I. Ryskin, *Zh. Eksp. Teor. Fiz.*, **94(6)**, 174 (1988) [*Sov. Phys. JETP* **67,** 1184 (1988)].

A Light Addressing Electrochromic Device

Xu Yanzhong, Qiu Muqing, Li Naipin, Cao Guangjun, Liu Sanqin
Dept. of Solid State Electronics, Huazhong University of Science and Technology,
Wuhan, Hubei 430074, P. R. China

Summary:

A novel electrochromic device structure is proposed in the study of the electrochromic information memory properties. A high resistive amorphous silicon dot is intercalated between the working electrochromic layer and the electronic electrode. Focused light beam can be applied to switch on the high resistive dot, hence the memory spots are addressed in two dimensional space.

Key Words: Electrochromic, Amorphous Silicon, Memory

I. Introduction

Electrochromism has been widely researched for electronic information displays and other small area applications in historical[1-3]. In recent years, the research interest is turned to the large area applications such as smart windows because the electrochromic technology has yielded to the liquid crystal technology in the case of a major commercial success, the digital watch, during the same period. A direct result of such research is the commercially available small size (under $0.5 \times 0.5m^2$) products of smart window glass. A much deeper understanding of the various electrochromic materials is been establishing by correlating the micro-structure with the energy band structure, albeit the quantitative work is very complex and difficult.[4-6]

Many electrochromic materials show an open circuit color memory ability, that is the electrochomic film changes its optical transmission under the external pulse drive, and remains the optical property for a long time in the state of external open circuit. This characteristic makes the electrochromic film a possible application in the information memory. Although this property has not been thoroughly studied, the successful experience in research of smart windows offers fundamental in establishing such research. In principal, this open circuit memory ability is determined by the mechanism of electrochromic film, whose color change corresponds to the injection of ions and electrons or extraction of them. When the external drive is removed, the ions and electrons will stay in or outside the film to keep the electro-neutrality, which is much like the case of the electron traps in the semiconductor programmable memory devices. The electrochromic device, however, is an electro-optical device, both very complex electronic connection and optical apparatus are involved in the practical application. An initial direct consideration for the electrochromism's application in information memory can be a complex large scale matrix of individual electrochromic units, which causes a complex problem in the electronic connection and isolation. For this reason, the research on the electrochromic information memory should be initialized from a simple device structure like compact disk or optical magnetic memory.

Hydrogenated amorphous silicon has been widely applied in energy conversion, large scale panel display and light detector. The conductivity of the high resistive amorphous silicon film can vary as high as 10^{5-6} under the light irradiation, hence it can be applied as a light sensitive switch in controlling the information read and write in the elecrochromic memory.

A proposal of electrochromic information memory of light addressing is presented by combining the features of electrochromic film and high resistive amorphous silicon film, which simplifies the electronic connection.

II. Device Construction

2.1 Electrochromic windows

In the current state of technology, several device designs exist and are represented by a variety of commercial and prototype devices. The electr0ochromic materials can be organic, inorganic or polymeric. The typical electrochromic layer in a device is amorphous or a sufficiently loose packing tungsten oxide WO_3, iridium oxide IrO_2 or nickel oxide NiO film, which facilitates mobiles of ions in it. A generalized schematic of a solid state electrochromic device is shown in fig.1. For transmissive devices the electrical conductor layers are usually highly doped semiconductors, such as SnO_2:F and In_2O_3:Sn. The second

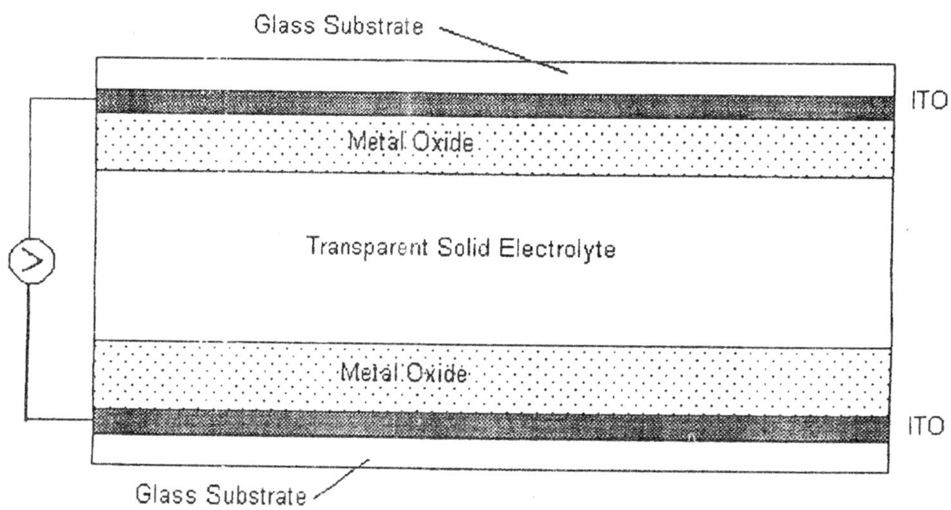

Fig.1 Schematic cross-section of a generalized solid state elelctrochromic device

layer is the electrochromic layer. Usually this layer ranges from 0.05-0.3um upwards. The third layer is the electrolyte. This layer exhibits ionic conductivity but little electronic conduction. The fourth layer is a counter electrode. When an electrolyte is used , with sufficient ion storage, or with a redox couple, the counter electrode is not needed.[7] The last layer is another electronic conductor. It is usually believed that the following reaction is corresponding to the coloring and bleach of the electrochromic film.

$$MeO_n + xI^+ - xe^- = IxMeOn \qquad (1)$$

Here Me is a metal atom, I^+ is a singly charged small ion such as H^+ or Li^+, e^- is an electron, and n depends on the particular type of oxide.

2.2 Electrochromic Device With a - Si

Amorphous silicon has been widely applied in the solar cells, panel display and optical detectors. A high resistive amorphous silicon layer can be modulated its resistivity by focused light beam in the extent of conductivity increase as high as 10^{5-6} times, hence it can be used as the switch. In the device the irradiated spot controls the selected information writing. A schematic of this novel device is shown in fig. 2.

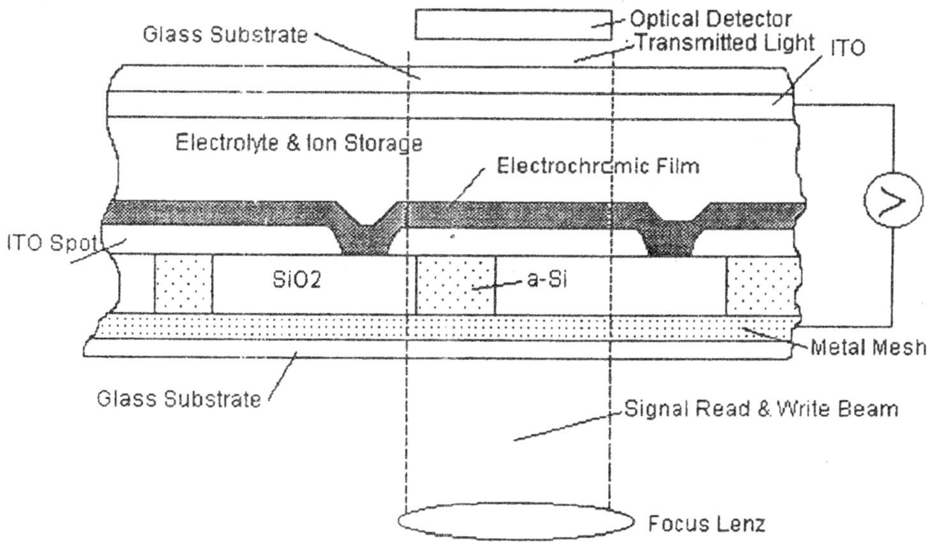

Fig.2 Cross Section of Electrochromic Device with a-Si Switch

Here the ITO layer prepared into individual small spots, so do the a-Si but in a smaller size, which will change its conductivity under light irradiation operating like a switch. When there is a writing operation, the external drive is simultaneously supplied, then the addressed spot changes its optical transmission according to the applied external voltage. Because the electrochromic film's response time is usually slow, this procedure needs several ten milliseconds. While in the case of read operation, the external drive is removed, the electrochromic film maintains the optical transmittance, the passed light beam heats the light detector over on the other side of the device, a signal is quickly read out. This procedure needs much shorter time. Due to the optical property, the stored information can be in continuos instead of conventionally binary for the CD-ROM, or magnetic media.

III. Experimental

As an example, nickel oxide film has been prepared. The optical transmittance and the response of the device have been tested and shown in fig.3 , fig.4. In the transmission curve, it is obvious that the optical transmittance increases rapidly with a negative external driving voltage, and decreases with an opposite bias. Then it is easy to modify the stored information by adjusting the external bias. In the fig. 4, the current response of the device is presented. the upper peak corresponds to a bleaching procedure, the lower to the coloring. The bleaching speed is quicker than the coloring from the tested curve.

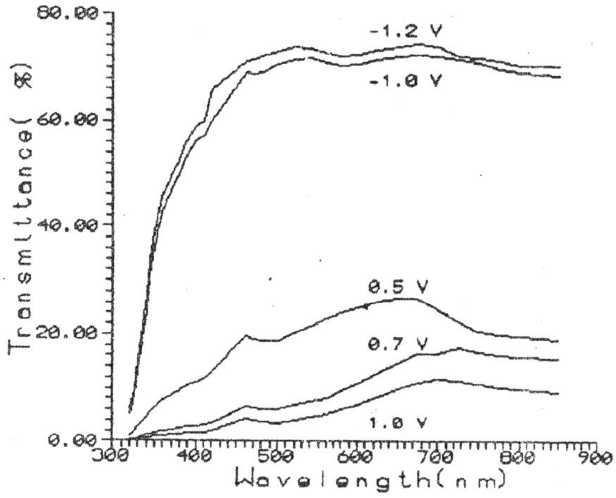

Fig.3 The Transmission of the device

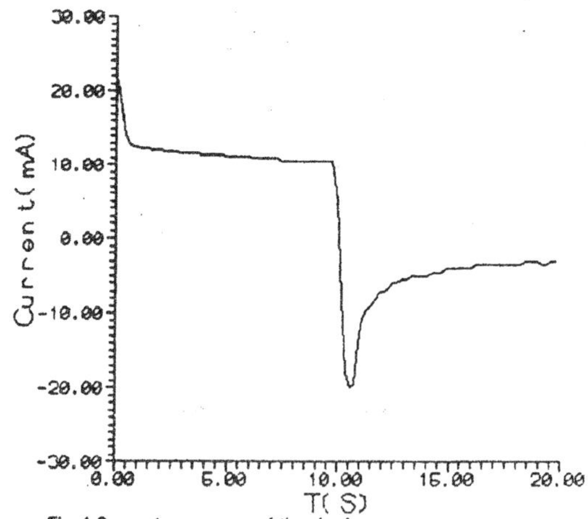

Fig.4 Current response of the device

IV. Conclusion

A novel electrochromic device for information memory is presented in the report. The amorphous silicon is used as the light addressing switch, controlling selection of the appropriate unit. This device structure can be easily fabricated in the disk form, and be applied as a mass memory media.

Reference

[1]Carl M. Lampert, Solar Energy Materials and Solar Cells 32(1994)307-321

[2]Richard S. Crandall and Brian W. Faughnan, Applied Physics Letters, Vol. 28, No2, 15 January 1976

[3]Kuo-chaun Ho, David E. Singleton and Charles B. Greenberg, J. Electrochem. Soc., Vol. 127,No.12, December (1990)3858-3864

[4]Xu Yanzhong et al, J. Phys. III France 5 (1995) 1491-1499,

[5]C.G.Granqvist, Solar Energy Material and Solar Cells, Vol. 32, No.4, April(1994)369-382

[6]Xu Yanzhong, Solar Energy Materials and Solar Cells, accepted

[7]Z. Crnjak Orel, B. OrelJ. of Materials Science Vol.30 (1995)2284-2290

$ZnO/CdS/CuGaSe_2$ Single Crystal Solar Cells.

A.Nateprov[1],S.Radautsan[1], J.H.Schon[2],K.Kloc[2],H.Riazi-Nejad[2] and E.Bucher[2]

[1] Institute of Applied Physics Academy of Science of Moldova,Kishinev,Moldova
[2] Universitat Konstanz,Konstanz,Germany

Abstract

The $ZnO/CdS/CuGaSe_2$ single crystal solar cells with thermaly evaporated CdS layers have been prepared. The cells with undoped and In-doped oxygen free buffer layers CdS had a low open-circuit voltages in comparision with chemicaly deposited CdS buffer layers. The strong oxygen influence on open-circuit voltage of devices was establised. It is shown that using of oxygen during ZnO films preparetion by rf-sputtering lead to high open circuit voltages for devices with thermal evaporated CdS buffer layers.

Introduction

A chalcopyrite semiconductors $CuGa(In)S(Se)_2$ have proven their suitability for use in photovoltaic devices. The more development device has efficiency 17.1%[1]. One of the chalcopyrites $CuGaSe_2$ with an energy gap of 1.68eV and a high optical absorption coefficient is a favourable candidate for use as a top cell in tandem solar cells. Recently the high efficence solar cell with $CuGaSe_2$ single crystal have been fabricated [2]. It is clear that for high efficient convertors thin films photovoltaic cells are preferable tham single crystal devices. However investigation single crystal devices is important becouse cell perfomance depends not only on properties of the bulk material but also on grain boundaries too.
The role of oxidation for performace of $CuGa(In)S(Se)_2$ based solar cells is well known[5]. Practically all high-efficiency chalcopyrite-based solar cells are contained chemical bath deposition CdS buffer layer. At the same time it is well known that chemically deposited CdS films are not a pure CdS and included a large amount of $Cd(OH)_2$ and/or CdO[3]. Moreover heat treatment in air are necessary to improve performance. The influence of oxidation for perfomance of $CuGa(In)S(Se)_2$ policrystaline thin film solar cells was investigated before [4]. At the same time the comparison between properties of $ZnO/CdS/CuGaSe_2$ single crystal solar cells with thermal evaporated CdS and chemically deposited CdS buffer layers have been not analyse yet. As a first step the results for $ZnO/CdS/CuGaSe_2$ single crystal solar cells are presented.

$CuGaSe_2$ single crystal as absorber

The single crystals $CuGaSe_2$ was grown by chemical vapor phase transport using iodine as the transport agent. The crystals was p-type conductivity. Without intentional doping the carrier concentration was approximately $10^{23}m^{-3}$ and Hall-mobility about 0.004 m^2/Vs at room temperature. Some of as-grown crystals were chemically polished in a solution of bromine and methanol and washed in deionized water before using. Practically the parameters of single crystals was as in [2] when the highest-to-date efficience for $CuGaSe_2$ single crystal solar cell device was obtained.

CdS films as buffer layer

The CdS films were prepared by thermal evaporation CdS powder (CdS-99.99). The „hot wall" method have been used and condition of process typicaly was a next: temperature of sustrate -150^{0}C, temperature of wall 550^{0}C, temperature of source CdS 400-450^{0}C. During evaporation vacuum was 10^5-10^6 Torr. The films were cooling after deposition in argon nitrogen gases mixture. Under these condition the CdS with different thickness have been prepared on glass substrate and on $CuGaSe_2$ single crystal. The composition of films have been verified by X-ray analises. In all cases the hexagonal (wurtzite) phase CdS have been observed. The films attain a high degree of (002) preferred orientation on glass substrate. For $CuGaSe_2$ substrate the main

0-7803-3223-7/96/$5,00

orientation CdS films was not so pronounsed as for glass substrate and preferred orientation was (102). The interesting fact was established when the as-grown (unetched) single crystal $CuGaSe_2$ was used as substrate. The surfase of crystal was decorated by microcrystals CdS (Fig.1). These small microcrystals grows on CdS film which cover $CuGaSe_2$ surface.

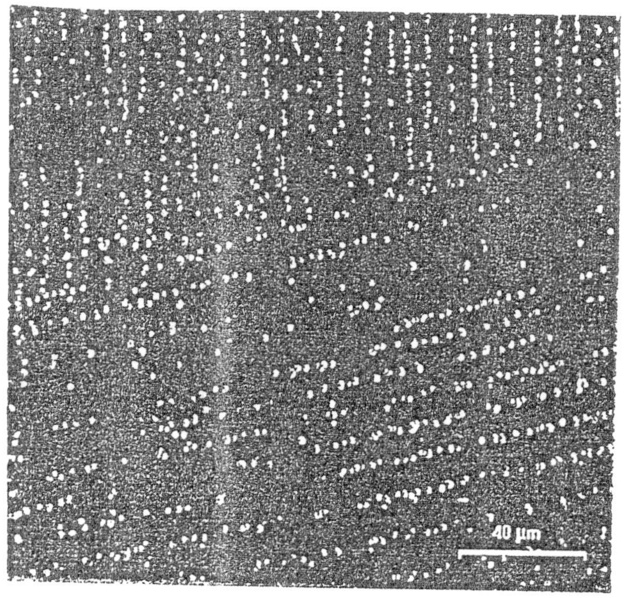

Fig1.The surface of $CuGaSe_2$ single crystal decorated by CdS microcrystas.

Microcrystals were concentrated at the places where $CuGaSe_2$ surface has a steps. The next important conclusion follows from this experimental result. Mobility nuclei of CdS on surface substrate is so high and as result the films CdS has no holes.It is important for devices because often thermaly evaporated CdS layers have columnar structure with small pinholes [3]. CdS layers were also deposited on glass substrate by chemical buth deposition method for comparison with thermaly evaporated one. The mixture of four solutions have been used for preparetion: 1) 0.015M cadmium chloride, 2) 0.043M ammonia chloride, 3) 2M ammonia and 4) 0.4M thiourea. A procedure was the same as in [2].

Optical transmission measurements were carry out on CdS films on a glass substrates at room temperature. Transmission curves for thermaly evaporated films are shown on Fig2.

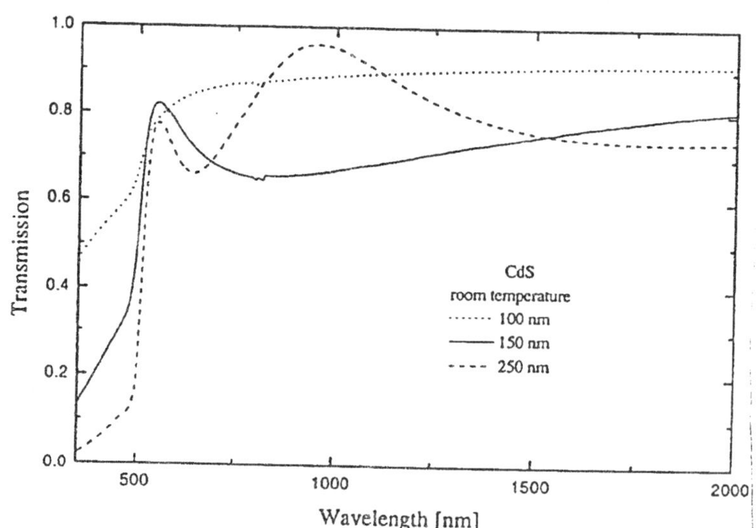

Fig.2 Transmission of thermaly evaporated CdS films.

Transmission measurements confirmed high quality of films, transmission is about 80% (96% at maxima of interfierence) at short wavelength edge of transmission is sharp. The thiner films exhibited transmission curves spread towards short wavelength tham more thicher. Transmission curves for CdS thermaly evaporated and CdS chemicaly deposited films with the same thickness are shown on Fig3. The important feature of picture is a approximately equality of transmission coefficients for both films. This is in contradiction with [3] where transmission in short wavelength ragion for chemically deposited film was higher.

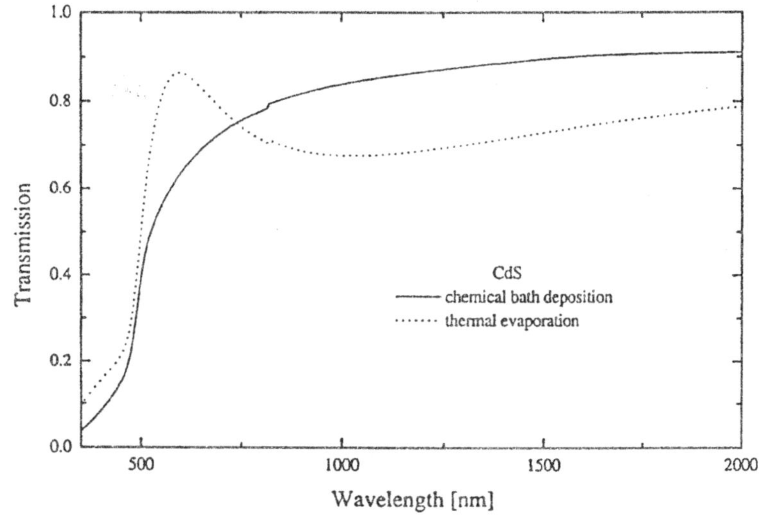

Fig3. The transmission of thermaly evaporated and chemically deposited CdS films.

Device preparation and properties

Ohmic back contacts wre prepared by evaporation of pure gold onto the rear (opposite sensitive) side of single crystal $CuGaSe_2$. Low resistivity conductive transparent film of ZnO was deposited onto CdS film by rf-magnetron sputtering. Gold wires were electricaly contacted to the ZnO and gold spots by conductivity silver paste.

Carrent voltag measurements were perfomed on devices before and after device annealing.

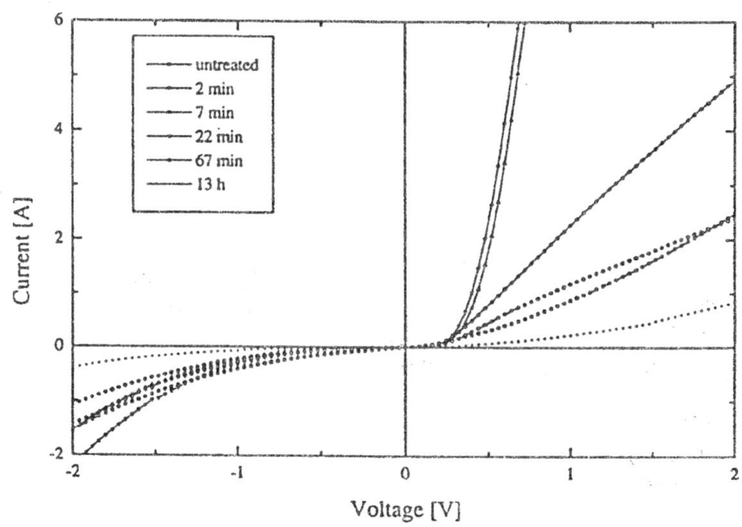

Fig.4. j-V (durk) curves of devices Zno/CdS/$CuGeSe_2$.

Figure 4 shows dark I-V characteristics of Au/$CuGaSe_2$/CdS/Zno cell with thermal evaporated CdS buffer layer before and after annealing on air with different times. The temperature of annealing was 200^0C usual temperature of heat treatments for solar cell based on $CuGaSe_2$. The well pronounced influence of annealing is clear seen. Short time (about 10 minutes) annealing improves diode quality. For more long times annealing the aggragation diode characteristics is obvious fact. Usualy the improvement diode characteristics after short time annealig connected with CdS-$CuGaSe_2$ interface quality and often have been seen before [4]. The aggravation diod characterristics during long time annealing is connected with copper diffusion into a buffer CdS layer.

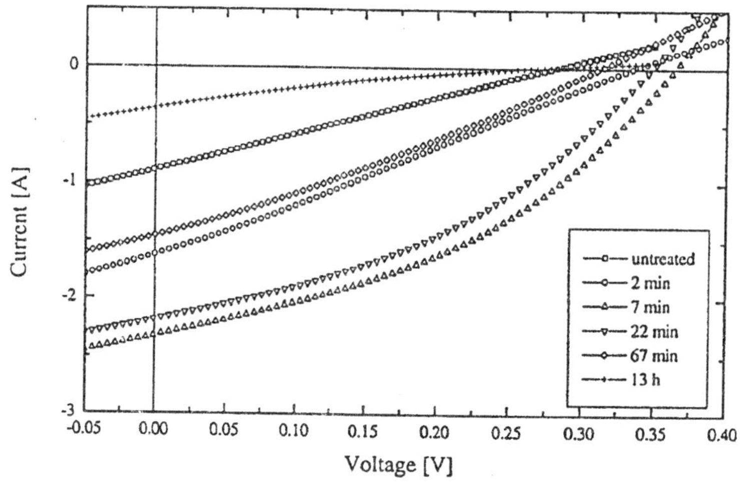

Fig.5. j-V (under illumination) curves of device.

Figure 5 shows development of the light I-V characteristics of a device. The influece of annealing is as for dark characteristics. The more essential differences our ZnO/CdS/$CuGaSe_2$ structures with thermal evaporated CdS layer are low values of V_{oc} and J_{sc} in comparison with the values in [2]. These values are low as for structures with undoped CdS buffer layer as well as for In doped. Because main difference between our buffer layers and buffer layers which was in [2] is absence of oxygen in our thermaly evaporated CdS layer it is reasonable to connect these values V_{oc} and J_{sc} with influence of oxygen. The next fact confirmed this conclusion. When we used oxygen during the ZnO-films sputtering the V_{oc} reached up to 630mV insteaad of usualy observed 0.35mV. This value V_{oc} is equal the V_{oc} value in [2] for cells with chemically deposited CdS buffer layers before annealing.

Thus for achievement high values of V_{oc} in ZnO/CdS/$CuGaSe_2$ solar cells oxygen availability is necessary.

References

[1] J.R.Tuttle, T.A.Berens, S.E.Asher, M.A.Conteras, K.R.Ramanathan, A.L.Tennant, R.Bhattacharya, J.Keane and R.Noufi. Proceedings 13th European Photovoltaic Solar Energy Conferece, Nice, France, 1995 (in press).

[2] M.Saad, W.Simon, K.Friemelt, H.Riazi-Nejad, E.Bucher and M.Ch.Lux-Steiner. Proceedeng 13th European Photovoltaic Solar Energi Conference, Nice, France, 1995 (in press)

[3] Tokio Nakada, Nobukata Okano, Voshlyuki Tanaka, Hiroyuki Fukuda and Akio Kunioka. Proceedings First WCPEC, 5-9 Dec., Waikoloa, Hawaii, 1994 (95-98).

[4] R.Klenk, R.Menner, D.Schmid, D.Cahen and H.W.Schock.Conf.Rec.22nd IEEE Photov.Spec.Conf, Las Vegas,1991 (1071-1074).

[5] T.Walter, D.Hariskos, R.Herberholz, V.Nadenau, R.Schaffler, H.W.Schock. Proceedings 13th European Photovoltaic Solar Energy Conference, Nice, France, 1995 (in press).

RADIATIVE RECOMBINATION AND SHUBNIKOV-DE HAAS EFFECT IN $Cd_xHg_{1-x}Se$ (X<0.5)

E.I.Gheorghita, G.S.Gudima, F.M.Munteanu
Institute of Applied Physics, Chisinau, Moldova

SUMMARY

In this paper Shubnikov - de Haas effect and radiative recombination spectra in the temperature range 2.1-66 K are investigated, experimentally observed peculiarities of radiative recombination spectra are explained, and the main parameters of the energy spectrum of $Cd_xHg_{1-x}Se$ solid solutions are determined.

1. INTRODUCTION

As it is known, in wide concentration ranges HgSe and CdSe form $Cd_xHg_{1-x}Se$ solid solutions with the structure of zinc blende at $0<x<0.77$ and of wurtzite at $x>0.81$. As far as $Cd_xHg_{1-x}Se$ with zinc blende structure are narrow gap semiconductors, then they are of practical interest in optoelectronics as an alternative material to $Cd_xHg_{1-x}Te$. Photodetectors made on $Cd_xHg_{1-x}Se$ can be used in the range 4 -14 μm . In contrast to $Cd_xHg_{1-x}Te$, $Cd_xHg_{1-x}Se$ solid solutions have a number of significant peculiarities in the energy structure /1/ requiring a detailed investigation.

2. RESULTS AND DISCUSSION

2.1. RADIATIVE RECOMBINATION

Fig.1 shows typical photoluminescence spectra $Cd_xHg_{1-x}Se$ solid solutions at the magnetic field absence. The photoluminescence spectrum (PL) consists of three bands conditionally designated as A, B, C. The most intensive is the shortwave band A. At the energetic distance 2.2 *meV* less intensive band B is placed. The long-wave band C having its maximum E =257 *meV* at 4.2 K is placed at the distance ~7.5 *meV* from the shortwave band A. In the external magnetic field all the radiative bands shift in significantly into the shortwave region of the spectrum. Besides, in the magnetic field the intensivity of the band A decreases, and that of the band B increases. PL spectrum alters strongly at the temperature rise. Already at T~6 K the long-wave band C practically disappears, and only two bands - A and B remain in the spectrum. A sharp shift of the band A into the high-energy side in the range T=6.3-13 K by ~2.5 *meV* can testify to the change of the radiative recombination channel: A band transformation from PL band connected with recombination via impurity centers into the recombination band conditioned by the "conduction band - valence band" transitions. At the extrapolation of the experimental temperature dependence of A band position we have obtained the value of the forbidden band width of the studied alloy at 4.2 K Eg=267*meV*. Using semiempirical formula Eg=Eg(x,T) /2/ we have specified the composition for the given compound x=0.291, that is in agreement with the results of the roentgen-spectral analysis.

0-7803-3223-7/96/$5,00

The square-law dependence of the radiation intensivity on the pumping level, the energetic position of A band and its transformation in the temperature field allow to consider that the given radiation band is conditioned by excitons connected with the neutral donors. The energetic position of the radiative recombination band of the donor band exciton is determined by the formula:

$$h\nu = E_g - (E_b + E_{ex}) \qquad (1)$$

where E_b is the energy of the donor bond exciton; E_{ex} is the energy of the ground exciton state; $h\nu$ is the energetic position of the radiation maximum for the considered band. Adopting from the experiment that $h\nu$=264.5 *meV* at 4.2 K, we obtain E_b-E_{ex}=2.5 *meV*. It is known /3/ that the energy of the exciton bond with the neutral donor E_b=0.33 E_d. E_d - energy of the donor ground state. The exciton ground state energy according to /2/ in $m^*_h >> m^*_e$ approximation is expressed by $E_{ex}=E_d(1+m^*_e/m^*_h)^{-1}$. Under the mentioned conditions it follows from (1):

$$0,33E_d + E_d\left(1+m^*_e/m^*_h\right)^{-1} = 2,5 \quad (meV) \qquad (2)$$

at m^*_e=0,02m_0 ; m^*_h=0.48m_0. The calculation by the formula (2) gives E_d=1.9 *meV*, and by the formula (1) for E_g=267 *meV*. The obtained value for the energetic depth of the donor lying corresponds to the thermal energy of ionization experimentally observed by us.

The maximum of B band ($h\nu_B$=262 *meV*) is approximately 5 *meV* distant from the band edge. By its energetic position this line cannot be attributed neither to the donor-valence band recombination, not to the exciton recombination bond with neutral or ionized donors. We believe that the observed line B by its energetic position and behavior is conditioned by the recombination of the excitons localized on the composition fluctuations /4/. Theoretical calculations conducted according to /5/ confirm the above said. In the calculations the following parameters were adopted: α=1.76 *eV* /2/; N=1.3*10^{28} m^{-3}; m^*_h=0.48m_0; m^*_e=0,02m_0 x=0.291.

The long-wave band C ($h\nu_C$=257 *meV*) is approximately 10 *meV* distant from the band edge. This band almost completely disappears at T=6-8 K. That is why this band is connected with the radiative transitions from the donor level into the low acceptor level E_A=8 *meV*.

2.2. SHUBNIKOV OSCILLATIONS

Shubnikov oscillations in $Cd_xHg_{1-x}Se$ were earlier investigated (up to x=0.2) in /6/. The expansion of the range of investigation of oscillation effects in these crystals at x>0.2 is of interest.

The oscillation curves registered at various orientations of the magnetic field in $Cd_xHg_{1-x}Se$ solid solutions at x>0.2 are characterized by frequencies differing in size up to 25% (almost spherical Fermi surface of electrons). The ratio of the maximum electron cross-section $S_{\parallel}$ of the isoenergetic surface to the minimum one $S_{\perp}$; $\eta=S_{\parallel}/S_{\perp}$=(1.23±0.03) in the range $0,2\le x\le 0,5$ practically does not depend on the composition.

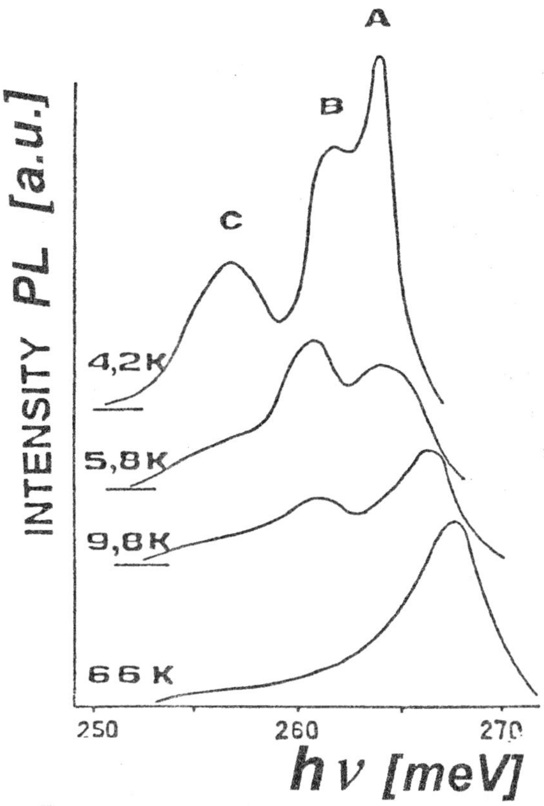

Fig. 1. Photoluminescece spectrum of $Cd_xHg_{1-x}Se$

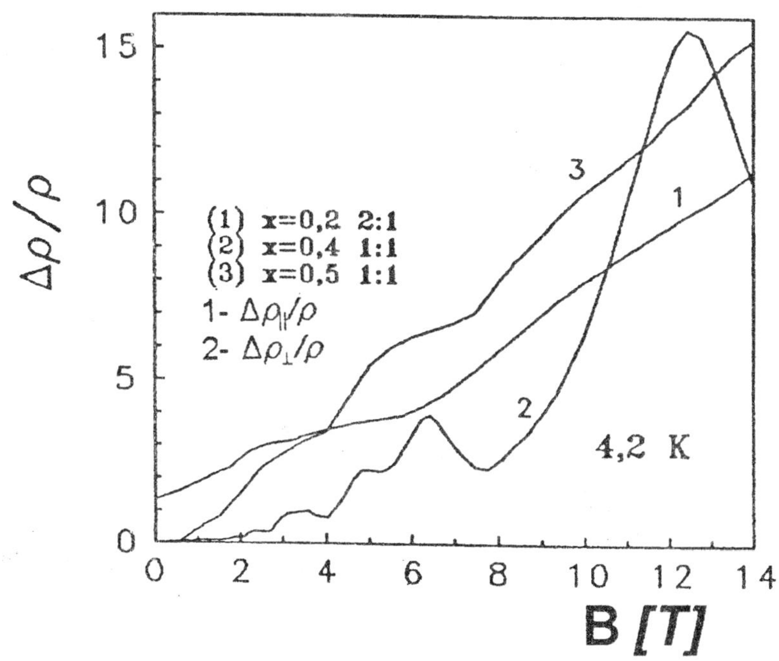

Fig.2. The quantum oscillations of magnetoresistance in $Cd_xHg_{1-x}Se$

The electron concentrations calculated by the Hall coefficient saturation in strong magnetic fields $R_{H\infty}=-(eN_H)^{-1}$, and from the Shubnikov oscillation frequencies coincide in size with accuracy up to 5%.

From the temperature and magnetic field dependences of the Shubnikov oscillation amplitude the cyclotron masses of electrons in alloys with range $0,2\leq x\leq 0,5$ were estimated. The obtained data testify to the fact that ($m^*/(1-m^*/m_0)$) dependence on $N^{2/3}$ in this concentration range is the linear function according to the law of the charge carrier dispersion. The Dingle temperature T_D characterizing non-thermal broadening of Landau level for electrons in alloys $Cd_xHg_{1-x}Se$ ($0,2\leq x\leq 0,5$) T_D ~2.6-4.2 K.

The quantum oscillations of the transversal magnetoresistance on some samples have a clearly expressed structure connected with spin-splitting of the latter Landau levels. The most clearly it is seen on $Cd_{0,4}Hg_{0,6}Se$ sample. One can judge the character of the electron spin-splitting near the large cross-section of the isoenergetic surface by the oscillation maxima position.

It was found that the ratio of the spin ($\Delta\varepsilon_S$) to the orbit ($\Delta\varepsilon_0$) splittings of Landau levels ($\gamma=\Delta\varepsilon_S/\Delta\varepsilon_0=g(m^*/2m_0)$) $\gamma\approx 0.16$ (extrapolation of the linear dependence $n(B^{-1})=0$ gives semi-integral numerical value on the axis of ordinates, the latter maximum is placed close to this line).

For Kane band at $Eg\neq 0$ $\gamma=\Delta/2\Delta+3E_g+3E_F$), that is why at known E_g and E_F one can estimate the spin-orbit splitting value Δ. For $Cd_xHg_{1-x}Se$ alloys it was found that $\Delta\approx 0.33\,eV$ ($E_g=0.46eV$, $E_F=0.013eV$), that is some what less than in HgSe(0.45 *eV*) and CdSe (D=0.42*eV*).

3. CONCLUSIONS

It was found that:

- PL spectrum of $Cd_xHg_{1-x}Se$ solid solutions at 4.2 K consists of three energetic bands which in the external magnetic field shift insignificantly into the shortwave region of the spectrum;

- the Fermi surface of electrons has almost spherical form, the electron concentrations calculated by the Hall coefficient saturation in strong magnetic fields and from Shubnikov oscillation frequencies coincide in value with the accuracy up to 5%;

-the ratio of the spin-orbit splitting of Landau levels in $Cd_xHg_{1-x}Se$ alloys ($0.2<x<0.5$) is $\gamma\approx 0.16$.

REFERENCES

/1/. M.Singhetal, J.Phys.C., **15**, N33, p.6731-6739 (1982).
/2/. C.J.Summers and J.G.Broerman, Phys.Rev. (b), **21**, N2, p.559-573 (1980).
/3/. M.A. Lampert, Phys.Rev.Lett.,**1**, p.450-453 (1958).
/4/. V.I.Ivanov-Omski et al., Sol.St.Comm., **46**, N1, p.25-28 (1983).
/5/. S.D. Baranovskii, A.L.Efros, Fiz.tech.polup., **12**,N12, p.2233-2237 (1978)
/6/. J.Stakiewicz,W.Giriat, and W.Dobrowolski, Phys.Stat.Sol. (b), 61, p.267-276 (1974)

VIBRATIONAL MODES OF $CuAl_{1-x}Ga_xSe_2$ SOLID SOLUTION

N.N. Syrbu [a], M. Bogdanash [a], V.E.Tezlevan [b]

[a]Technical University of Moldova, 277012, Stefan cel Mare, Bul.168, Chisinau, Moldova [b]Institute of Applied Physics, Moldavian Academy of Sciences,277028, Grossul str.5, Chisinau, Moldova

SUMMARY

In the present paper new results of Raman scattering of $CuAl_{1-x}Ga_xSe_2$ crystals by the unpolarized light at 77K are presented. IR reflectivity spectra of this solid solution have been investigated for polarizations E ll c and $E \perp c$ at 300K. Reflectivity spectra contours were calculated and phonon parameters and dielectric constant were determined.

1. INTRODUCTION.

The $CuAlSe_2$, $CuGaSe_2$ and $CuAl_{1-x}Ga_xSe_2$ crystals are perspective materials for being used in nonlinear optics and as solar energy converters. There are some investigations of Raman spectra of $CuAlSe_2$ and $CuGaSe_2$ at 300K [1-3]. IR reflectivity spectra of these materials were presented in [4]. The solid solution $CuAl_{1-x}Ga_xSe_2$ has not been investigated earlier.

The $CuAlSe_2$ and $CuGaSe_2$ compounds have a chalcopyrite structure and belong to the $I42\text{-}D_{2d}^{12}$ spase group. The unit cell contains two formulae units. with 21 optical vibrational modes. The vibrational modes in the Brilloin zone center are

$$\Gamma_{opt} = 1A_1 + 2A_2 + 3B_1 + 3B_2 + 6E$$

The A_2 modes are neither IR nor Raman active. Three vibrational modes B_2 are IR active in the Ellc polarization and six E modes are for $E \perp c$. Besides, the modes A_1, B_2 and E are Raman active.

$CuAl_{1-x}Ga_xSe_2$ crystals have been grown from the gas phase. 15 x 25 mm^2 plates cut from ingots were used for measurings. We investigated Raman scattering by using a double spectrometer DFS-32 with a 5 Å/mm line dispersion. Lasers with 5145, 4880 and 4765 Å lines were used as an excitation source. The IR reflectivity spectra were measured by vacuum spectrometer KSDI-82 (50-250 cm^{-1}) and SPECORD M-80 (200-4000 cm^{-1}) with 0.2 cm^{-1} and 1 cm^{-1} spectral breadth of the slit, respectively.

2. RAMAN SCATTERING OF $CuAl_{1-x}Ga_xSe_2$ SOLID SOLUTION CRYSTALS.

Raman scattering spectra of $CuAlSe_2$, $CuGaSe_2$ and $CuAl_{1-x}Ga_xSe_2$ (x=0.2; 0.3; 0.5 and 0.6) solid solutions are presented in Fig. 1. For compounds of corresponding materials $CuAlSe_2$ (x=0.0) and $CuGaSe_2$ (x=1.0), frequencies of the E, B_2, B_1 and A_1 vibrational modes are observed. For $CuAlSe_2$, these modes are revealed in the range 82 - 407 cm^{-1}, and for $CuGaSe_2$ - in the range 59 - 280 cm^{-1}. In spectra of mixed compounds, there are vibrational mode components of both $CuGaSe_2$ and $CuAlSe_2$ crystals. The vibrational mode distinction of component corresponding to $CuAlSe_2$ is observed with confidence in the range of high frequencies (ν>280 cm^{-1}). In this range $CuGaSe_2$ crystals have not vibrational modes. As X is increasing, that is with the compound change from $CuAlSe_2$ to $CuGaSe_2$, the intensity of high-frequency vibrational modes decreases.These modes are noticed up to x=0.5, and for larger X values, the scattering lines are not practically revealed (Fig.1).

The dependence of the most vibrational mode frequencies on the composition of $CuAl_{1-x}Ga_xSe_2$ is linear. The A_1 vibrational mode frequency is the same for all solid solutions.

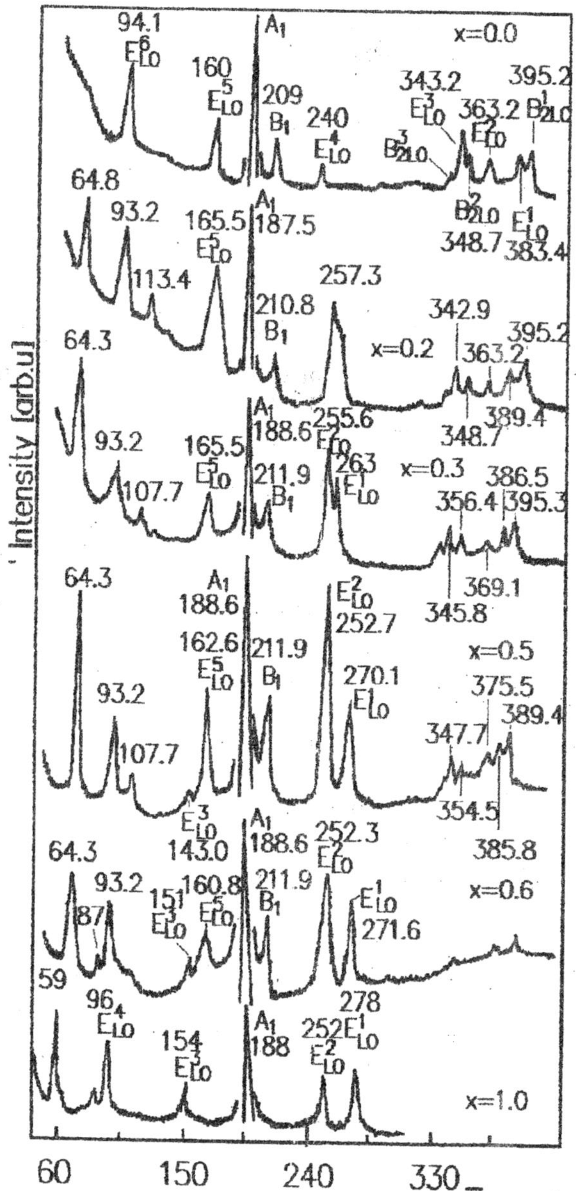

Fig. 1. Raman scattering of $CuAl_{1-x}Ga_xSe_2$ crystals for $CuAl_{1-x}Ga_xSe_2$ x=0.2; 0.3; 0.5 and 0.6.

The vibrational modes E_{LO} (154 cm^{-1}) observed in $CuGaSe_2$ are hardly distinguished in solid solution up to x=0.3. We have found high frequency modes at 278 and 252 cm^{-1} in $CuGaSe_2$ crystals which are due to $E_{LO}^1(B_2^1)$ and $E_{LO}^2(B_2^2)$ phonons, respectively. This is consistent with the results of paper [2]. The frequency of the E_1 mode decreases considerably in solid solution $CuAl_{1-x}Ga_xSe_2$ as X increases (Fig.1). The frequencies of the rest vibrational modes do not change as X increase. As X decreases, the vibrational mode frequencies, corresponding to the B_2^1 and E^1 oscillators of $CuAlSe_2$, decrease too.

3. INFRARED (IR) REFLECTIVITY SPECTRA OF $CuAl_{1-x}Ga_xSe_2$ SOLID SOLUTION CRYSTALS.

Reflectivity spectra of $CuAl_{1-x}Ga_xSe_2$ (x=1.0; 0.8; 0.5, 0.2 and 0.0) solid solution for the polarization E ∥ c are presented in Fig.2. The compounds X=1.0 and X=0.0 correspond to $CuGaSe_2$ and $CuAlSe_2$ crystals. For high-frequency vibrational modes B_2 caused by $CuGaSe_2$ compounds as well for polarization E ⊥ c the decrease of peak intensities is observed when X is decreasing. The intensity of the B_2^1 mode due to $CuAlSe_2$ here increases. This regularity is clearly observed for all vibrational modes of B_2 symmetry (Fig.2). The value of longitudinal-transversal splitting also decreases when X is decreasing for the modes due to $CuGaSe_2$ and increases for the modes due to $CuAlSe_2$. Frequencies of longitudinal-transversal vibrational modes change with the change of the compound.

Reflectivity spectra for all X values (x=1.0; 0.8; 0.5, 0.2 and 0.0) for the polarization E ⊥ c are presented in Fig.3. As mentioned above in both crystals for this polarization six vibrational modes of E symmetry have been found. In X=0.8; 0.5 and 0.2 compounds there are modes belonging to both crystals. As X is decreasing, the intensity of the vibrational mode due to $CuGaSe_2$ decreases too (the peak marks 1Ga) and the intensity of the vibrational mode E_2 due to

$CuAlSe_2$ (the peak marks 2Al) increases (Fig.3). The analogous regularity is practically observed for all vibrational modes. Simultaneously with this removal of the frequencies both LO and TO vibrational modes due to $CuGaSe_2$ decreases as X decreases. The increase of the longitudinal-transversal splitting of vibrational modes due to $CuAlSe_2$ components takes place simultaneosly (as X decreases).

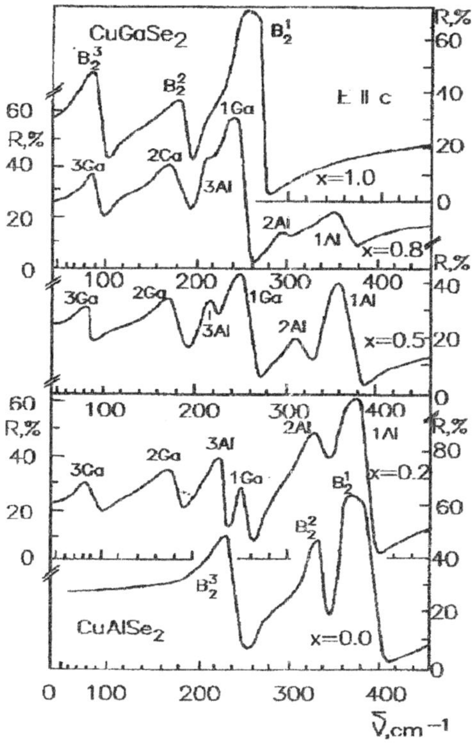

Fig. 2. Reflectivity spectra of $CuAl_{1-x}Ga_xSe_2$ crystals for the E II c polarization.

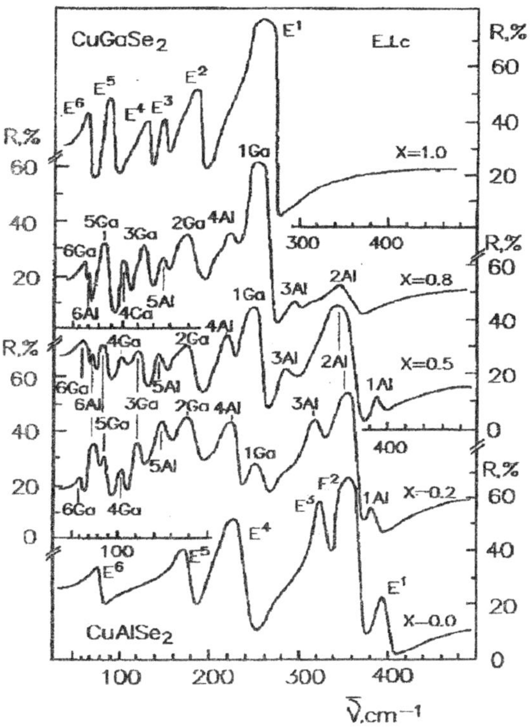

Fig. 3. Reflectivity spectra of $CuAl_{1-x}Ga_xSe_2$ crystals for the E ⊥ c polarization

Reflectivity spectrum contours are calculated from the multy-oscillator model of the dispersion relations for both polarizations. The calculations are made proceeding from the condition that for polarization E ⊥ c twelve vibrational modes are revealed, where six modes are due to $CuGaSe_2$ compound and six modes are due to $CuAlSe_2$ compound and for polarization E II c six ones are revealed, where three modes are due to $CuGaSe_2$ and three are due to $CuAlSe_2$. The reflectivity coefficient value in the range 3000-4000cm^{-1} has been used to determine the high-frequency dielectric constants ε_∞ for each compound. These values are 5.13 ($CuGaSe_2$), 6.78 (X=0.8), 6.48 (X=0.5), 6.12 (X=0.2), 6.85 ($CuAlSe_2$) for the polarization E ⊥ c and 4.20 ($CuGaSe_2$), 6.12 (X=0.8), 5.44 (X=0.5), 6.12 (X=0.2), 4.82 ($CuAlSe_2$) for the polarization E II c. The damping factor and the oscilator strengths for each vibrational mode of $CuGaSe_2$ and $CuAlSe_2$ components in solid solutions have been determined proceeding from calculations of reflectivity contours. The calculation of reflectivity contours shows also that the damping factor changes as X changes. As it was expected modes in the region of compounds close to X=0.5 possess the most damping factors. In fact, solid solutions have less perfect cristalline structure than pure crystals. As it is seen oscillator strengths caused by $CuAlSe_2$ components fall when X increases, i.e. as $CuGaSe_2$ components decrease.

REFERRENCES

1. W.H. Koschel, V. Hohler, A. Romber, I. Baars, Solid St. Com. 13, 1011, (1973).
2. W.H. Koschel, M. Bettini, Phys. St. Sol. (b) 72, 729, (1975).
3. I.V. Bondar, A.G. Karoza, G.F. Smirnova, Phys. St. Sol. (b) 84, k65, (1977).
4. A.M. Andriesh, N.N. Syrbu, M.S. Iovu, V.E. Tazlavan, Phys. St. Sol. (b) 187, 83, (1995).

DIMORPHITE BASED UV DETECTORS

D.TSIULYANU, G.GOLBAN, E.KOLOMEYKO,
N.GUMENIUC, O.MELNIC and S.MARIAN

Department of Physics, Technical University,
Datchia str.41, Kishinev, 277060, Moldova.

Photoelectrical properties of artificial dimorphite allow the fabrication of device structures sensitive to ultraviolet radiation. Photoconductivity spectra of these sensitive structures depend on technology of the film preparation. The result for fabricated of UV sensors are presented including experimental results for UV-dosimeter.

1. Introduction

Dimorphite, a yellow-orange mineral discovered for the first time near Vesuvius (Naples) is known since the middle of last century /1/. The chemical formula of the mineral has been discussed during more than one hundred years, and now is determined as As_4S_3 . Dimorphite can also be obtained artificially by melting of corresponding components in vacuum.

Being one of the fourth known chemical compounds of the As-S system /2/ dimorphite exhibits typical semiconductor properties with characteristic peculiarities. In particular, optical absorption of dimorphite thin films /3/ exhibits a rather sharp edge with shape and position depending on condition of preparation. Present work is devoted to the investigation of photoconductivity of artificial dimorphite thin films, including fabrication and test of UV- sensitive structures. Application of dimorphite based devise structures as sensors of UV dosimeters is given.

2. Experiment

Ingots of polycrystalline dimorphite were obtained by the melt-quenching method with pure As and S in evacuated quartz ampoules. Thin films were obtained by vacuum sublimation or evaporation of synthesized dimorphite on glassy or quartz substrates. Velocity of films growth and substrate temperature have been controlled during the deposition.

Photoconductivity was studied in specimens Al - As_4S_3 -Al with sandwich configuration and semitransparent upper aluminum electrodes. Spectral measurements were carried out by using equipment with monochromator MDR-12 controlled by computer EC-1841. A halogen lamb has been used as a source of illumination in all tests. Calibration of fabricated radiometer- dosimeter was performed by using vacuum thermoelement VTh-8 (Carl Zeiss, JENA).

3. Results

a)Photoelectrical measurements.

Following characteristics have been studied.

1. Ratio of specific resistance $K= \rho/\rho_d$ in dark and under illumination with integral UV light. The level of illumination and applied voltage were contact.
2. K vs applied voltage.
3. Spectral distribution of photo conductivity.

0-7803-3223-7/96/$5,00

Photoelectrical measurements have been carried out using specimens obtained with different technological conditions and with beforehand given requirements to level of vacuum, velocity of growth and substrate temperature. Any regularities in variation of K with condition of the sample preparation have not been found. This parameter is in general determined by dark resistance which can vary in limits of 10^{14} -10^{15} Ohm cm. On the other hand, K strongly depends on applied voltage. Fig.1 show this dependence at room temperature under UV illumination with λ = 250-400 nm and at light intensity 1.34 10^{-4} W/cm^2. The increase of K is observed with saturation at field strength E=2.2 10^4 V/cm.

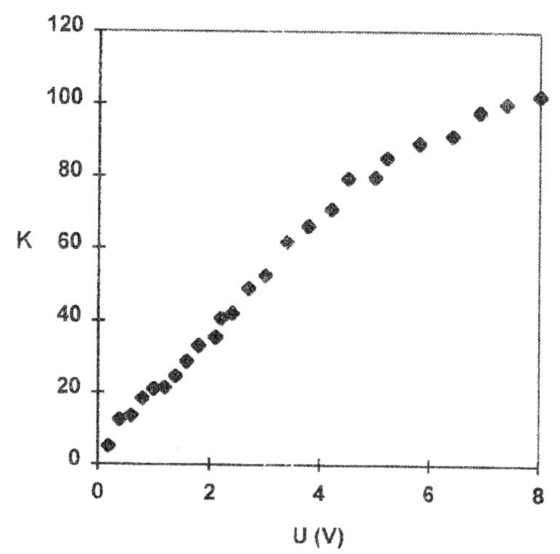

Fig.1. Ratio ρ/ρ_d vs applied voltage.

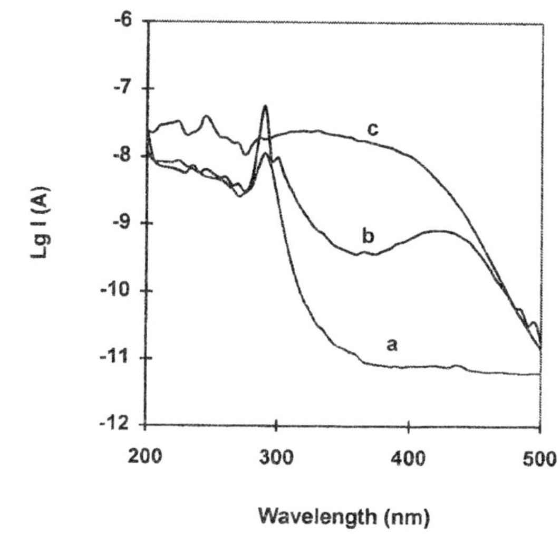

Fig.2. Photocurent vs wavelength. Velocity of film growth decreses in order a-b-c.

Nevertheless application of such large value of field strength is not reasonably because of danger of sample breakdown.

Fig.2 shows the dependence of photocurrent in dimorphite based device structures on wavelength for specimens made with different velocity of growth. It can be seen that at high velocities of film growth (fig.2, curve a) this dependence has the only maximum in UV region of spectra at λ =290 nm. The "red" edge of the maximum is sharp and photocurrent decreases with three orders of magnitude.

On lower velocities of growth the maximum of photoconductivity (see fig.2,curve b) in visible region of spectra appears. The position and width of the additional maximum of photoconductivity depend on the technology of film the preparation, first of all on the velocity of growth and substrate temperature. At very small velocities of growth, the maximum of photoconductivity at λ = 420 nm increases, and can reach the value of the principal maximum situated at λ = 290 nm. Fig.2, (curve c) shows the dependence of the photocurrent on wavelength for a specimen made with velocity of growth v~0,2 μm/h. It can be seen that the maximum in question joins with principal ones and practically the unselective spectral distribution of photoresponse in region at 200-400 nm occurs.

b)Detector structures and behavior under illumination.

The schematic representation of the UV detector fabricated in present work is shown in fig.3. It consists of fotosensitive structures fixed onto plastic base with two contact legs. In order to avoid completely the influence of visible light because of small tail of sensitivity in this region, the UV filter is placed in front of the the photosensitive structure. The basic detector structure can be considered both as photoresistor or as photosensitive capacitor. So, there were two possibilities to built UV dosimeter: using both variation of resistivity or capacity of the structure under illumination.

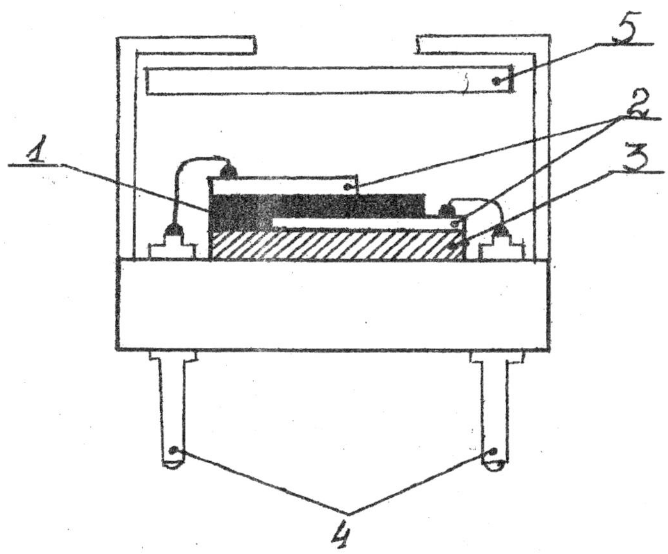

Fig. 3 UV detector: 1.- Dimorphite film; 2.- Electrods; 3.- Substrate; 4.- Contact legs; 5.- UV filter.

Building up both types of UV- dosimeter we have found the capacitances devices to be more suitable. In such version the detector is a relaxation oscillator with a frequency that depends on UV intensity. Fig.4 plots the oscillation frequency in arbitrary units against the UV irradiance. It can be seen that this dependence is very close to linear, therefore the counter can be used to register the total dose received by the detector. The respective dosimeter circuit has been elaborated and tested for long term UV radiation. Fabricated UV detector and dosimeter operates over the radiance range 10^{-1}- 10^{2} W m^{-2}, dose range 10^{2}- 10^{6} J m^{-2}. Other parameters are: supply voltage from 4,5 to 10 V and power requirement 5 10^{-3} W.

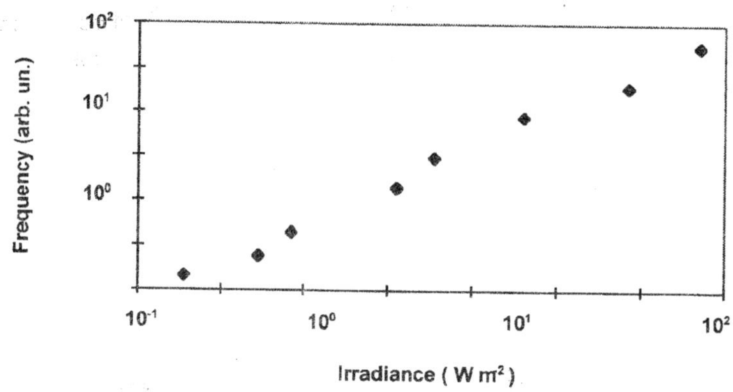

Fig. 4 Oscillator frequency vs intensity of UV radiation

4. Discussion

The appearance of two maxima of photoconductivity can be due to microheterogenity of films. Therefore X-ray analyses of synthesized dimorphite and thin films on its bases have been carried out. Both known phases α and β of dimorphite /1/ have been found in powder material but judging by intensities of X-ray diffraction peaks, the portion of β-dimorphite was larger. This situation essentially was changed for thin films. The films growth by sublimation in vacuum contain α-dimorphite in larger quantities,but the ratio of concentration of α and β dimorphite can be easily changed by variation of velocity of growth. The fact is important that device structures based on artificial dimorphite can be sensitive to UV radiation both selective with maximum at $\lambda = 290$ or unselective depending on technology of the film preparation. The peak of biological activity (erythema, DNA damage) is in the region from 280 to 320 nm peaking at 297 nm /4/. Therefore dimorphite based sensitive structure may be adequate for a low cost UV-sensor for biological and medical applications.

5. Conclusion

Thin films of artificial dimorphite exhibit photoelectronical sensivity in UV region of spectra and thus allow to be used in fabrication of UV detectors. Both selective and unselective UV sensors can be fabricated by variation of technology of the film preparation.

This paper has presented an exemple of such device with dosimeter operating over a wide range of light intensities.

References

/1/ A.Scacci,Memorie Geologiche sulla Campania,Accad.Scienze, Napoli,1850, p.84.

/2/ G.Z.Vinogradova, Glassy Formation and Phase Equilibrium in Chalcogenide Systems, Izd. Nauka, Moscow, 1984. (in Russian).

/3/ D.Tsiulyanu, G.Golban and O.Melnic, CAS'93, 571.

/4/ L.Koller, Ultraviolet Radiation, Wiley, New York, 1965.

The Dependence of the Photomagnetoelectric Effect upon the Impurity Distributions of the Diffused Junctions. Abrupt junctions

Eugen Lakatos

BANEASA S.A.

32, Erou Iancu Nicolae, Bucharest 72996, Romania

ABSTRACT

This paper describes the dependence of the photomagnetoelectric effect upon the impurity distributions of the abrupt diffused junctions, in constant conditions of the applied magnetic field and luminous radiation. The calculated photomagnetic voltages depending on surface concentrations and substrate impurity concentrations are exhibit.

1. INTRODUCTION

In the paper /1/ we discuss about the dependence of the photomagnetoelectric effect (PMEE) upon the impurity distributions of the diffused junctions (linearly graded junctions). Because in the modeling of the phenomenons which describe the functioning of the p-n junctions, we use two approximations (linearly graded junction and abrupt junction), this paper describes the dependence of the PMEE upon the impurity distributions of the diffused abrupt junctions.

2. THEORY

Starting from the theoreticals hypothesiss showed in the paper /1/, in constant condition of the applied magnetic field and luminous radiation, the photomagnetic voltage varies function of the built-in potential, V_{bi}, according to (dependent on the direction of the applied magnetic field)

$$\pm V_{pm} = A - V_{bi} \tag{1}$$

where

0-7803-3223-7/96/\$5,00

$$A = \frac{kT}{q} \ln \frac{p_n}{p_{no}} + V_p$$

(in constant condition of luminous radiation and applied magnetic field);

$p_n = \Delta p_n + p_{no}$;

Δp_n - excess generated holes concentration;

p_{no} - constant holes concentration in the case of the thermal equilibrium;

V_p - phototension;

q - electronic charge;

kT/q - thermal voltage at 300 K.

For the abrupt junction, in the relation (1), V_{bi} is replaced by (complete ionization case)

$$V_{bi} = \frac{kT}{q} \ln\left(\frac{N_A N_D}{n_i^2}\right) \tag{2}$$

If we consider the "box" approximation of the diffused impurity distribution (Fig.1), the relation (2) becomes (p^+-n abrupt junction)

$$V_{bi} = \frac{kT}{q} \ln\left(\frac{C_S C_B}{n_i^2}\right) = f(C_B)\big|_{C_S = ct.} \tag{3}$$

Therefore, for the abrupt junction, we have the following functions

$$A\_ + V_{pm} = f(C_B)\big|_{C_S = ct.} \tag{4}$$

The calculated photomagnetic voltages (until the constant A), depending on substrate impurity concentrations, for three values of the surface concentrations (10^{19}, 10^{20}, 10^{21} cm^{-3}) are exhibit in Fig.2.

The calculations have been made for the nondegenerate semiconductors case. Because for the p^+-n junction is valid the relationship

$$qV_{bi} = E_g - (E_c - F_n) \tag{5}$$

where

E_g - energy bandgap

E_c - bottom of conduction band

F_n - Fermi energy level (for n type),

we had in view to execute the relationships

$$\frac{E_g}{q} - V_{bi} \geq \frac{kT}{q} = 0.0259V \tag{6}$$

$$V_{bi} - \frac{E_g}{q} < \frac{5kT}{q} = 0.13V \tag{7}$$

The values C_B, C_S which for the relationships aren't carry out, haven't been taken into consideration.

Note

The condition for the total degenerate semiconductors is

$$\frac{F_n - E_c}{q} > \frac{5kT}{q} = 0.13V \tag{8}$$

The condition for the nondegenerate semiconductors is (n type semiconductor)

$$\frac{E_c - F_n}{q} \geq \frac{kT}{q} = 0.0259V \tag{9}$$

3. CONCLUSIONS

We calculate and we plot the photomagnetic voltages (until the constant A), depending on substrate impurity concentrations, for three values of the surface concentrations (10^{19}, 10^{20}, 10^{21} cm^{-3}). It has been found that the photomagnetic voltage is directly proportional to substrate impurity concentration and to surface concentration.

REFERENCES

/1/ E.LAKATOS, CAS'94 Proceedings, vol.1, p.195 , 1994.

/2/ E.LAKATOS, Doctoral Thesys, University of Bucharest, Faculty of Physics, 1995.

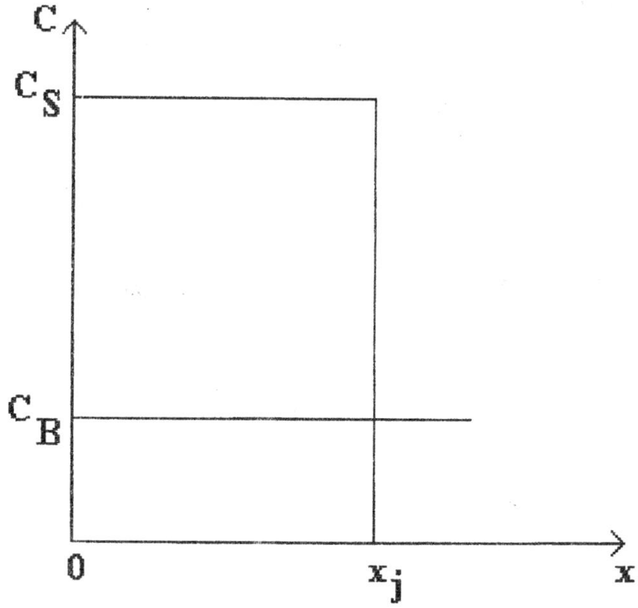

Fig.1. "Box" approximation of the diffused impurity distribution

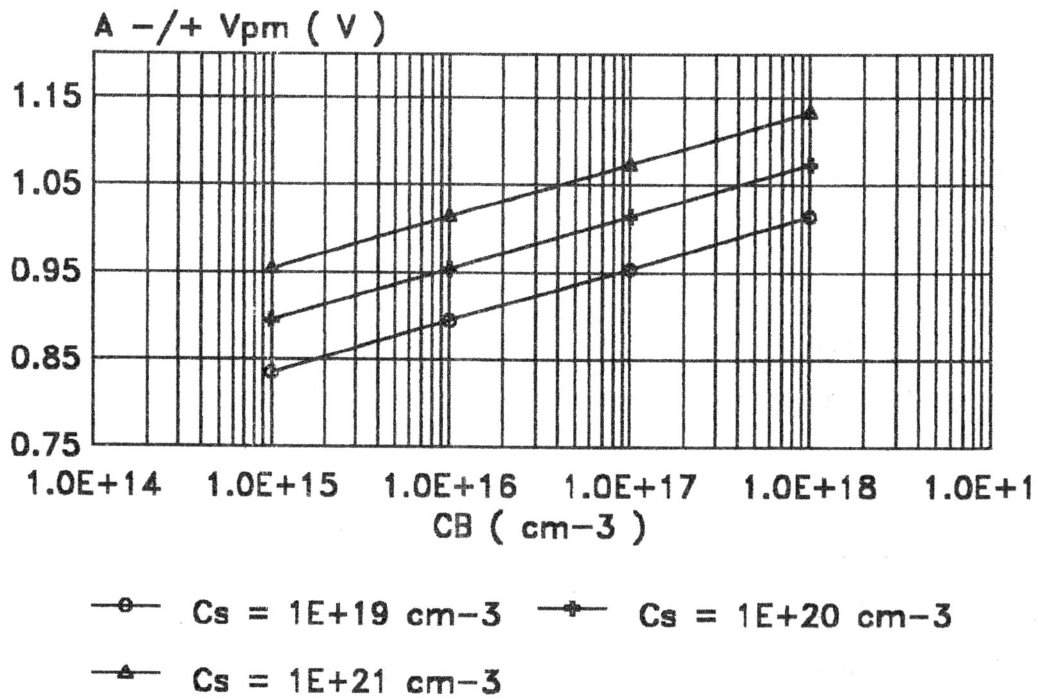

Fig.2. Abrupt junction

OPTIMIZATION OF N-TYPE InSb PHOTOCONDUCTIVE DETECTORS PERFORMANCE

C. E. A. Grigorescu, S. A. Manea[1], M. F. Lazarescu[1], C. Logofatu[2], T. Necsoiu, T. Botila[1], I. Munteanu[3]
Inst. of Optoelectronics, Bucharest, PO Box MG - 22, 76 900 Romania
[1] Institute for Physics and Technology of Materials, Bucharest, Romania
[2] Institute for Physics and Technology of Radiation Devices, Bucharest, Romania
[3] University of Bucharest, Faculty of Physics, Bucharest, Romania

Summary: The aim of this paper is to study the influence of power dissipation on the performance of n - type InSb photoconductive detectors for 3 - 5 μm atmospheric window. By providing mainly a convective heat transfer among the elements of the device the power dissipation has been increased till 10 W/cm^2. An improvement in the spectral detectivity is observed. No enhancement in the total noise is remarked.

1 Introduction

N - type InSb photoconductive (PC) detector elements have usually spectral detectivity (D^*_λ) value smaller than for p - type material, when operated at liquid nitrogen temperature [1]. This conclusion follows from the difference in the transport properties and recombination mechanisms involved in n - type and, respectively, p - type InSb PC cooled elements [2].

An enhancement of the spectral detectivity (D^*_λ) would be obtained in n - type case of a reasonable purity level provided that the power dissipation of the device could be increased [1]. Our work is aimed to investigate the changes in the performance of n - type InSb PC element detector by acting on the power dissipation and by improving the heat transfer among the elements of the system.

2. Basic physical considerations

The dependence of electron and hole lifetimes on purity at the operating temperature of the detector is important to proper detector design , because the lifetime values affect the photosignal, the response time and the noise too. In n - type InSb of moderate to high purity, at 84 K the lifetimes are equal [3], having a value of 8×10^{-7} sec., assumed to be representative

0-7803-3223-7/96/$5,00

for 77 K too. At this temperature the recombination mechanism is the low level Shockley - Read type[1]. Because photoconductive detectors are limited by thermal noise and/ or generation - recombination noise, the choice of a proper expression for noise is strictly necessary for determining the detector performance at the operating temperature. The spectral responsivity of n - type InSb detector as a function of the power dissipation at 77K [1] is given by:

$$R_\lambda = \frac{\lambda \tau_n P_D^{1/2}}{hcwt^{3/2} e^{1/2} \mu_n^{1/2} n_0^{3/2}} \tag{1}$$

where λ is either the cutoff or the peak wavelength value, τ_n and μ_n are, respectively, the electron lifetime and mobility, e is the electrical charge, n_0 - the equilibrium electron concentration, h - Planck's constant, c - speed of light, P_D - the power dissipation value, w - the width of the sample and t - its thickness. By considering the appropriate expressions for the g - r noise [4] and for the thermal noise [1], the spectral detectivity can be written as:

$$D_\lambda^* = \frac{\lambda}{2hcn_i} \left(\frac{\tau_n n_0}{t} \right)^{1/2} (1+\Gamma)^{-1/2} \tag{2}$$

where n_i is the intrinsic carrier concentration at liquid nitrogen temperature ($10^{10} cm^{-3}$). As it is observed from Eq. (2), the spectral detectivity of n - type InSb detectors increases if Γ diminishes, i.e. if increasing the bias current i_b on the detector, provided that the power dissipation threshold of the device is not exceeded. The expressions of Γ and D^*_λ in terms of power dissipation per unit area P_D are given by [1]:

$$\Gamma = \frac{kTn_0^3 t}{n_i^2 \tau_n P_D} \tag{3}$$

$$D_\lambda^* \approx \frac{\lambda \tau_n}{2hcn_0 t} \left(\frac{P_D}{kT} \right)^{1/2} \tag{4}$$

3. PC detector fabrication and characterization

InSb single crystals grown by the Czochralski method, along <110> axis [5, 6], with a majority carrier concentration of 8×10^{13} cm^{-3} and a mobility of 3×10^5 $cm^2/$ V sec at liquid nitrogen temperature were cut into rectangular slabs having 2 mm in length, 1 mm in width and 20 μm in thickness [7, 8]. Cylindrical sapphire supports were provided for the detector elements as prepared and the systems have been introduced in glass dewars with sapphire windows, antireflecting coated. A copper radiation shield was used, provided with a band pass filter in order to decrease the background radiation temperature bellow 300K. Both the sapphire support and the end part of the radiation shield were in direct contact with the liquid nitrogen used as a coolant, providing a convective heat transfer within the system. A temperature value of 85 K at the InSb/sapphire interface and of 170 K at the front surface of the band pass filter have been calculated from the theoretical model [9], assuming thermal equilibrium between the filter and incident radiation and using the "thin wall" approximation for the conductive heat transfer between the copper shield and the glass housing of the detector. Spectral responsivity measurements have been performed for the detector at different power dissipation values, ranging from 8.56×10^{-3} to 10.02 W/cm^2. In Fig. 1 the experimental results for $R_{4.9}$, $D^*_{4.9}$, and the calculated values obtained by using Eq.(1) and Eq. (4), respectively, are shown. Fig. 2 shows the variation of the total noise V_N versus the applied bias current.

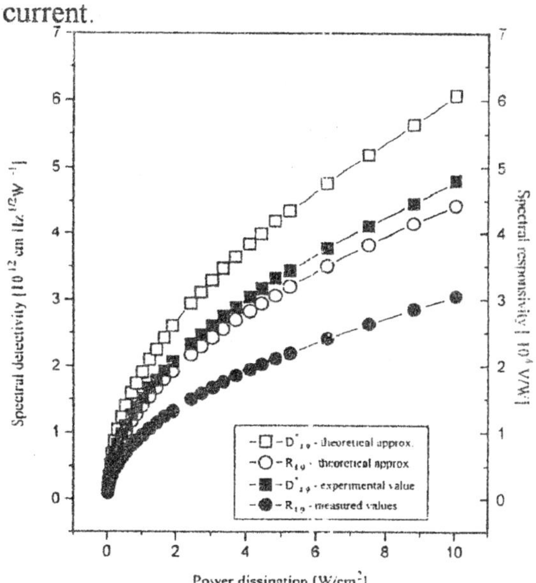

Fig. 1 R_λ and D^*_λ versus power dissipation

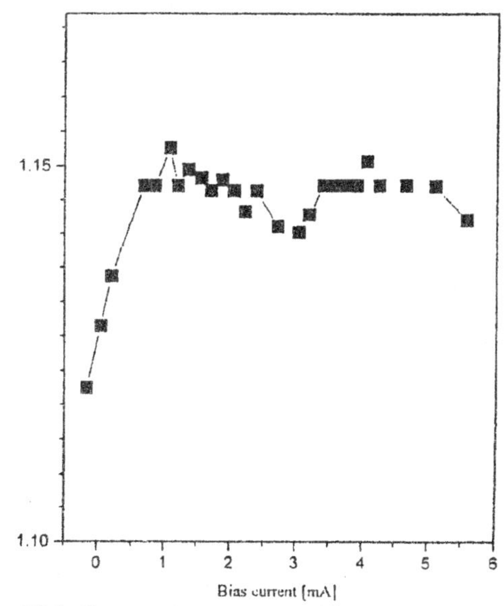

Fig. 2 Total noise versus bias current

The spectral detectivity $D^*_{4.9}$ has been obtained from Eq. (4) in the theoretical approximation and for the experimental values the following equation was applied:

$$D^*_\lambda = \frac{R_\lambda (A_D \Delta f)^{1/2}}{V_N} \qquad (5)$$

were V_N is the total noise, A_D is the detector area and R_λ is given by:

$$R_j = \frac{V_S}{P} \qquad (6)$$

using the experimental values for V_S (the signal value at λ_{peak} = 4.9 μm) and P - the corresponding emittance power of the IR source (0.8×10^{-6} W). It is observed that for high power dissipation the theoretically predicted values and the experimental ones do not differ spectacularly. The cooled radiation shield enabled the experimental value to be closer to the theoretical one, by reducing the background contribution to the noise.

4. Conclusions

The convective heat transfer provided for the detector by the combination of sapphire support and copper radiation shield in direct contact with the liquid nitrogen enables an increasing of the power dissipation under operating conditions of the device. Both the experimental data and the theoretical approximation for the spectral responsivity and respectively for the spectral detectivity at λ_{peak} = 4.9 μm show that for higher power dissipation an enhancement is correspondingly obtained in the performance for the detector made from n -type InSb of high purity.

REFERENCES

[1] P. W. Kruse, in Semiconductor and Semimetals, R.K.Willardson and Albert C. Beer, Vol. 5, 1970, p. 48 - 52

[2] G. K. Wertheim, Phys. Rev., 104, (1956), pp. 663 - 664

[3] R. A. Laff and H. Y. Fan, Phys. Rev.,121 (1961), p. 53

[4] K. M. Van Vliet, Appl. Optics, 6 (1967) p. 1145

[5] K. F. Hulme and J. B. Mullin, Solid - State Electron., 5 (1962), p. 211

[6] J. B. Mullin, personal communication, Bucharest, March 1993

[7] C. E. A. Grigorescu, S. A. Manea, E. Elena, T. Botila, M. F. Lazarescu, SPIE, Proceeding ROMOPTO ' 94,Vol 2461 (1995), p. 483

[8] C. E. A. Grigorescu, M. F. Lazarescu, S. A. Manea, E. Elena, Romanian Patent No. 110708, (1996)

[9] C. Logofatu, paper in press

MEASUREMENT OF PERFORMANCES OF UNCOOLED HgCdTe PHOTODETECTORS IN HETERODYNE REGIME OF OPERATION

*N. Iftimia, **I. Spanulescu, ***I. Iftimia
*Army Research and Development Institute, O.P. Bucharest-Bragadiru, Fax: 7804863
**Hyperion Research and Development Institute, Bucharest, Romania
***Technical Military Academy, Bucharest, B-dul G. Cosbuc. No. 81-83, Sect. 5

SUMMARY

1. Introduction.

An ordinary infrared detector than may be dominated by thermal generation or amplifier noise can serve as nearly theoretically perfect receiver when used in a heterodyne mode.

This is a consequence of the gain inherent in the non-linear mixing of the coherent laser signal and a local oscillator radiation in the detector [1].

As long as the photon noise, due to local oscillator radiation, exceeds all sources of noise, the detector will be capable of sensing a few photons per measurement time interval.

The two fundamental parameters of interest for heterodyne regime of operation are bandwidth and sensitivity. The bandwidth of heterodyne detector and so the bandwidth of heterodyne receiver is a function of many parameters, as: the minority carrier mobility and carrier generation outside the space charge region, the junction resistance and capacitance, the preamplifier bandwidth.

Heterodyne sensitivity of detector is usually defined in terms of noise equivalent power $(NEP)_{SL}$, which is defined in [2] for signal noise limited case:

$$(NEP)_{SL} = h\upsilon \; B / \eta_{eff} \tag{1}$$

where: $h\upsilon$ - energy of the incident photons;

B - bandwidth of preamplifier;

η_{eff} - heterodyne effective quantum efficiency.

The noise equivalent power is the power that must follow the detector to cause S/N to be equal to one.

The effective quantum efficiency of detector in heterodyne regime is reduced to the effect of incomplete domination of a local oscillator noise, because the shot noise can not be neglected [3].

This expression is given by:

$$\eta_{eff} = (M-1)\eta \; / M \tag{2}$$

where M is the ratio of detector noise powers with and without local oscillator power present.

2. Advanced techniques that increase the performances of HgCdTe uncooled detectors.

The photodetectors that work at near-room-temperature are essentially sub BLIP (background-limited-operation) detectors. Their performances can be increased by using new technologies and methods.

A number of concepts to improve performance of NRTPD (near-room-temperature photodetectors) have been proposed, including the optimisation of classical devices for high temperature operation: photoconductors, photodiodes, photoelectromagnetic and Dember detectors.

0-7803-3223-7/96/$5,00 © 1996 IEEE

The performance of NRTPD can be improved by use of an optical resonant cavity and optical immersion of a detector to a high refraction index hemispherical lens [4]. Another way is to use of nonequilibrum phenomena at the heterojunction contacts [5].

The optical immersion of detectors to high refraction index hemispherical or hyperhemispherical lens reduces the image size by a factor dependent on n (the refraction coefficient of lens). This size is reduced by a factor equal to n^2 if the detector is placed in the centre of hemispherical lens, or by n^4, if the detector is placed at the first Weierstrass point of a hyperhemispherical lens, extended beyond the centre of the sphere. With optical immersion a large detector is replaced by a smaller one that has the equivalent apparent optical area.

The detectivity of optical immersion detectors is given by [6].

$$D^* = \frac{\lambda}{hc}\left(\frac{A_0}{A_e}\right)^{1/2}\frac{\eta}{t^{1/2}}[2(G+R)]^{1/2} \tag{3}$$

where: A_0 - the optical area;

A_e - the electrical area;

η - the quantum efficiency;

t - the thickness of detector;

G,R, - the generation and recombination rate.

This equation shows that the detectivity at a given wavelength can be maximised if: - the ratio of optical area to electrical one is high, the ratio of quantum efficiency to square root of thickness is high and the sum of generation rate and recombination one is low.

The immersion increases normalised detectivity by a factor of n in the case of hemispherical immersion and by n^2 in the case of hyperhemispherical immersion [7]. Optical immersion offers the additional advantage of reducing bias power dissipation by factor of n^2 or n^4 in the case of hemi-and hyperhemispherical immersion, respectively.

Application of an optical resonant cavity increases the voltage responsivity and detectivity [7]. The effective absorption of radiation in thin photoconductors can be increased using the interference effect to set up the resonance optical cavity. In the simplest resonant structure interference occurs between the waves reflected at the rear highly reflective surface of semiconductor.

For effective interference the two surfaces have to be sufficiently flat. The optimised resonant - cavity 10,6 um photoconductor exhibits detectivity enhanced by a factor of aproximately 2.5 [7].

3. Heterodyne operation of uncooled HgCdTe photodetectors

The laboratory arrangement for testing the heterodyne operation of HgCdTe uncooled photodetectors is shown in fig.1.

The laser is a CW - CO2 laser with frequency stabilisation circuit, to ensure that the power output is maintained within the top of a single line P20. The laser power and wavelength are continuously monitored. The output beam of laser is split into a "local oscillator" and a "signal" path, the latter one passes through A/O modulator where a carrier of 60 MHz is applied.

The "zero" order beam is stopped in an absorber and the "first" order, modulated beam is slightly deflected. After attenuation, the beams are recombined and aligned coaxially before being focused onto the detector.

The adjustind of this arrangement is very critical for the proper measurements, because any misalignment can reduce the detector efficiency and hence the minimum noise equivalent power.

The power in each beam was measured with a RM - 6600 Radiometer and a very sensitive RK - 570 C probe. For the detectors with 1*1 mm^2 sensitive area, the local oscillator power was kept at maximum 1W/peak, to protect the detector. The power of "signal" beam was controlled by means of calibrated attenuators. The pulse repetition frequency of carrier RF signal could be varied in the range of 1 - 5 kHz. The heterodyne signal was amplified in a preamplifier with 40 dB gain and 5 MHz bandwidth. The signal to noise ratio was analysed with and Hewlett Packard spectrum analyser.

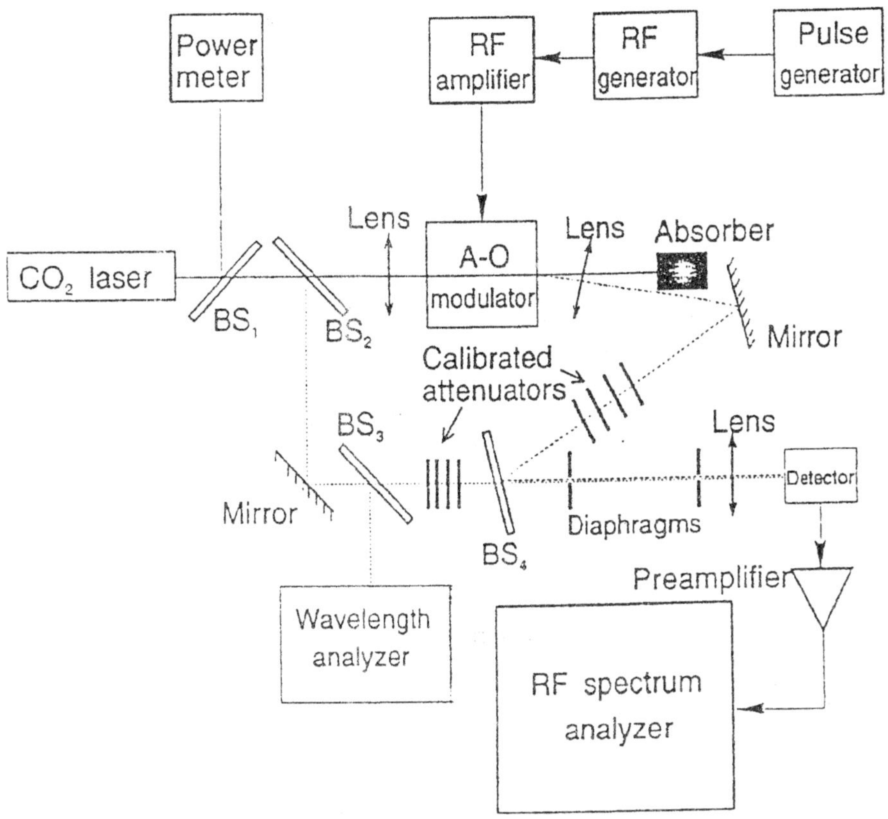

Fig.1 Optical scheme of heterodyne measuring system

First, a comparation between the sensitivities of detectors with and without optical immersion has been done.

Without optical immersion, the voltage sensitivity of P.E.M. detectors is smaller than photoresitors. Ours results show, that immersion lens increases the sensitivity of this detector to the level higher that of typical photoresistor R 005.

After this, the heterodyne operation of detectors was tested. Varying the power of a "signal" beam, the heterodyne signal to noise ratio (SNR) was measured, and then extrapolated to the point in which its value is equal to 1 (NEP). The dependence of SNR on the local oscillator power was also tested to find the optimum operation conditions.

The measured values of NEP for both detectors (the average value of five measurements in each experimental point) are plotted in fig.2. The NEP attained its minimum value for a local oscillator power of approximately 150 mW for the P.E.M., optical immersion detector and of approximately 500 mW for the photoresistive detector.

We can observe that the results are two orders of magnitude lower than the theoretical limit. Among several possible reasons we can indicate the smaller detectivity for near room temperature operation and not very high accuracy in setting of the system. At the other side, the shown level of NEP is high enough for application that don't need a very high sensitivity (LIDAR measurement using topografic targets).

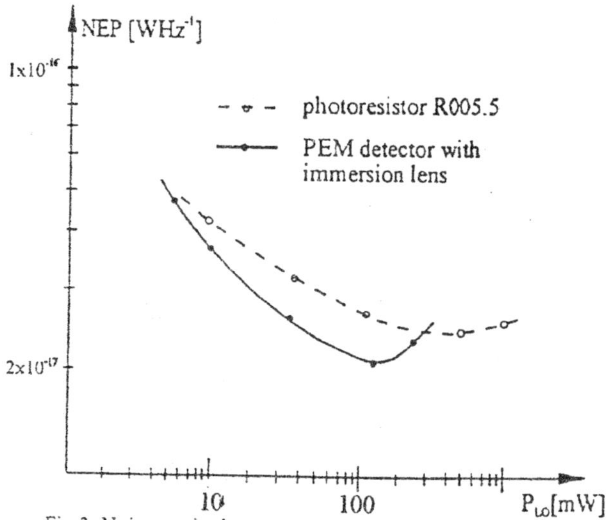

Fig.2 Noise equivalent power as function of L.O.power

4. Conclusion.

An photoresistor has been compared with an optical immersed photoelectromagnetic detector with the same electrical area. Better results were obtained with the last detector.

For the other P.E.M detector with smaller area ($0.1*0.1$ mm^2), the results of heterodyne operation will be published in the near future.

References:

[1]. Henry P.M., Powers J.K., Rowe R.L. and Morris H.B., -" HgCdTe Photodiodes for Heterodyne Application "SPIE - Proceedings, Vol. 663 - Laser Radar Technology and Applications (1986)
[2]. Kingston R. H.,"Detection of Optical and Infrared Radiation", Springer Series in Optical Sciences.
[3]. Foord R., Jones R., Vaughan J.M. and Willetts D.V., Applied Optics, Vol. 22, No. 23, (1983)
[4]. Jozwikowski K. , Novak M. and Piotrowski J. , Infrared Physics, 24, pg. 371, (1984)
[5]. Grudzien M. and Piotrowski J., Infrared Physics, 29, pg. 261, (1989)
[6]. Rogalski A., Opto - Electronics Review. No. 3, pg. 95, Warsaw, Poland (1993)
[7]. Piotrowski J. , Gallus W. and Grudzien M., Infrared Physics, Vol. 31, No. 1, (1991).

Photodetectors on base of AlGaAs/GaAs heterostructures grown by liquid phase epitaxy using rare-earth additions

E.F.Venger, G.N.Semenova, T.G.Kryshtab, A.V.Svitelskiy, P. M. Lytvyn and S.I.Krukovskii*.

Institute of Semiconductor Physics NASU, pr.Nauki 45, Kyiv 252650, UKRAINE.
*Carat Scientific Prod. Ass., Lviv 290031, UKRAINE.

Summary

The $Al_xGa_{1-x}As/GaAs$ heterostructures represent a potentially useful material for photodetectors with high efficiency in spectral range 0.4÷0.9 μm. To produce the thermostable detectors with high radiation hardness on the base of liquid phase epitaxy (LPE) the physical and chemical foundation of such epilayers formation was developed. The abilities of LPE method were improved due to utilization of multicomponent solution-melts with rare-earth additions. The horizontal sliding-boat step-cooled LPE technique with controlled composition of gaseous medium and additions of Yb in Bi and Ga-Bi-Al solution-melts was used as the base of techology. As a result of our investigation the quickly responsive photodiodes with improved technical, exploitative and economical parameters were produced.

1. Introduction

At present it is well known that A3B5 heterostructures and photodetectors with essentially different photosensitivity for different spectral range could be produced by metal organic chemical vapor deposition and molecular beam epitaxy technologies. The liquid phase epitaxy method can also be used as the most useful for production of 0.4÷0.9μm spectral range photodiodes. Low cost, possibility of the controllable epilayers deposition with thickness from ~100 Å to tens μm and wide-range doping are the advantages of this method.

The aim of this work is to test the achievments having been got by several authors in using of LPE technology of the production of quickly responsive PIN photodetectors in required spectral range for especially considering the possible space applications. In thas it was necessary to pay attention to the development of the technology of thermostable and having good radiation hardness epilayers. It was essential for this work to use multicomponent melts (epilayers, grown from Bi melts are more thermostable, due to the absence of Ga inclusions) and the rare-earth additions in the melt for i-layer preparation. It has been found that the presence of rare-earth elements during growth from the melt leads to efficient gettering of background impurities [1-4], change of the stoichiometry, and quality of epilayers. Our preliminary investigations have proved the influence of rare-earth on the increase of the radiation resistance of the diodes [1].

2. Epitaxial layers growth and characterization

On the first stage of our work epitaxial GaAs and AlGaAs layers were grown on semi-insulating (100) GaAs substrates at optimum technology conditions. Then multillayer heterostructures for photodiodes were prepared.GaAs and AlGaAs epilayers with thickness about 0.1÷12 μm were grown from Ga, Bi, Ga-Bi-Al melts by LPE . Epilayers were nominally undoped. Various amounts of Yb (N^L_{Yb} up to 1×10^{-2} at.%) we added to the melts.The AlGaAs films were grown with 5 °C supercooling. Such ΔT prevented the GaAs from homogeneous

0-7803-3223-7/96/$5,00

crystallization in the melt (AlAs is more refractory than GaAs). Composition of AlGaAswas regulated by the Al content in the melt and by the value of ΔT.

The grown films were characterized by various methods. Carrier concentration, resistivity and mobility were obtained using Hall effect measurements at 300 and 77 K. Near band edge PL spectra excited by Ar^+ or He-Ne laser were recorded at 4.2, 77 and 300 K. We used high resolution X-ray diffraction measurements on the double-crystal diffractometer with X-ray source Cu $K_{\alpha 1}$ such as (004) reflection rocking curves, measurements of the integrated intensity (R) of X-ray diffraction for Bragg geometry and measurements of R for X-ray (002) quasiforbidden reflections (XFR) to determine the stoichiometry.

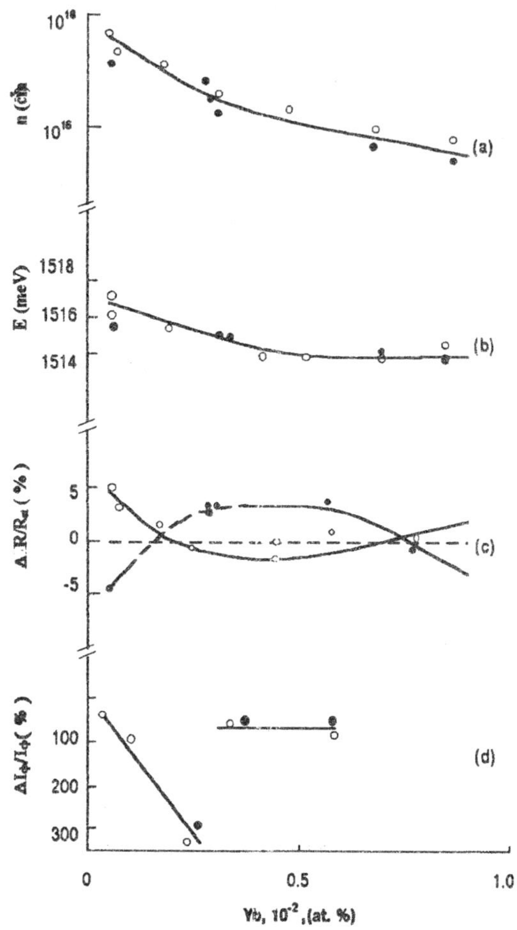

Fig.1. The 300 K electron concentration of GaAs layer (a), 4.2 K bound exciton energy (b), ratio $\Delta R/R_{st}$ for (002) reflection (c), ratio $\Delta I_\Phi / I_\Phi = (I_0 - I_\Phi)/ I_\Phi \times 100\%$ as a function of Yb concentration in Ga melt (d): • - for Ga-rich films, o- for As-rich films.

3. Experimental results and discussion:materials and devices

3.1. Materials

Experimental results shown that:

a) The electron concentration (see Fig.1,a) in the layers decreased from about 5×10^{17} cm^{-3} ($N^L_{Yb} = 0$) (this value was determined by background contamination and was so high in our technology conditions) to $< 1\times10^{15}$ cm^{-3} for $N^L_{Yb} \sim 10^{-2}$ at.% without a conversion to p-type. Compensation coefficient i) $K = N^+_D / N^-_A$ (N^+_D, N^-_A - the density of ionized donors and acceptors respectively)and separate values ii) N_D and N_A were calculated from measured mobility and n at 77 K. We obtained the increase of K with the increase of N^L_{Yb} and concluded to i) and ii) that the gettering properties of the Yb in the melt were restricted by n-type impurities at any growth conditions.

b) As shown in Fig.1,b, the position of bound exciton line I_1 shifted to the lower energy (from 1.518 to 1.514 eV) and, the full width at half maximum (FWHM) of its spectra decreased to 2.5 meV with increasing of the Yb addition in the melt for As-rich films. It may be explained by incorporation of Yb into Ga-sublattice as single atoms. The value of the shift corresponded to $\varepsilon \sim 10^{-4}$ which could be attributed to different atomic radii of Ga and Yb ($r_{Ga} << r_{Yb}$). For Ga-rich films the shift of the bound exciton line ($h\nu$ =1.5144 eV, 4.2 K) was essential.

c) GaAs films grown from Ga-Bi solution were As-rich. With the increasing of the value of Yb in the melt the $\Delta R / R_{st}$ (R_{st} corresponds to quasi-forbidden reflection intensity for the liquid horizontal Brigman method) decreased, passed through the minimum at optimal Yb concentration (N^L_{Yb} = 0.0048 at.%) leading to the best electrical properties ($n \leq 1\times10^{15}$ cm^{-2}, μ ~ 6000 cm^2 /Vs, 300 K).

The stoichiometry decreased (increased) with increasing amount of Yb for As-rich (Ga-rich) GaAs films, passed through the minimum (maximum) for the optimal Yb concentration, correspondingly to epilayers stoichiometry. At high concentrations of Yb (up to 0.01 at.%) insignificant deviation from stoichiometry was different for As- and Ga-rich films.The above deviation from stoichiometry for As-rich films may be explained by incorporation of Yb into Ga sublattice as single atoms as well as its influence on the thermodynamic parameters of the melt as "catalyst" [4]. For Ga-rich films these deviations can be ascribed to Yb atoms incorporated into the films just as impurity on Ga sites which are occupied by host atoms.

Nature of residual acceptors (from PL spectra) for films, which were grown from Bi melt were not similar, but concepts developed for mechanism of formation of single crystalline films grown by LPE using REEs may still be used interchangeably.

Original AlGaAs epilayers were grown on GaAs substrates with amount of Al (~ 0.023 at.%) to the Ga melt. Composition of alloy corresponded to x = 0.032. Very high purity $Al_{0.03}Ga_{0.97}As$ epilayers were grown by adding of the optimum amount of the Yb (~ 0.0048 at.%) and have the hole concentration $p \sim 10^{14}$ cm^{-3} and μ = 450 cm^2 /V s, 300 K.

Fig. 1. shows the ratios of the integrated intensities of the near band edge photoluminescence spectra at 77 K of as-grown samples (I_0) and γ-irradiated (I_Φ). The ratios of the intensities I_0 /I_Φ , where I_Φ corresponds to a flux of about 8×10^{17} cm^{-2}, are plotted as a function of the amount of Yb in the Ga or Bi melts. From Fig.1 one can see that the effect of the γ-ray irradiation is the decrease in the PL intensities. Two features are noted. First, for $N^L_{Yb} \leq 0.34\times10^{-2}$ at % the decrease in the I_0 /I_Φ is greater than that in the undoped samples. Second, for N^L_{Yb} >0.34×10^{-2} at.% the ratio I_0 /I_Φ does not depend on the rare-earth concentration and is about 30. As shown in Fig.1, the films purity and the crystal quality become high with the increase of the amount of Yb in the melt (0.34×10^{-2} at.% <N^L_{Yb}), whereas the radiation resistance does not decrease. The observed effect is independent from the growth conditions (either As-rich or Ga-rich epilayers) but was controlled by the rare-earth concentration and generally could be attributed to the creation of non-radiative centers by γ-ray irradiation. However, the irradiation does not affect equally the PL spectra of the as-grown samples and the samples with rare-earth additions.

To explain the obtained results we have examined the interaction of radiation-induced native defects with rare-earth atoms (or coagulates). Some of the simple native defects - vacancies and interstitials created by the room temperature irradiation are expected to be mobile and thus may interact (annihilate) with rare-earth complexes or in internal fields around them. A similar situation might also be applicable to AlGaAs epilayers. This rare-earth-impurity defect interaction could play an important role in the increasing of radiation resistance in AlGaAs epilayers.

3.2. AlGaAs/GaAs heterostructure photodiodes

The epitaxial layers for PIN photodiodes were grown by LPE on (100) oriented n^+ GaAs:Sn ($n = 2\times10^{17} \div 2\times10^{18}$ cm^{-3}) substrates. The cross-section of the integrated structure is shown in the inset to Fig.2. The structure consists of the following parts: a about 7 μm thick n-$Al_xGa_{1-x}As$ ($x = 0.2\div0.35$, $n \leq 1\times10^{18}$ cm^{-3}) base layer; a 1÷3 μm thick i(ν)-$Al_xGa_{1-x}As$ (x =

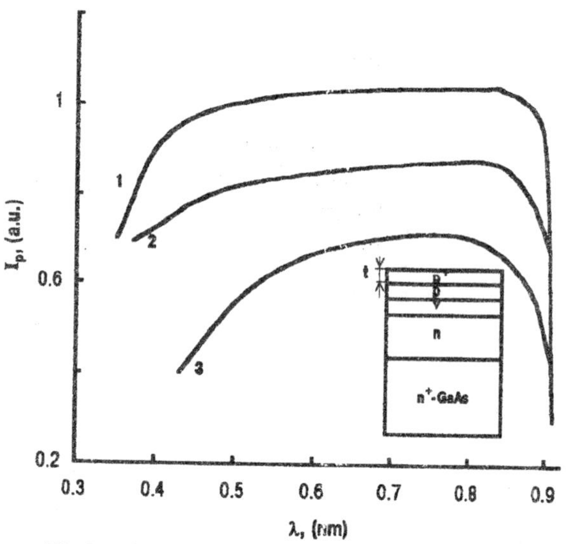

Fig.2. Photoresponse spectra of AlGaAs photodiodes with different wide-gap window thickness: 1- 300 Å, 2- 700 Å and 3- 1 μm. Basic photodiode structure is shown in the insert.

$0.05 \div 0.1$, $n \leq 1\times10^{15}$ cm^{-3}); a $0.25 \div 1.5$ μm thick p-$Al_xGa_{1-x}As$ ($x = 0.3 \div 0.8$, $p \approx 5\times10^{17} \div 1\times10^{18}$ cm^{-3}) window layer and a 0.5 μm-thick p -$Al_xGa_{1-x}As$ ($p \geq 10^{18}$ cm^{-3}) cap layer.

Both the optimal growth conditions, predetermined by composition of the Ga-Bi-Al solution-melt, and the rare-earth concentration in the melt ($N^L_{Yb} \approx 0.05$ at.%) for obtaining i-layer (ν-layer, actually) were preliminary found. The room-temperature spectra of the photoresponce in the 0.4÷0.9 μm region of the samples grown by LPE method with rare-earth addition are shown on Fig.2. The difference between 1, 2 and 3 spectral curves is due to the difference in the wideband window layer thickness *t*. The intensity of the photoresponce is correlated with the value of thickness *t*.

The following characteristics were achieved for low-cost photodetectors grown by the high productive LPE technology with rare-earth addition in the Ga-Bi-Al melts:

- spectral sensitivity range $\Delta\lambda = 0.4\text{-}0.9$ μm;
- responsivity on the peak of spectral responce wavelength $S_{0.63} = 0.4$ A/W;
- rise and decay time of the transfer characteristics $\tau = 5\times10^{-10}$ s;
- operation voltage $u = 5 \div 10$ V;
- dark current $I_d = 1\times10^{-6}$ A;
- effective area of the responcive element $A_{eff} = 1 \div 50$ mm^2;
- capacitance $C = 5 \div 10$ pF.

The thermostability and good radiation hardness of these detectors were higher than those of photodiodes grown from Ga-melts without rare-earth additions.

5. Conclusion

AlGaAs/GaAs based heterostructures for photodiodes application were grown by the specially developed LPE method. The optimal growth conditions, solution melt composition and rare-earth concentration in the melt for obtaining thermostable heteroepilayers with high radiation hardness were found.

6. References

1.G.N.Semenova, T.G.Kryshtab, V.P.Klad'ko, S.I.Krukovskii, A.V.Svitelskiy. *Proceed. of V Intern.Conf. in Physics and Technology of Thin Films*, vol 2, p.28, Ivano-Frankovsk, Ukraine, 1995 [in Russian].

2.M.-C. Wu and C.-M. Chiu. *J.Appl.Phys.*, Vol.73(1), pp.468-470, January 1993.

3. B.Podor, L.Csontos, K.Somogyi and D.Vignaud. *Acta Physica Polonica A*, Vol. 87(2), pp.465-468, February 1995.

4.G.N.Semenova, T.G.Kryshtab, V.P.Klad'ko, S.I.Krukovskii and A.V.Svitelskiy. *Ukrainian Phys.J.*, Vol.40(10), pp.1101-1106, October 1995 [in Ukrainian].

INDOOR PV DEVICES TESTING

Carmen Ilie, Nicolae Olariu, Călin Oros
State University "Valachia" Târgoviște, Energy - Environement Research Department
236 Calea Domnească, 0200 Târgoviște, Romania ; Tel +40-45-611904 ;
Fax +40-45-631197 ; E-Mail nolariu@nira.univt-net.ro

SUMMARY

The paper presents a solar simulator for photovoltaic modules characterisation. We chose for making a "steady state" simulator with halogen lamps verifying if it is adequate for PV modules testing. For the spectral matching of solar spectrum and simulator's spectrum we have developed a method of adequate correction there were considered both lamp spectral distribution and PV module spectral response.The evaluation to simulator's performation consisted on uniformity's study of irradiance at test surface plane and temporal stability of irradiance.

I. INTRODUCTION

There are known two types of solar simulators for PV performance evaluation: the "steady state" simulator consisting halogen lamps for PV cells and small modules and second one the pulsed type. This type consist of one or more long-arc xenon flash lamps and is utilised for large modules and arrays that need an uniform irradiation on large areas. The advantage of the pulsed type consists in the negligibile heat input to the test cells, that allows them to remain uniformly at the ambient temperature and the measuring to be easily and accurately done. Unfortunately the pulsed simulators an very expensive regarding "steady state" simulators.

II. EXPERIMENTAL MODEL. CHARACTERIZATION AND EXPERIMENTAL RESULTS

The solar simulator must corespond to the international standards requirings elaboreted by Joint Research Centre - Ispra - Italy. These standards divide solar simulators into three classes of precision (A, B, C) according to their spectral matching, uniformity of irradiance at test surface plane and temporal stability of irradiance. The specified tolerances are listed in Table 1:

Parameter	Class A	Class B	Class C
Spectral matching	0.75 - 1.25	0.6 - 1.4	0.4 - 2.0
Non-uniformity of light	< 2%	< 5%	< 10%
Light instability	< 2%	< 5%	< 10%

Table 1. Simulator requirements

0-7803-3223-7/96/$5,00

The main advantage of measurements with solar simulators is the high reproductibility which can be achieved and which is required in particular in the context of module type testing in accordance with Standard Test Conditions (S.T.C.) :

- Irradiance in module surface plane : 1000 W/ m^2
- Solar spectral irradiance distribution : AM 1.5 global
- Cell temperature : 25^0 C

The accomplished solar simulator uses 100 W "Sylvania" halogen lamps, the lightened surface is (50 cm # 150 cm).
The technical characteristics of the halogen lamps used are presented in Table 2:

General information	
Suplly Voltage	220 V
Lamp Rating	100 W
Mechanical Data	
Maximal Overal Length	136 mm
Diameter	122 mm
Reflector	aluminium
Base type	E27
Lens	clear
Coil	axial
Average life	2500 hrs
Ilumination Characteristics	
Peak Luminous Intensity	3500 cd
Colour Temperature	2900 K
Half Peak Angle (nominal)	30^0

Artificial light is characterized mainly by irradiance distribution and by temporal stability. In figure 1 it is compared solar reference spectrum with halogen lamp spectrum:

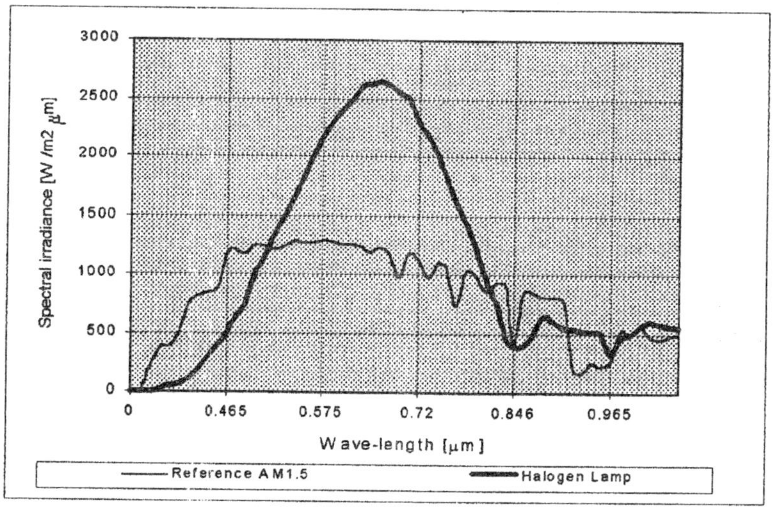

Figure 1: Reference AM 1.5 spectrum in comparison with halogen lamp spectrum.

In order to ensure a proper evaluation of the spectral matching there were considered both lamp spectral distribution and PV module spectral response.

It was made a mobile lamps construction, being possible to vary the irradiance at test surface plane(figure 2):

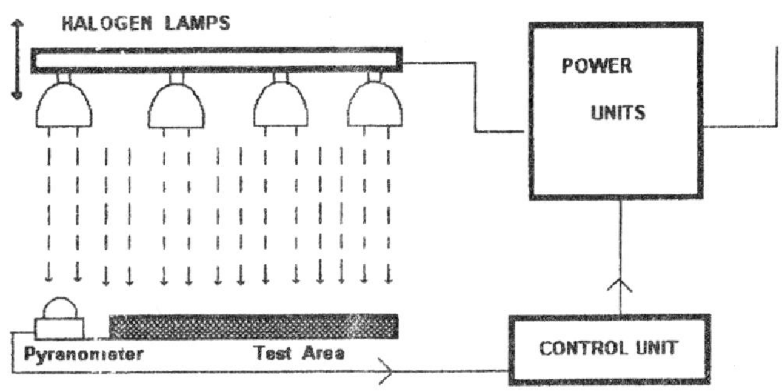

Figure 2: Block circuit diagram of the simulator unit.

The radiation sensor at the test surface plane serves the active light regulation of the simulator.It is coupled with the control unit and ensures the temporal stability of the irradiance. It also serves the compensation of the ageing effects of the lamps.

The lightening of a certain surface comes from the solar simulator by over laping the light of several lamps. For the uniform lightening of a (50 cm # 150 cm) surface there were used eight halogen lamps.Their characteristics are in Table 3.

The results of the measurements regarding the irradiance at module surface plane uniformity are ploted in figure 3.

III. CONCLUSIONS

The designed halogen lamps solar simulator is from the precision class B.In order to improve it, we aim to analyse different light dispersion optical systems for better uniformity and to analyse spectral adjustment (lamp emission - element absorbtion) for elaborating adequate corrections.

For the temperature stability during the I-V characteristic measurement we intend to create a thermostatal structure so that the temperature at module surface plane should not be influenced by infrared radiations absorbtion.

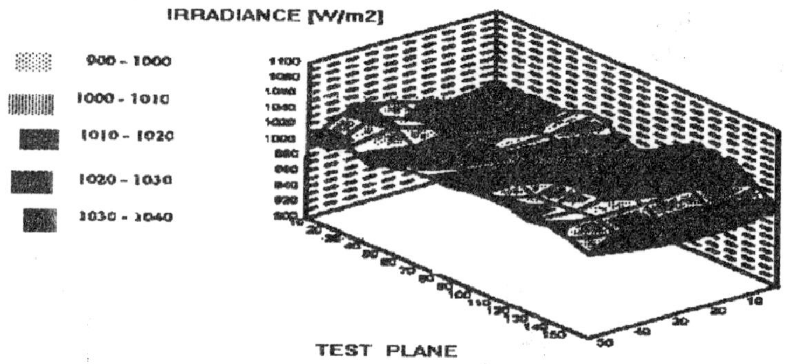

Figure 3: Uniformity of irradiance

REFERENCES

1. J.C. Larne - Proc. 2nd E.C.Photovoltaic Solar Energy Conf. - pg. 477(1979)
2. K. Bogus, J.C. Larne, R.L. Crabb - Proc. 1st E.C. Photovoltaic Solar Energy Conf. - Luxembourg 1977, pg.745
3. ERDA/NASA 1022/ 77 /16 - Revised Terrestrial Photovoltaic Measurement Procedures
4. "Standard procedures for terrestrial photovoltaic performance measurements" - K. Krebs - Joint Research Centre Ispra Establishment - Italy, No. 101/2/ 1981
5. IEC / TC 82/ WG2,82 (Central Office) 16 - "Solar Simulator Performance Requirements" - 1986
6. F. Nagamine, R. Shimokawa - "Calibration of Pyramometers for the Photovoltaic Device Field" - Japanese J. of Appl. Phys. - 29 (1990) - pg. 516-521.

PHOTOCONDUCTIVITY, LUMINESCENCE AND OPTICAL ABSORPTION OF WS_2 CRYSTALS.

E. Arushanov[a,b], E.Bucher[b], K.Friemelt[b], O.Kulikova[a], L.Kulyuk[a], A.Nateprov[a], A.Siminel[a]

[a] Institute of Applied Physics, Academy of Sciences of Moldova
[b] Faculty of Physics, University of Konstanz, P.O.Box 5560, D-78434

SUMMARY

For the first time the photoluminescence of the tungsten disulphide 2H-polytype single crystals have been observed and the photoconductivity spectra in the region of indirect transition have been investigated. The analysis of the experimental data has been performed taking into account the band structure of 2H - WS_2 layered crystals.

INTRODUCTION

The layered transition metal dichalcogenides are rather attractive systems because they present many unusual physical and chemical properties which are not commonly encountered in three-dimensional materials.

Tungsten disulphide is a semiconductor which exhibits some favourable properties as a solar cell base material [1].

SAMPLES AND EXPERIMENT

The tungsten disulphide 2H-polytype single crystals have been grown by the gas transport method using the iodine as a transport agent.

Photoconductivity (PC) measurements were performed using the standard computer assisted lock-in method in the wave length range of 400-1200 nm in the temperature interval from 80 to 300 K. A pyroelectric detector was used as a sensitive element for measuring of the absorption coefficient.

The steady-state photoluminescence (PL) spectra have been measured in the temperature range 2-120 K. The excitation was provided by Ar^+-(λ_{ex}=514 nm) and He-Ne (λ_{ex}=633 nm) lasers.

RESULTS AND DISCUSSION

The investigation of optical absorption spectra has shown availability of a direct forbidden band (Eg=1.74eV, T=300 K) and indirect one (Eg^{ind}=1.35eV), that in a good agreement with Eg values, determined from photochemical potential spectra [2].

0-7803-3223-7/96/$5,00 © 1996 IEEE

The photoconductivity spectra of 2H-WS_2 are shown in fig.1. The high photosensitivity has been observed in a ratherwide spectral range.

The shape of PC spectra finds out three brightly expressed sites in the energy region: $h\nu < 1.4$ eV (I), 1.4 eV $< h\nu <$ 1.8 eV (II) and $h\nu>1.8$ eV (III), each of which has a complex structure.

A feature which is evident of the spectra (area III) in the whole temperature interval is several dips in the photocurrent which correspond to the positions of the A and B exciton absorption maxima [3].

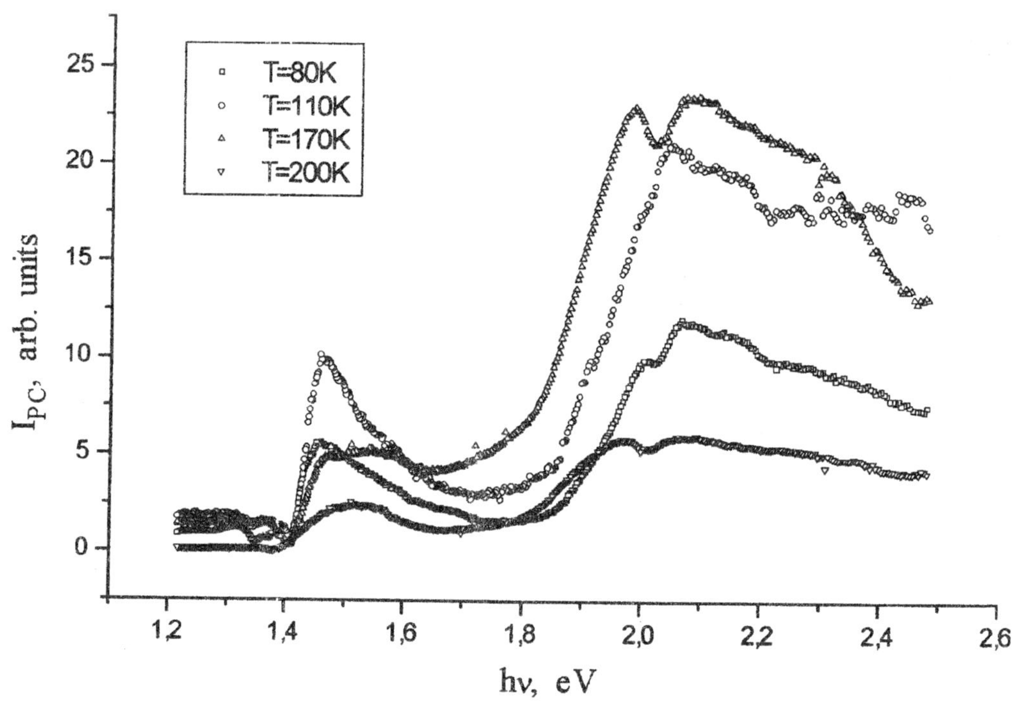

Fig.1 Spectrum of photoconductivity at different temperatures

The position of edge short-wave peaks II and III corresponds to characteristic sites of the optical absorption spectrum, connected with indirect (II) and direct (III) interband transitions.

Taking into account that the PC signal is proportional to the absorption coefficient in the region of the fundamental absorption edge caused by direct optical transitions [4], we have also evaluated the width of the direct gap of tungsten disulphide from PC: Eg = 1.86eV (80 K) and 1.76eV (300K), that is in a good agreement with the data obtained from absorption measurements.

The temperature dependence of a spectrum PC in all intervals has a complex character (fig. 2). Peaks of the area III $h\nu$=1.98 eV and $h\nu$=2.06eV, with the monotonous increase in the whole temperature interval, are connected with direct interband transitions.

The peak hν=1.46 eV, the appropriate area of indirect interband transitions, has a minimum of intensity at T~200 K and sharp maximum at T=250 K.

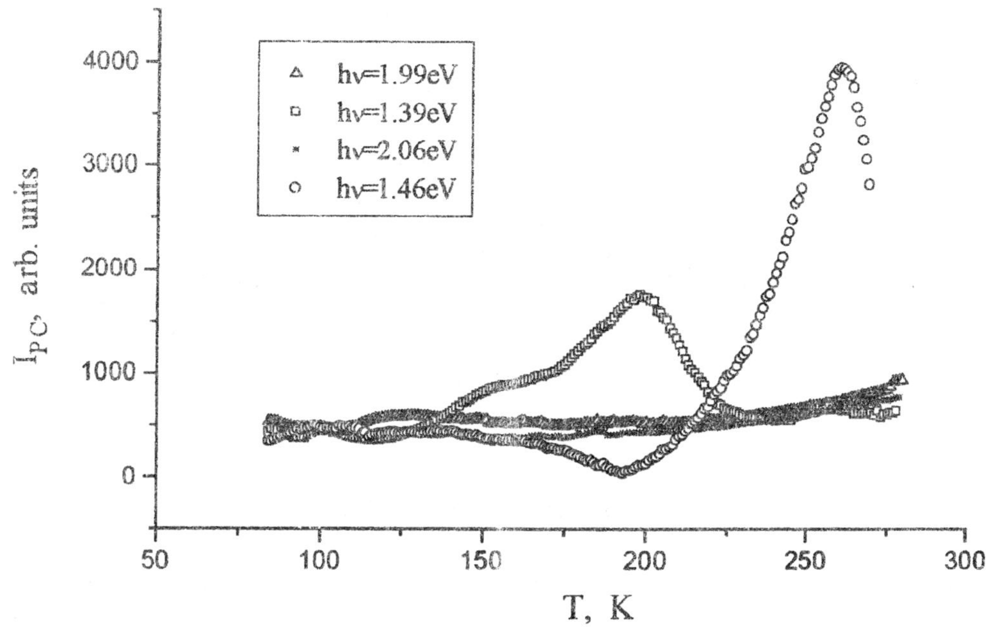

Fig.2 Temperature dependences of photoconductivity at different exitation

Spectrum PL has been observed in the range of energy of I region (spectrum PC), i.e. at $h\nu < E_g^{ind}$ (Fig. 3).

At low temperature PL spectra of 2H-WS_2 crystals are formed by the superposition of up to ten narrow (halfwidth~5 meV) components, some of them being equidistant. The position of the spectral maximum $h\nu_{PL}^{max}\cong 1.33$ eV does not practically depend on temperature. The distance between the equidistant components is about 13 meV, and is likely to correspond to the optical phonon energy.

The temperature dependence of the PL intensity exhibits the maximum at T≅50 K. Further T increase leads to a rapid PL quenching, and at 120 K its intensity is over 10^3 times weaker than at 50 K.

In PL spectra the availability of an asymmetric band with a sharp long-wave edge located in the energy region of the indirect gap $h\nu_{PL}^{max}\cong$ 1.35eV is worth noting. A model of localized levels allowing to explain the observed PL temperature activation as well as peculiarities of PC

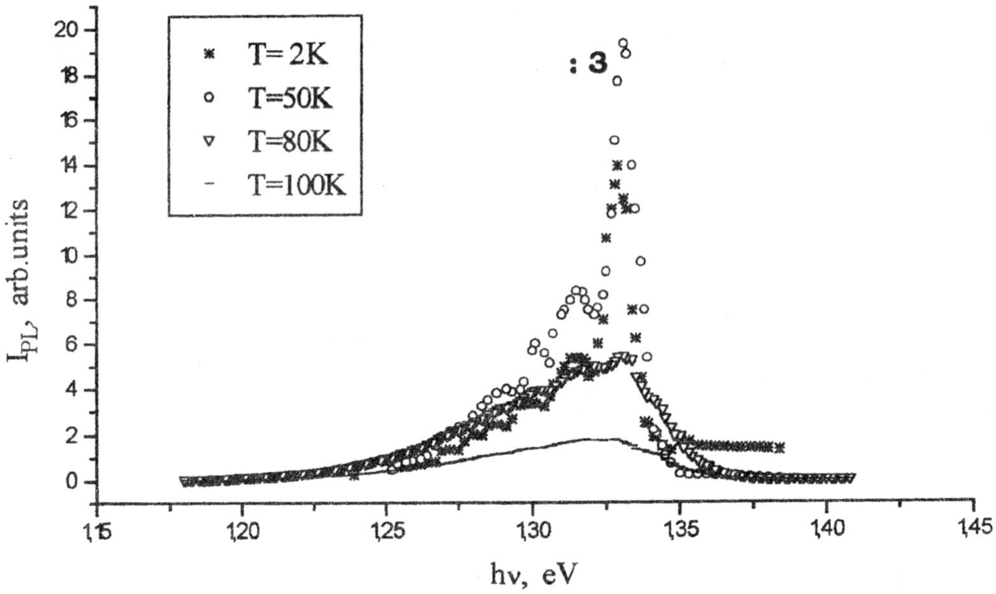

Fig.3 Spectrum of photoluminescence
for different temperature

temperature dependence at hν<Eg^{ind} is proposed. The asymmetrical shape the 1.45 eV PC spectral band is described in the assumption of direct photo-transitions under the conditions when the extreme of the conduction and valence bands are shifted into the K-space [5].

REFERENCES

1. A.Aruchamy, Photoelectrochemistry and Photovoltaics of Layered Semiconductors, in: Phys. and Chem. of Mat. with Low-Dimensional Structures, Kluver Acad. Pub.(Dordrecht, 1992)
2. K.K.Kamt, B.A.Parkinson, J.Phys.Chem., 86, 4631982)
3. J.A.Wilson,A.D.Yoffe, Adv.Phys.,18, 193 (1969)
4. W.Horing, H.Neumann, B.Schumann, G.Kuhn, Phys. St.Sol.(b), N1, K57, (1978)
5. C.Sourisseou, F.Cruege, M.Fouassier, M.Alba, Chem. Phys. 150, 281 (1991)

Temperature Study of the Carrier Lifetime in Solar Cells with one dominant Trap Energy Level.

Dragos Dobrescu and Adrian Rusu

University "Politehnica" Bucharest, Romania

Summary

Based on experimental observation, a new mathematical model is made for explaining the carrier lifetime versus temperature dependence in silicon solar cells manufactured on p-type material with one dominant acceptor trap energy level, situated in the lower part of the forbidden band. This model explains the behaviour of the lifetime tacking into account the SRH model for the lifetime and the physical parameters temperature dependencies.

1. Introduction

Several characterizations made for solar cells for spatial applications [1], [2] have proven that, in the temperature range 80K - 320K, the carrier lifetime values increases as the temperature is growing. The importance of the lifetime versus temperature dependence study lies in the fact that some important physical parameters can be extracted from these curves. These parameters are the energy level, the capture section and the activation energy of the trap. In this paper, after the measurements were performed on several samples of silicon solar cells, the observed dependencies of the lifetime versus the temperature are explained and a model for a solar cell having only one dominant bulk acceptor trap with a shallow energy level is disscused.

2. Experimental results

Characterizations were performed on silicon solar cells samples having different active layer doping concentration (N_A is 10^{16}, 1.2×10^{15} and 8×10^{14} cm^{-3} respectively for the three type of samples denominated A, B and C in this paper), with different electrical methods [3], [4], in order to determine from the carrier lifetime versus the temperature dependence the physical parameters for the traps in the bulk material. In Fig. 1 are represented typical curves obtained using the diffusion capacitance method in the temperature range between 80 and 320K, for all the three samples above mentioned.

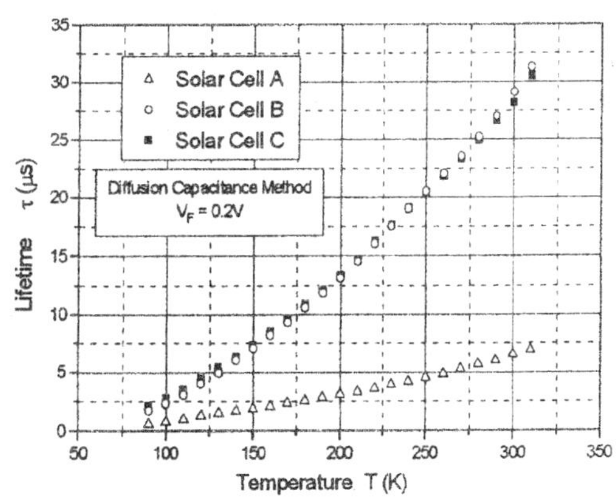

Fig. 1
Typical dependence of the lifetime versus temperature for the studied solar cells.

0-7803-3223-7/96/$5,00

It can be seen that the lifetime increases when the temperature grows, having not a linear dependence. Also in these samples, a dominant energy trap level was found using EDMR and DLTS techniques [4]. The trap is Mn and has an energy 0.23eV above the energy of the valence band. For such an energy level, which is far from the middle of the forbidden gap, using physical explanations, a new mathematical model is described as follows.

3. Mathematical model

This analyse is made for p-type silicon, with one dominant trap shallow energy level, considering a SRH recombination mechanism. If a small traps concentration is considered, it results that changing electron and holes population at the level of the traps doesn't effect the electron and holes excess concentrations. Using these hypothesis, the following expression for the SRH lifetime can be written [5]:

$$\tau_{SRH} = \frac{\tau_p(n_0 + n_T + \Delta n)}{n_0 + p_0 + \Delta n} + \frac{\tau_n(p_0 + p_T + \Delta n)}{n_0 + p_0 + \Delta n} \qquad (1)$$

In the small level injection conditions, the equilibrium concentrations for holes p_0 and the p-type trap density p_T are much greater than the excess charge carrier concentration Δn:

$$p_0, p_T >> \Delta n \qquad (2)$$

Using (2) the expression in (1) can be simplified:

$$\tau_{SRH} = \frac{\tau_p(n_0 + n_T)}{n_0 + p_0} + \frac{\tau_n(p_0 + p_T)}{n_0 + p_0} \qquad (3)$$

For the p-type silicon the hole concentration is much greater than the electron one:

$$p_0 >> n_0 \qquad (4)$$

Modifying the relationship (3) with the observation in (4) results:

$$\tau_{SRH} = \frac{\tau_p(n_0 + n_T)}{p_0} + \frac{\tau_n(p_0 + p_T)}{p_0} \approx \tau_p \frac{n_T}{p_0} + \tau_n \frac{p_0 + p_T}{p_0} \qquad (5)$$

The variables in equation (5) have the following expressions:

$$\tau_n = \frac{1}{\sigma_n v_{th} N_T}, \quad \tau_p = \frac{1}{\sigma_p v_{th} N_T},$$

$$n_T = n_i \exp(\frac{E_T - E_i}{kT}) = N_C \exp(\frac{E_T - E_C}{kT}), \qquad (6)$$

$$p_T = n_i \exp(-\frac{E_T - E_i}{kT}) = N_V \exp(\frac{E_V - E_T}{kT})$$

where there been used the following symbols:

N_T - the trap density, σ_n - the electrons cross section, σ_p - the holes cross section,

v_{th} - the thermal velocity, n_T - the n-type trap density, p_T - the p-type trap density,

n_i - the intrinsic concentration, N_C - the conduction band concentration,

N_V - the valence band concentration, E_T - the energy level of the traps,

E_C - the energy level of the conduction band, E_V - the energy level of the valence band,

E_i - the intrinsic energy level.

Using the temperature dependencies presented in (6) the expression in (5) can be studied for the case with the Fermi energy level E_F greater than the shallow trap energy level E_T as it was stated in the experimental results chapter. This case appears when the temperature of the semiconductor increases. Supposing that all the impurity atoms are ionised for temperatures above the ionisation temperature, in conditions of thermal cuasi-equilibrium the concentrations p_0 and n_0 are constant. Due to the rising temperature, the p_T concentration is getting greater than p_0, but in the same time the concentration n_T remains much lower than p_T. Using these observations the expression in (5) can be simplified to:

$$\tau_{SRH} \approx \tau_n \frac{p_T}{p_0} \tag{7}$$

For temperatures greater than the ionisation temperature for acceptor impurity $p_0 = N_A$, and if the relationship (6) is introduced in (7) the following relationship is obtained:

$$\tau_{SRH} \approx \tau_n \frac{N_V}{N_A} \exp\left(\frac{E_V - E_T}{kT}\right) \tag{8}$$

Taking into account that N_V increases proportional with $T^{3/2}$ and neglecting the weak dependence of the thermal velocity v_{th} versus the temperature, the following relationship results:

$$\tau_{SRH} \sim T^{3/2} \exp\left(-\frac{E_T - E_V}{kT}\right) \tag{9}$$

In (9) it can be seen that the charge carrier lifetime increases very strongly with the temperature. This kind of variation is strictly in concordance with the measurements performed on several silicon solar cells as it will proven in this paper.

If this analysis is continued it can be seen that for higher temperatures the semiconductor enters in the region with an intrinsic behaviour, in this case the following equation can be written:

$$n_0 = p_0 = \sqrt{N_C N_V} \exp\left(-\frac{E_C - E_V}{kT}\right) \tag{10}$$

If the values for p_0 and n_0 from the previous relationship are introduced in (3) the following dependence for the carrier lifetime versus the temperature is obtained:

$$\tau_{SRH} = \frac{\tau_p}{2}\left(1 + \sqrt{\frac{N_C}{N_V}} \exp\left(\frac{E_T - \frac{1}{2}(E_C + E_V)}{kT}\right)\right) + \frac{\tau_n}{2}\left(1 + \sqrt{\frac{N_V}{N_C}} \exp\left(\frac{E_T - \frac{1}{2}(E_C + E_V)}{kT}\right)\right) \tag{11}$$

From the relationship (11) it may be seen that the lifetime values are decreasing when the temperature increases. If the temperature increases at even higher level the exponential terms in equation (11) can be approximated with 1. In conclusion at very high temperatures the lifetime becomes constant again like at very small temperatures, but heaving a different value as it can be seen from the next relationship:

$$\tau_{SRH} = \frac{\tau_p}{2}(1+\sqrt{\frac{N_C}{N_V}}) + \frac{\tau_n}{2}(1+\sqrt{\frac{N_V}{N_C}}) \approx \tau_p + \tau_n \quad (12)$$

In Fig. 2 are described all the temperature regions presented above. The temperatures T_1, T_2 and T_3 from this figure are the temperatures at which the Fermi energy level reaches respectively: the dopant acceptor energy level, the trap energy level and the intrinsic energy level. Finally a good agreement may be seen if we compare the model region between T_2 and T_3 with the measurements presented in Fig. 1.

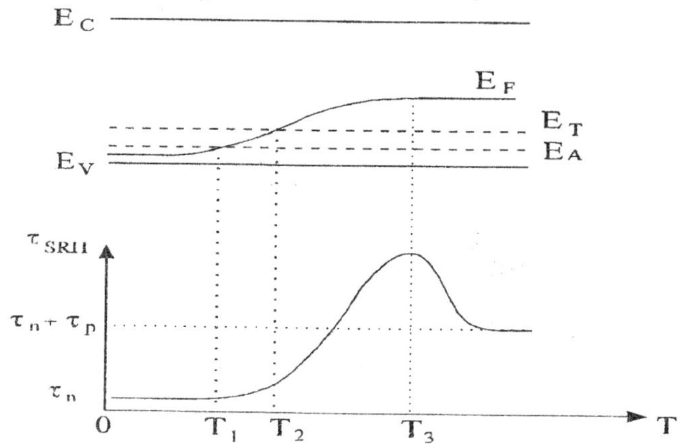

Fig. 2
The lifetime versus temperature dependence in p-type silicon with one acceptor energy level E_A and single trap shallow energy level E_T.

4. Conclusions

Based on experimental results obtained for silicon solar cells and considering the physical aspects of the temperature dependence of the carrier lifetime in p-type silicon with one shallow dominant trap energy level, a new model was developed. A good agreement between the model and the experimental measurements can be seen.

References

[1] D. Dobrescu, "Contribution to the Study and Characterization of Silicon Solar Cells with Spatial Applications", PhD Thesis, Bucharest, p. 110-115, 1996.

[2] D. Dobrescu, A. Rusu, "Minority Carrier Lifetime Measurements in Silicon Solar Cells", CAS'94, Sinaia, Romania, Proceedings, Vol. 2, p. 681-684, 1994.

[3] D. Dobrescu, A. Rusu, D. Stievenard, "Minority Carrier Lifetime Electrical Measure Methods used for Silicon Solar Cells Characterization", Rev. Roum. Sci. Tech.-Electrotech. & Energ., Vol. 40, 1, 1995, Ed. Academiei, p. 75-84, 1995.

[4] D. Dobrescu, A. Rusu, D. Stievenard, "A Complex Electrical Characterization of the Solar Cells for Spatial Application", Proceedings 20th International Conference on Microelectronics, MIEL'95, Vol. 1, Nis, Serbia, p. 281-284, 1995.

[5] M. Draganescu, "Solid State Electronics", Ed. Tehnica, Bucharest, p. 429-434, 1972.

Session 20

PROCESSES & TECHNOLOGIES (Poster)

Computer Modeling of Selective Area Epitaxy with Organometallics

A. Mircea, D. Jahan, A. Ougazzaden, D. Delprat, L. Silvestre, G. Zimmermann,
France Telecom, Centre National d'Etudes des Télécommunications, Laboratoire de Bagneux, France
and
A. Manolescu, A.M. Manolescu, "Politehnica" University of Bucarest, Romania

1. Introduction

Selective Area Epitaxy with Organometallics (SAEOM) enjoys presently a wide interest. This is due to the capabilities that this technique offers in the area of optoelectronic device integration. Typically, SAEOM is used in conjunction with very thin ("quantum") layer structures to easily achieve controlled variations of the optical absorption edge along an integrated optical wave guide, leading to such integrated devices as, e.g., enhanced optical coupling laser [1], laser plus electro-absorption modulator [2], etc. This process is rather typical for Organometallic Vapor Phase Epitaxy (OMVPE) and cannot be duplicated with other competitive techniques such as GSMBE (Gas Source Molecular Beam Epitaxy) or CBE (Chemical Beam Epitaxy). Computer modeling the epitaxial deposition process can be used in this context, both for pre-designing the process and for post checking its success. For such a tool to be more than a qualitative guide, it must be tailored carefully to suit the experimental conditions. It is the purpose of the present paper to show that indeed, and in spite of the process complexity, the predictions of the computer model can be in quantitative agreement with experiment. This work builds upon several previous publications, particularly [3-4,6-8]. Its field of application is basically in the technology of InGaAsP optoelectronic devices.

2. Mask Layout.

The SAEOM process makes use of dielectric masks (silicon nitride or oxide) deposited on the epitaxial substrate. The masks are typically pairs of rectangles, aligned to a main crystallographic orientation. This elementary pattern is usually repeated over an area of the substrate (fig.1). Sometimes the rectangles have flared ends in an effort to provide a smooth transition between the masked and unmasked regions [1]. After growth upon such a masked substrate, and provided that the operating conditions are such that no growth occurs over the masked areas, the growth rate is enhanced in the small openings (and more generally in the vicinity of the mask edges) as compared to the value it would normally have with an unmasked substrate. Since, in a well/barrier quantum structure, the optical absorption edge, determined by the direct transition between the first occupied electron quantum level and the first occupied hole quantum level in the wells, depends not only on the composition of the wells and barriers but also on their thickness, it varies with the position on the substrate and has its minimum energy (maximum wavelength) in the small openings between the pairs of rectangular masks.

3. Purpose of modeling.

It is the purpose of modeling to predict all the important features of the SAEOM process, given the nature and crystalline orientation of the substrate, the nature and geometry of the mask and the epitaxy operating parameters (temperature, pressure, reactants). Typically one would like to be able to accurately predict the thickness and composition distributions over the masked substrate area, both initially - when the thickness of the layer is negligible in comparison with all the other dimensional parameters - and later on as a function of growth duration.

4. A simple model.

The general features of SAEOM can be understood from a very basic model [9] shown schematically in fig.2. The diffusion equation (Fick's second law) is solved in a parallelepipedic box with periodic boundary conditions at the 'vertical' walls. The active molecules get into the box through the ceiling $z=h$ and deposit on the free substrate area at the bottom $z=0$. The boundary conditions at $z=0$ are such that there is no deposit at all on the masked area and at the same time all the molecules arriving at the free substrate area are immediately adsorbed and deposited.

In the general case the problem is three-dimensional, however, far from the mask end a two-dimensional treatment (mid-plane shown in fig.2) is appropriate.

This simple model is not suited for realistic calculations. Its main weakness lies in the fact that it cannot take into account the existing differences between the various active molecules which contribute to the growth. The single free parameter available for fitting the calculation to experiment in this model is h, the height of the 'stagnant layer'. Although this parameter is physically related to the thermal decomposition of the active molecules, so that we expect it to vary among depositing species, its influence is not sufficient, let alone basic for the understanding and correct simulation of ternary and quaternary alloy SAEOM.

5. Closed-form calculations with the simple model.

In spite of its above mentioned drastic limitations, the basic model yields qualitatively interesting results and is also useful as a testing tool for more sophisticated models. This is so since a) the simple model can be viewed as a particular case of the more complex models, and b) the simple model lends itself readily to very accurate, closed-form solutions which can be used as comparison and reference for calculations made with approximate (finite-difference, finite-element and Fourier transformation have been used) numerical methods.

Closed-form solutions are obtained in the two-dimensional case (plane xOz in fig. 2). The plane (x,z) is looked at as the existence domain of analytic functions of a complex variable $w=x+iz$. Such functions automatically satisfy the Laplace equation. The active molecule concentration $C(x,z)=C(w)$ is identified to the real part of

0-7803-3223-7/96/$5,00 © 1996 IEEE

such a function. An electrical analogy is immediate when we identify C(w) to an electrical potential, its gradient is then the analog of an electric field and the particle flux (D.grad C) at any point is the analog of the electric current density flowing through that point (this analogy holds also in the more general 3-D case of fig.2). The boundary conditions of fig.2 are such that the upper limit of the stagnant layer (z=h) and the unmasked parts of the substrate are the analogs of perfect electric conducting electrodes, while the masked regions on the substrate are the analogs of perfect insulators. Such problems have been extensively analyzed and solved in the case of electrical thin-film transducers, using conformal mapping techniques [10]. The solution involves four steps :

a) Find the Schwarz-Christoffel conformal mapping of the rectangular domain ($-c/2<=u<=+c/2$, $0<=v<=h$) in the plane $w=x+iz$ on to the upper $Y>0$ complex half-plane $Z=X+iY$. The solution is [11]:

$$w = c/2\, F[\sin^{-1}(Z),\mu]/K[\mu]$$

where :

$$F(\varphi,\mu) = \int(1-\mu \sin^2\theta)^{-1/2}d\theta\ ,\ \{0,\varphi\}$$

is the incomplete elliptic integral of the first kind, and

$$K(\mu) = F(\pi/2,\mu)$$

is the complete elliptic integral of the first kind and μ is calculated from the nonlinear equation

$$hK(\mu) = (c/2)K(1-\mu).$$

The range $0<\mu<1$ covers all possible (positive) values of $2h/c$ (fig.2).

The inverse functionality is expressed as

$$Z = \sin[\mathrm{am}(w.K(\mu)/(c/2),\mu)]\ (1)$$

where the "JacobiAmplitude" am is the inverse of F, that is, if $\omega=F(\varphi,\mu)$ then $\varphi=\mathrm{am}(\omega,\mu)$ [12].

In particular, the lower right corner $(c/2,0)$ of the rectangular domain in the w plane transforms to the point $(b_2=1,0)$ in the Z plane, while the upper right corner $(c/2,h)$ in the w plane transforms to the point $(a_3=\mu^{-1/2},0)$ in the Z plane. The points $(g/2,0)$ and $(g/2+m,0)$ in the w plane transform to the points $(b_1,0)$ and $(a_2,0)$ in the Z plane, where a_2 and b_1 are calculated from (1).

b) Find in the Z plane the complex field (gradC) which is the conformal mapping of the (constant) field for the case when there is no mask (m=0). It is expressed as follows:

$$E_{uniform}(Z)=E_{uniform}(w).dw/dz=(C_0/h)(c/2K(\mu))((1-Z^2)(1-\mu Z^2))^{-1/2}$$

c) Find the complex field in the Z plane for the nonuniform case using Mushelishvili's method [10]; this amounts to performing a few definite integrals along the real axis X. For the particular case depicted in fig.1, where for each elementary pattern on the substrate there are three symmetrically aligned unmasked portions $-c/2<x<-(g/2+m)$, $-g/2<x<g/2$ and $g/2+m<x<c/2$, the field is expressed as follows:

$$E_{masked}(Z)=(e_0+e_2.\ Z^2)((Z^2-b_1^2)(Z^2-a_2^2)(Z^2-b_2^2)(Z^2-a_3^2))^{-1/2}$$

The constants e_0, e_2 are obtained by imposing the conditions that the potential (or concentration) differences across the masked segments, obtained by integration along the x axis from b_1 to a_2 and from b_2 to a_3, are equal to zero and C_0 respectively. This leads to a system of two linear equations for the two unknowns e_0, e_2. The coefficients are obtained by numerical calculation of four integrals.

d) Return to the original w plane using $E_{masked}(w(Z))= E_{masked}(Z).dw/dZ$.

The entire calculation is a matter of tens of seconds using "Mathematica" on a personal computer and a complete exploration of a range of geometrical parameters can be done in a very short time. In all cases, at the masked/unmasked interface the field is singular (goes to infinity).

6. Using the simple model.

One of the central questions in SAEOM is to be able to predict r_c , the thickness enhancement ratio in the center of the gap (for unmasked growth $r_c=1$).In fig.3 we compare a particularly impressive set of published experimental data [1] for InP growth with the corresponding simple model calculation. There is good qualitative agreement; the quantitative aspect is not perfect, for one of the openings (w=12 μm) the fit is remarkable. For the other opening in the figure (w=40 μm) the experimental values are approximately 15 percent higher than the calculated ones, we shall see below that this is quite an extreme case.

7. The limits of the simple model.

In fig.4, the thickness profiles from two SAEOM growth experiments of InP and GaAs from TMIn and from TMGa, at atmospheric pressure, using the same geometry, are shown. It is apparent that the GaAs profile is very much smoother than the InP one. Similar differences are observed between growths carried out at different pressures. There is no way to reproduce this behaviour with the simple model, therefore, the need for a more comprehensive one is evident. At least one additional parameter is required which would take care of changes in the pressure or in the deposited species.

8. Full Surface Interaction model (FSIM).

The model depicted in fig.5 [3,6] accounts for the adsorption - desorption interaction of the active molecules with the solid surface, as well as for the migration of the adsorbed molecules before their eventual incorporation (epitaxial growth) or desorption. It adds as much as nine additional parameters (the diffusion coefficient in the gas phase and the adsorption - desorption - incorporation - migration coefficients over the free semiconductor and the masked surfaces respectively).

9. Reduced Surface Interaction model (RSIM).

The main difficulty when dealing with FSIM is in the choice of the parameters. Fortunately, most of these can be neglected in normal circumstances. It has been repeatedly shown [4, 15] that surface migration plays, at most, a secondary role, therefore it can be neglected without losing the essential features of SAEOM (it should be retained, however, in a more in-depth study). Moreover, if there is no deposit at all on the mask, one might also neglect the adsorption over the masked area. When the desorption (or evaporation) coefficient is also neglected, e.g. at not too high growth temperatures, only the adsorption coefficient of the unmasked substrate and the gas phase diffusion coefficient are left. It can be shown that the model of fig.2 can be used, provided that the boundary condition at the semiconductor surface is modified as follows :

$$D.dC/dz = a.C\ (z=0)$$

which amounts to adding to the model a single additional parameter, a/D (or the reciprocal $L_f=D/a$ which has the dimension of a length). The boundary condition at the masked surface remains as in fig.2. The case of fig.2 (the simple model) would apply if a were infinitely large or L_f=0. Actually, a is bound to be a fraction of the random thermal velocity of the active molecules in the gas phase:

$$a = \theta\, v_{th} = \theta\, (R_0.T/(2.\pi.M))^{1/2}$$

where M is the molecular mass of the active molecule and θ is a (non-dimensional) sticking coefficient, expected to approach unity. For InP growths carried out using TMIn in hydrogen, at atmospheric pressure (1 bar) and T=650°C, the diffusion coefficient D is typically 2 cm^2/s and v_{th} is typically 85 m/s so that we expect D/a to be at least 2.3 µm. With lighter organometallic molecules such as TMG, v_{th} is about 20 percent higher, D remains approximately the same so that we expect D/a to be at least 1.9 µm. At lower pressures, D/a is expected to vary in the same way as D, i.e. to increase as 1/p.

10. Using FSIM and RSIM.

We implemented the Full Surface Interaction, 3-D model using finite differences. The program is optimized for speed using variable space increments and was tested under DOS, OS/2 and Unix. A fast Pentium PC with 128 MO RAM can treat realistic 3-D cases in a reasonable time; most of the actual calculations were done in 2-D however. We checked both FSIM and RSIM against experimental data, both our own [13-14] and from the litterature [1,7]. The main conclusions to-date are as follows:

1) RSIM correctly predicts the thickness enhancement ratio r_c (fig.3) as well as the thickness profiles across the gaps, particularly for gap widths larger than 10 µm (fig.4).

2) The best fits for D/a, obtained from the shape of the thickness profiles (rather than from the thickness enhancement ratio at the gap center, see below), confirm the 1/p law of variation with pressure and yield for InP a most probable value of 3+-0.5 µm at atmospheric pressure, in good agreement with other investigations [4]. For GaAs from TMGa in H2, D/a is approximately four times larger: 12+-2 µm, which is quite different from [4]. We note that according to the physical meaning of this parameter, outlined above, this large value implies a surprisingly small sticking coefficient θ for TMGa.

3) Two useful rules-of-thumb concerning r_c, the thickness enhancement ratio at the gap center, can be derived from the numerical calculations : a) for any given semiconductor and not too small gap width, a broad optimum in the growth pressure exists which will yield the largest r_c. This occurs when D/a is approximately one-fourth of the gap width (e.g., the case w=40 µm in fig.3); b) when D/a is about the same as the gap width (e.g., the case w=12 µm in fig.3), r_c is about 15 percent below the maximum and is approximately equal to the value obtained from the simple model (D/a=0).

11. Conclusion

The simulation of SAEOM using 2-D conformal mapping can be very useful as a first, fast step towards a full understanding of the process. More detailed, 2-D and 3-D, simulation can be carried out with realistic results on a personal computer, provided that the openings are not too small as compared to a characteristic, growth pressure-dependent, vapour phase diffusion length D/a. For very small openings it is necessary to take into account surface migration and differential crystal plane growth rate effects.

12. References

[1] T.Takiguchi et al, Proc. ICMOVPE VIII, Cardiff, U.K., 1996 - to be published in Journal of Crystal Growth

[2] T.Tanbun-Ek et al, Proc. ICMOVPE VIII, Cardiff, U.K., 1996 - to be published in Journal of Crystal Growth

[3] D.E.Coronell and K.F.Jensen, J. Cryst. Growth 114 (1991) 581

[4] M.Gibbon et al, Semicond. Sci. Technol. 8 (1993) 998

[5] F. Scholz et al, J. Cryst. Growth 145 (1994) 242

[6] T. Fujii and M. Ekawa, J. Appl. Phys. 78 (9), 1 Nov 1995, 5373

[7] A.M.Jones et al, J. Electronic Materials 24 (1995), 1631

[8] M. Ida et al, J. Cryst. Growth 158 (1996) 437

[9] Y.D. Galeuchet and P. Roentgen, J. Cryst. Growth 107 (1991) 147

[10] A. Manolescu and A.M. Manolescu, Proc. Transducers '95-Eurosensors IX, Stockholm, Sweden 1995

[11] L. Volkovsky, G. Lunts, I. Aramanovitch, "Problems in the theory of functions of a complex variable", Mir Publishers, Moscow 1972, problem 1255

[12] S. Wolfram, "Mathematica", 2nd edition, Addison Wesley 1991, 785 (EllipticF, EllipticK) and 815 (JacobiAmplitude)

[13] G. Zimmermann et al, Proc. ICMOVPE VIII, Cardiff, U.K., 1996 - to be published in Journal of Crystal Growth

[14] A. Ougazzaden et al, Proc. ICMOVPE VIII, Cardiff, U.K., 1996 - to be published in Journal of Crystal Growth

[15] T. Fujii et al, J. Cryst. Growth 156 (1995) 59.

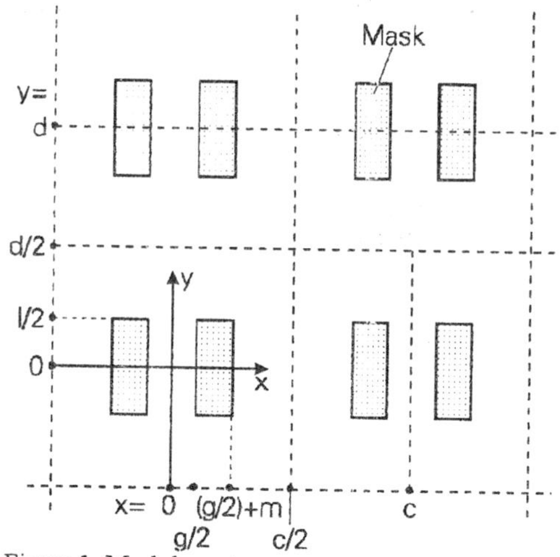

Figure 1. Mask layout

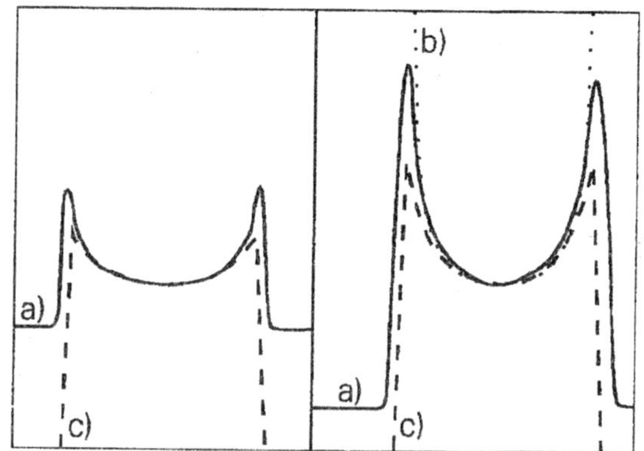

Figure 4. Thickness enhancement profiles. Left GaAs, right InP, same scale. a)experimental, b)simple model, c)RSI model

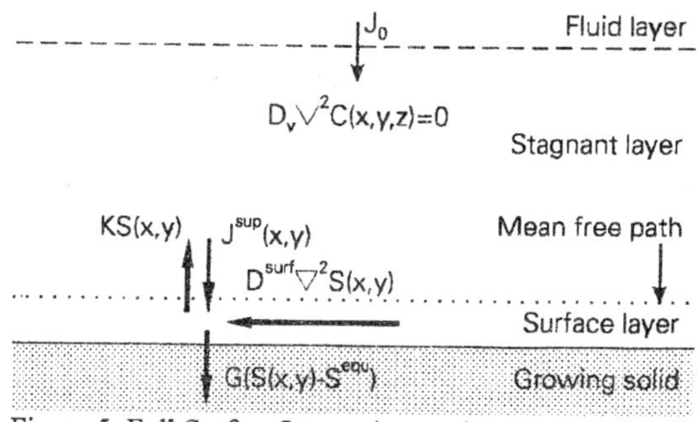

Figure 5. Full Surface Interaction model

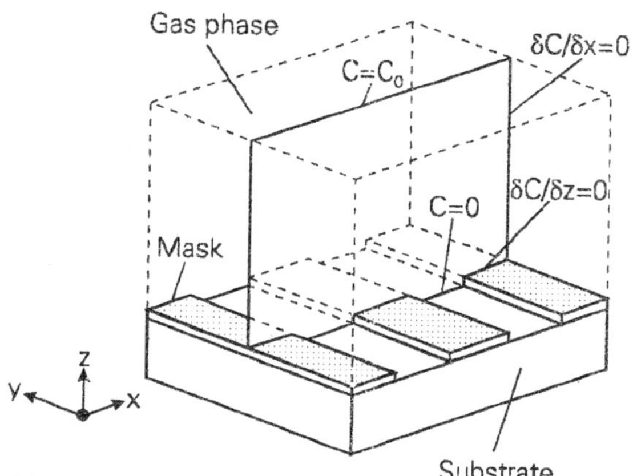

Figure 2. Simple model

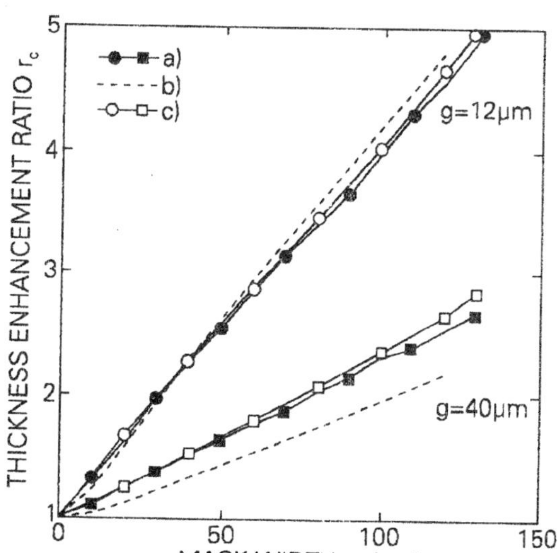

Figure 3. Growth enhancement ratio. a) measured [1], b) simple model, c) RSI model

Surface Pattern Formation and The Volatile Component Loss of Heat Treated Metallisations of InP

I. MOJZES*, B. KOVÁCS*, I. KUN**, L. MÁTÉ**, M. SCHUSZTER**, L. DOBOS***

* Technical University of Budapest, Department of Electronics Technology, H-1521 Budapest, Hungary

** Technical University of Budapest, Department of Mathematics and Computer Science, H-1521 Budapest, Hungary

*** Research Institute for Technical Physics of the Hungarian Academy of Sciences, H-1325 Budapest P. O. Box 76, Hungary

Summary

Palladium based metal systems can be used to make Ohmic contacts to $A^{III}B^{V}$ compound semiconductors. A covering layer of gold is advantageous from the point of view of bonding as well. Gold, palladium layers were studied on InP substrates. Samples were annealed in the vacuum chamber of a scanning electron microscope (SEM) and the volatile component loss (phosphorus) was monitored by a quadrupole mass spectrometer. The changes of the surface morphology were studied using the SEM images. It means that the change of surface morphology and the volatile component loss were monitored simultaneously. In the case of Pd/$A^{III}B^{V}$ samples a single characteristic peak due to the interaction taking place between the metallization and the substrate was observed on the volatile component loss vs. temperature curve. In this temperature range characteristic pattern formation can be detected. This surface morphology shows fractal behaviour. It is supposed that the giant volatile component loss originating from the interaction between the metallization and the substrate can be described using percolation theory.

1. Introduction

During the heat treatment of $A^{III}B^{V}$ compounds covered by several kinds of metal layers a large volatile component loss can be detected over the sample using the Evolved Gas Analysis technique [1]. Combining this technique with scanning electron microscopy, *in situ* investigations were carried out for monitoring both volatile component loss and surface morphology [2]. As a result of the interaction leading to the phase transition of the metal layer, accompanied the volatile component loss at a characteristic temperature, some typical pattern formation can be detected on the surface.

Gold based metallization in the $A^{III}B^{V}$ technology has some disadvantages (e.g., rough interface after annealing). The Pd based contacts could be more appropriate as a contacting metallization system [3-7]. To obtain the characteristic parameters of the surface morphology of Pd/InP and Au/InP structures the *in situ* technique originally described in [2] was applied.

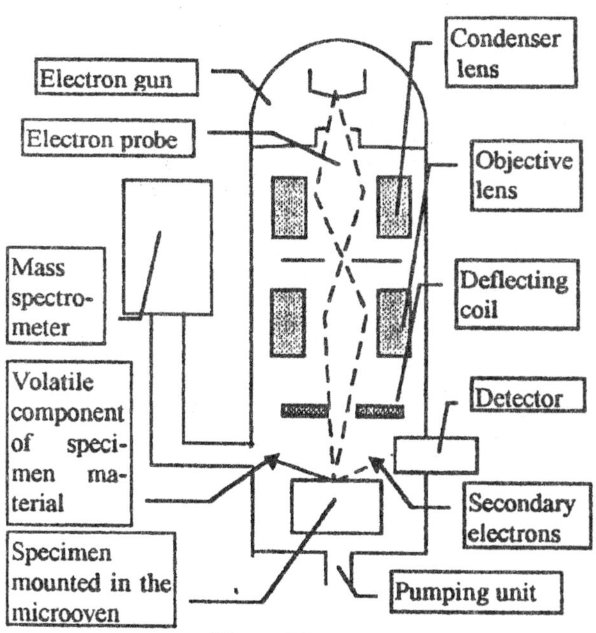

Fig. 1: The experimental setup

2. Experimental

50 nm thick Pd layers and 60 nm thick Au layers were evaporated onto chemically cleaned n^+-InP wafers.

The investigated samples were cut from single crystalline wafers. After the mechanical and chemical polishing the slices were chemically cleaned and etched. The applied procedure was: a., 5 minutes boiling in toluene, 3 minutes boiling in methanol, b., 15 seconds etching in $CH_3COOH : H_2O_2$ (3:1), 3x washing in distilled H_2O, c., 15 seconds etching in $HF : H_2O$ (1:20), rinsing in methanol, d., 5 minutes boiling in methanol, drying by stream of N_2. The metallization was evaporated in ultra high vacuum (UHV). Before the deposition the wafers were outgassed at 175°C and during the deposition the substrates were held at 75°C.

To study *in situ* the volatile component loss and the actual surface morphology the developed, combined

0-7803-3223-7/96/$5,00

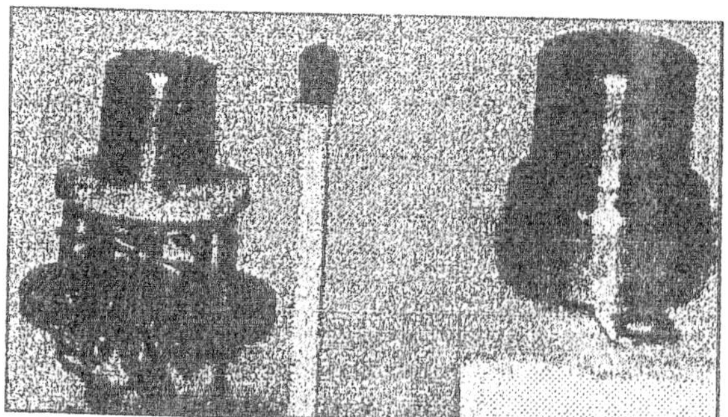

Fig. 2: The microoven used for the heating of the sample. The shield used for EMC is on the right hand side.

SEM (JEOL JSM-20) and mass spectrometer system (ATOMKI Q100) was used. The experimental setup is shown in Fig. 1. Applying this device the sample can be heated up to 630°C.

The samples were annealed applying the microoven shown in Fig. 2 using a heating rate of 30°C/min. The volatile component loss was monitored by the quadrupole mass spectrometer and the SEM images of the surface were taken during the heat treatments.

3. Results

Pd (50 nm) / InP and Au (60 nm) / InP samples were investigated. Comparing the volatile component loss vs. temperature curves of the investigated samples (Fig. 3.) it can be seen that the peaks of the phosphorous loss due to the interaction palladium with InP is rather sharp. It has a one-peak character in spite of the situation experienced in the case of gold-InP sample.

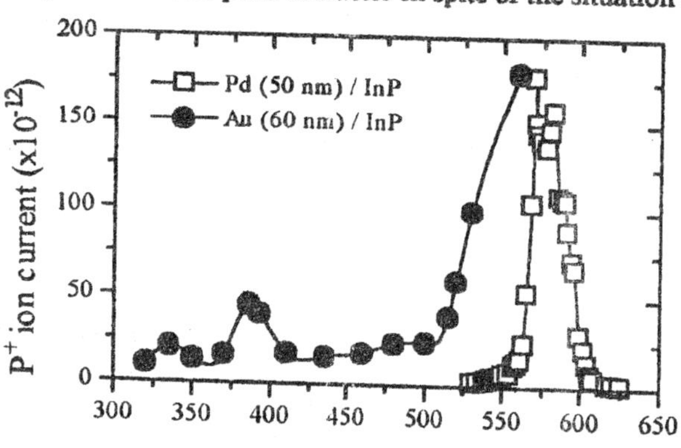

Fig. 3: Volatile component loss of metallized InP samples

The gold-InP interaction produces a two-peak character. The main peak of gold-InP samples lies in the same temperature range as for palladium-InP samples. The SEM micrographs of the samples taken at the temperatures of the giant volatile component loss shown in Fig. 4.a and Fig. 5.a.

As it can be seen on the micrographs the surface patters show apparent self-similarity at the so-called characteristic temperatures. Far below and far above the critical temperatures we got uniformly grey pictures. Photos about the pattern were digitised and analysed after storing in bit-map files where one of 256 grey shades can be registered in each byte.

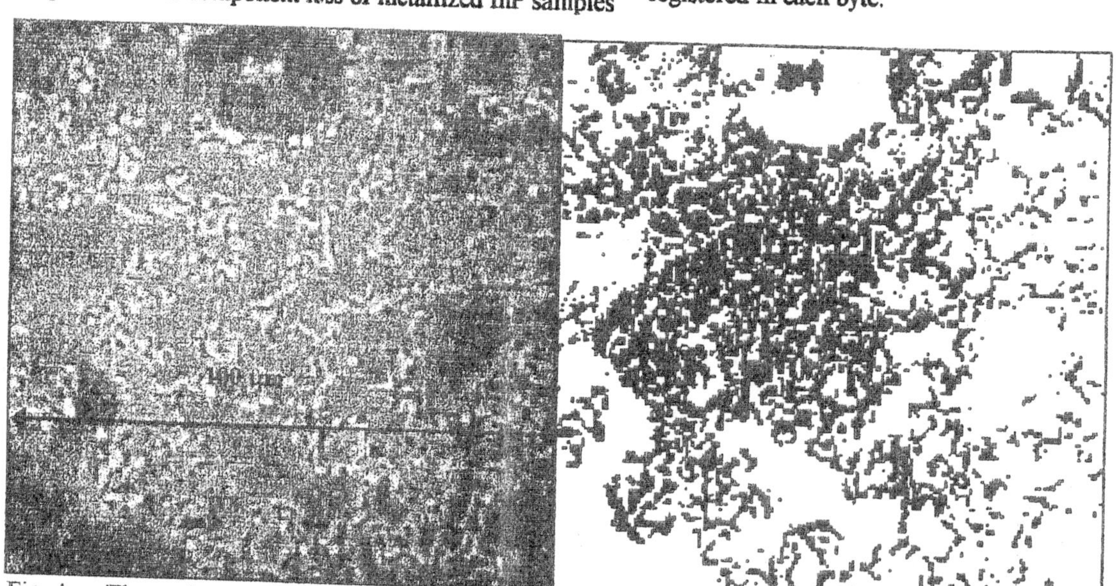

Fig. 4. a, The surface morphology of Pd (50 nm) / InP sample at 540°C, b, The digitised surface pattern used to evaluate the fractal site dimension.

The analysis was carried out on two samples, each corresponding to a matrix of 304 × 290 dots where a dot was considered as white under grey shade value 150 and as black above it (Fig. 4.b and Fig. 5.b). The spatial

correlation function of the pattern (density of points at the directed distance **r** from a starting point) was assumed to have the form $c(\mathbf{r}) \sim |\mathbf{r}|^{-\alpha}$ — where α is a constant, $0 < \alpha < 2$ — according to Ref.[10]. We denote by $M(L)$ the average mass in a box of linear size L, where mass is proportional to the integral of $c(\mathbf{r})$ on the box. It is easily seen that the assumption predicts $M(L) \sim L^D$ with $D = 2-\alpha$ i.e. self-similarity with scaling factor D. This means that the pattern has a fractal site dimension D. Starting from 1000 distinct points of the pattern, integration of $c(\mathbf{r})$ was approximated with simple summation in the boxes.
In good accordance with theory, as it is shown by the linear regression chart, $\ln M(L)$ is a linear function of $\ln L$ with slope D (Fig. 6). $D = 1.76\pm0.02$ was obtained for the fractal site dimension of the black pattern on the white background for Pd/InP (Fig. 4.a) and $D = 1.58\pm0.03$ for Au/InP (Fig. 5.b). This means that the pattern has a fractal character -- since the dimension D is significantly less then 2 — similarly to what was experienced in percolation theory [11] and in an experimental setting [12].

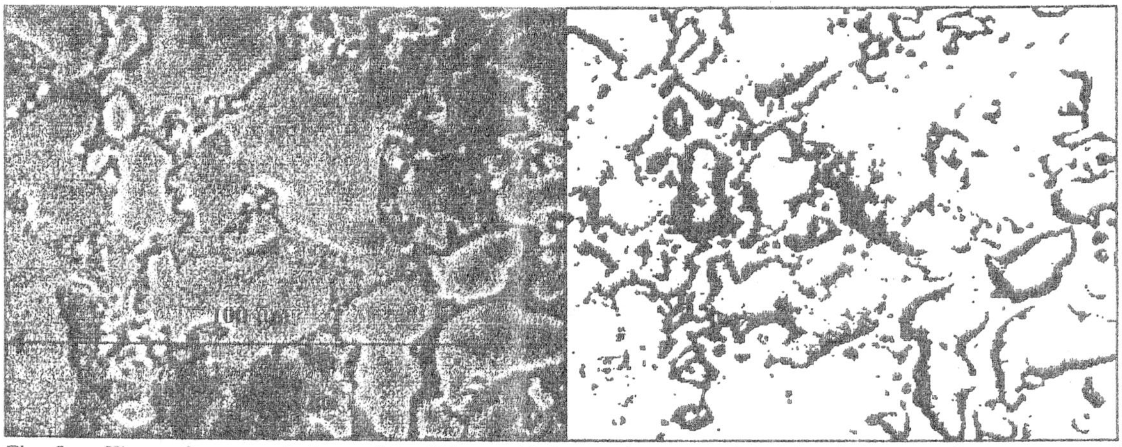

Fig. 5. a, The surface morphology of Au (60 nm) / InP sample at 570°C; b, The digitised surface pattern used to evaluate the fractal site dimension.

4. Discussion

Analysing our results we conclude that the simultaneous application of the EGA and SEM can lead to a valuable contribution in the investigation of the interaction of metallic layers with $A^{III}B^{V}$ materials. In our previous works the surface morphology was investigated only after the heat treatment "freezing" the sample at a given temperature [1]. Combining the two above mentioned techniques it became possible to take an image of the surface and to continue the heat treatment at the same vacuum cycle and to repeat it.

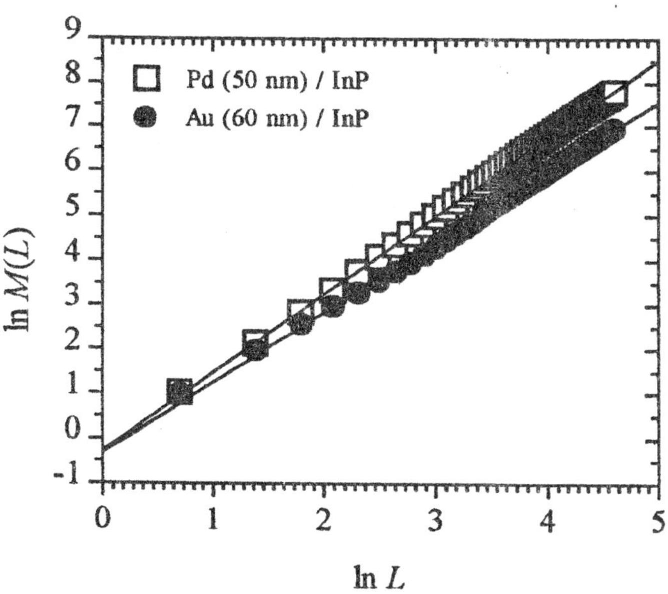

Fig. 6: Linear regression plots of $\ln M(L)$ vs. $\ln L$ for metallized InP samples.

The EGA results obtained in the present experimental arrangement are in good agreement with the results of former ex situ investigations [1,8].
According to references [3-8] it is well known that either the Pd/InP or the Au/InP system goes through some phase transition during annealing. These phase formations and transitions are investigated mostly after a constant temperature, long time heat treatment. Taking into account that nearly the same structure is obtained using either a long time, low temperature or a short time, high temperature heat treatment, it is obvious that phase transition temperatures, if any,

will differ in our experiment compared to those in literature.

We can assume that different heat treatment involves also different phase transition temperatures in our measuring system. Studying, e.g., the volatile component loss versus temperature curves of Pd/InP system, $Pd_2InP(II)$ phase decomposition [6] can be coupled with the peak of P loss. Along the same process, surface morphology has a fractal behaviour, as it was shown.

Comparing Fig. 4 and Fig. 5 the difference of surface morphology is obvious. However at the temperature characterised by the peak of P loss both surface show fractal character.

5. Conclusions

In the present work Au and Pd layers deposited onto the surface of InP were investigated using simultaneous EGA and SEM techniques. As a result of interaction of the thin metallization with the compound semiconductor a volatile component loss occurs leading to the formation of surface morphology showing fractal character. It should be stressed that these fractals are a result of interaction of different materials at given temperature range.

6. Acknowledgement

This work was supported by Hungarian National Scientific Foundation (OTKA) through Grants No. 1237, 1243, T4171 and T020596.

7. References

1. Mojzes, I., Sebestyén, T., Szigethy, D. (1982) Volatile component loss and contact resistance of metals on GaAs and GaP during annealing, *Solid-State Electronics* **25**, 449-460.
2. Dobos, L., Kovács, B., Mojzes, I., Malina, V., Pécz, B. and Karányi, J. (1992) The volatile component loss and the surface morphology of the gold-palladium metallizations to the compound semiconductor structures, in A. Katz, S. P. Murarka, Y. I. Nissim, J. M. E. Harper (eds.) *Advanced Metallization and Processing for Semiconductor Devices and Circuits - II, Mat. Res. Soc. Symp. Proc.* **260**, 641-646.
3. Braslau, N. J., (1981) Alloyed Ohmic Contacts to GaAs, *Vac. Sci. Technol.* **19**, 803-807.
4. Robinson, G.Y. (1985) Schottky diodes and Ohmic contacts for the III-V semiconductors, in C.W. Wilmsen (ed.) *Physics and Chemistry of III-V Compound Semiconductor Interfaces*, Plenum Press, New York, 73-163.
5. Kuan, T. S., Batson, P. E., Jackson, T. N., Rupprecht, H., Wilkie, E. L. (1983) Electron microscope studies of an alloyed Au/Ni/Au-Ge Ohmic contact to GaAs *J. Appl. Phys.* **54**, 6952-6957.
6 Ivey, D. G., Jian, Ping, Bruce, R. (1992) Reactions Between Pd Thin Films and InP, *J. Electronic Mat.* **21**, 831-839.
7. Caron-Popowich, R., Washburn, J., Sands, T., Kaplan, A. S., (1988) Phase formation in the Pd-InP system, *J. Appl. Phys.* **64**, 4909-4913.
8. Mojzes, I., Veresegyházy, R., Kovács, B., Gurbán, S., Pécz, B. and Radnóczi, G. (1985) Thermal dissociation of III-V compounds - Physico-chemical effects influencing the contact parameters, in J. Kassabov (ed.) *Physical Problems in Microelectronics*, World Scientific, 214-286.
9. Dobos, L., Karányi, J., Pécz, B., Kovács, B., Mojzes, I. and Malina, V. (1993) Palladium based contacts to GaAs and InP, *Per. Polytechnica Ser. El. Eng.* **37**, 21-30.
10. Vicsek, T. (1989) *Fractal growth phenomena*, World Scientific, Singapore.
11. Stauffer, D. (1979) *Phys. Rep.* **54**, 1.
12 Kapitulnik, A., Deutscher, G. (1982) Percolation Characteristics in Discontinuous Thin Films of Pb, *Phys. Rev. Lett.* **49**, 1444 - 1447.

TIN DIOXIDE SOL-GEL DERIVED THIN FILMS DEPOSITED ON POROUS SILICON

C. Cobianu[1], C. Savaniu[1], O. Buiu[1], D. Dascalu[1],
M. Zaharescu[2], C. Parlog[2], A. van den Berg[3] and B. Pecz[4]

[1]. Institute of Microtechnology, P.O. Box 27-17, 77550 Bucharest, Romania,
[2]. Institute of Physical-Chemistry of the Romanian Academy, Splaiul Independentei 202, Bucharest, Romania
[3]. MESA Research Institute, University of Twente, Faculty of Electrical, Engineering, P.O. Box 217, 7500 AE Enschede, The Netherlands,
[4]. Res. Inst. Tech. Phys., H-1325 Budapest, P.O. Box 76, Hungary.

SUMMARY

SnO_2 and SnO_2:Sb sol-gel derived thin films have been prepared from tin (II) ethylhexanoate and tin (IV) ethoxide precursors (the later not reported before) in order to be used for gas sensing applications where porous silicon is used as a substrate. Transparent, crack-free and adherent layers were obtained on different types of substrates (Si, SiO_2/Si). By energy dispersive X-ray spectroscopy (EDXS) performed on the cross section of the porosified silicon (PS) coupled with XTEM analysis, the penetration of the SnO_2 sol-phase in the nanometric pores followed by the SnO_2 consolidation film in the pores of the PS, during subsequent annealing at 500°C has been experimentally proved.

1. INTRODUCTION

Tin oxide semiconducting layers incorporating small amounts of noble metal (Pt, Pd) and metal oxides dopants (Sb, In, Bi, etc.), prepared by thick and thin film technology have been widely used for gas sensing applications where conductivity change is used to detect reducing gases /1/. Recently, /2-4/ a trend towards sol-gel synthesis of the SnO_2 thin films has been observed due to the method capability to tailor complex compositions and to control the porosity of the films /5/. This last property, determining the high specific area of the layer seems to be quite attractive for promoting increased sensing properties of the SnO_2 based sensors /6/. Based on this generic idea, it is interesting to investigate if there is any influence due to a porous substrate (in our case, porous silicon) on the SnO_2 sensing properties. As an initial step, it is necessary to study the capability of the sol-gel method to generate a SnO_2 (liquid) sol phase able to penetrate in the pores of the porous silicon (PS) and to determine by subsequent annealing a SnO_2 solid film on the walls of the PS.

It is the purpose of this paper to prove by coupled measurements of cross-section (X)TEM and EDXS performed on the cross section of the porous silicon that the tin oxide is present inside the pores of PS and to show thus the feasibility of sol-gel method for this task. Further technological improvements concerning increased wettability of the solution in conjunction with surface substrate preparation are investigated in order to enhance the penetrability of the SnO_2 in the pores.

2. EXPERIMENTAL PROCEDURE

The synthesis of the undoped and Sb doped SnO_2 sol-phase consists of the following precursors: tin (IV) ethoxide, prepared in our laboratory and tin (II) ethylhexanoate (Sigma-Aldrich) used as SnO_2 sources, $SbCl_3$ (J.T. Baker) used as a Sb source while absolute ethylic

0-7803-3223-7/96/$5,00

(Merck) and n-buthylic alcohols were used as solvents. The composition of the initial solutions, the pH of the sol, its long term stability and the thermal treatment conditions for the film firing and consolidation are presented in the Table I.

Table I. Solution composition (in molar ratios), sol stability, sol pH and SnO_2 film formation.

Preparation conditions	tin (II) ethoxide	tin (IV) ethylhexanoate
ROH/SnO_2 (molar ratio)	18.46	17.39
H_2O/SnO_2 (molar ratio)	0	2.02
$SbCl_3/SnO_2$ (molar ratio)	0.1	0
pH of the sol	1.4	4.63
Sol stability (days)	>30	>7
Annealing temperature (°C)	450-600	500-850
Annealing time in oxygen ambient (min.)	30	30

The SnO_2 thin solid films were obtained by spin casting of 0.2 μm filtered solution on three types of substrates: Si, SiO_2 /Si and PS/Si freshly cleaned before gel deposition. Uniform, transparent and crack-free layers were obtained and easily checked on the first two substrates at the end of the thermal sequence consisting of drying (150°C), firing and crystallization (450-850°C). The porous silicon substrates (about 2 μm in depth) were obtained by standard anodization process on highly boron doped silicon substrates with resistivity of about 1mΩ-cm and (100) crystallographic orientation. The structural analysis of the SnO_2 films deposited on different substrates has been performed by XRD using a HZG diffractometer (CuKα radiation). The PS morphology was observed by XTEM analysis performed at RITP Budapest on a Philips CM 20 TEM instrument equipped with NORAN EDS X-ray detector for composition investigation on the cross section of the PS.

3. RESULTS AND DISCUSSION

Starting from tin (IV) ethoxide, ten consecutive undoped SnO_2 layers were successively deposited on Si substrates. Each layer was dried and thermally annealed separately in order to avoid cracking. The thickness of each SnO_2 spin coated layer was about 25 nm while the refractive index was 2.1 in agreement with the literature data (4). Similar SnO_2 films were deposited from tin (IV) ethylhexanoate where the film thickness/spin-casting was about 50 nm. The Sb doped SnO_2 films obtained from tin (IV) ethoxide and $SbCl_3$ presented a lower thickness per spin-casting, of about 15 nm at a similar solution concentration (see Table I).

The XTEM image (500K magnification) of the PS silicon used as a substrate for SnO_2 sol-gel deposition is presented in Fig. 1. The contrast between the porosified silicon (upper side) and monocrystalline silicon (dark, at the bottom of Fig. 1) is quite evident. The very small dimension of the pores is associated with the high boron doping level of the silicon wafer. At this level of pore dimensions is not possible to "see" by direct observation (by means of the differences in the image contrast) the SnO_2 film on the walls of the PS

The XRD analysis indicated that the crystallization of SnO_2 started at annealing temperature of 450°C, the crystalline structure being of rutile type after a thermal annealing of 850°C for 30 min. Electron diffraction (ED) pattern of the SnO_2 deposited on PS/Si substrates and annealed at 500°C is presented in Fig. 2. Practically, the same image was obtained for PS/Si substrate before

SnO_2 deposition indicating that either the SnO_2 inside the pores is too thin or the layer crystallization is incipient, or both of them.

In order to analytically investigate the presence of the Sn atoms in the pores of the PS/Si substrates on which SnO_2 from tin (IV) ethoxide was deposited, the energy dispersive X-ray (EDX) spectroscopy was performed on the cross section of the porosified silicon with a small dimension beam focused in the middle side of the porosified silicon (about 1 μm below the surface level). The EDX spectrum is presented in Fig. 3. The diagram shows the presence of two peaks associated to the Sn atoms whose concentration appear to be higher than 1 at% and which are well above the detection limit of the equipment. The oxygen present in the spectrum can be assigned to both native silicon dioxide and tin dioxide existing in the pores according to Sn detected. These starting results confirm the ability of the polymeric sol with solid "particles" of molecular dimensions to penetrate the nanometric pores of PS/Si.

4. CONCLUSIONS

Transparent, crack free, undoped and Sb-doped SnO_2 films were prepared by sol-gel method from tin (IV) ethoxide (SnO_2 synthesis from this precursor was not reported before) with $SbCl_3$ as a doping source or from tin (II) ethylhexanoate. The layers were deposited on Si, SiO_2/Si and porous silicon/Si substrates. Based on XTEM/EDX analysis, the capability of the SnO_2 sol-phase (from tin (IV) ethoxide) to penetrate (as deep as 1 μm) in the nanometric pores of the PS has been evidenced.

REFERENCES

1. V. Ambrazeviciene, A. Galdikas, S. Grebinskij, A. Mironas and H. Tvardauskas, Sensors and Actuators B, 17 (1993) 27-33.
2. C.D. Feng, Y. Shimizu, and M. Egashira, J. Electrochem. Soc., 141, (1994), 220-225.
3. Y. Takahashi and Y. Wada, J. Electrochem. Soc.,137, (1990), 267-272.
4. S.S. Park, H. Zheng and J.D. Mackenzie, Materials Letters, 22 (1995), 175-180.
5. C.J. Brinker and G.W. Scherer, "Sol-Gel Science" The Physics and Chemistry of Sol-Gel Processing (Acad. Press, San Diego, 1990).
6. B. Gautheron, M. Labeau, G. Delabougliser and U. Schmatz, Sensors and Actuators B, 15-16 (1993), 357-362.

Acknowledgement

This research has been performed in the frame work of the European COPERNICUS program (PORSIS project, CP 94-0963), coordinated and financed by European Union.

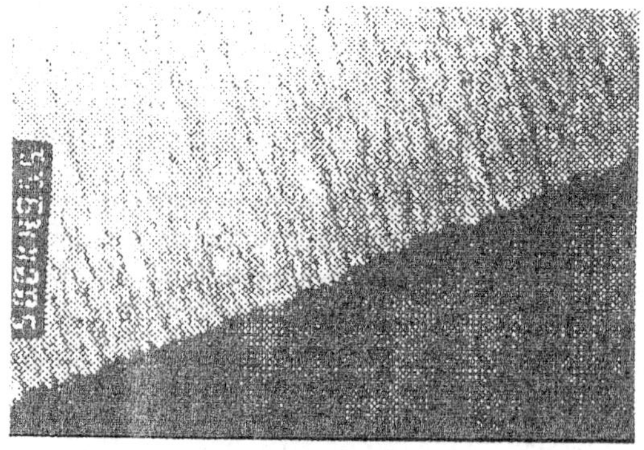

Fig. 1- A cross sectional view of SnO_2 on porous silicon

Fig. 2- ED pattern of SnO_2/ porous Si

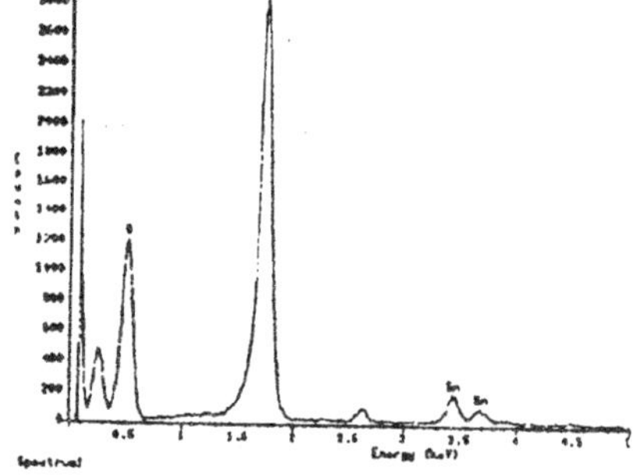

Fig. 3- EDS spectra of SnO_2 deposited on Si/porous Si
(the middle part of the porous Si layer)

EXPERIMENTAL AND THEORETICAL INVESTIGATION OF PbTe GROWTH BY MBE ON (111) BaF_2 SUBSTRATES.

K.I.Andronik, O.Yu.Keloglu, V.A.Mazur, E.A.Zasavitski
Institute of Applied Physics, 277028 Chişinău, Moldova

Lead telluride films have been grown on cleaved BaF_2 substrates by MBE method. Growth process was monitored by RHEED system and the crystal structure of the obtained films was investigated by Θ-2Θ and grazing angle X-ray tests. Optimal technological regimes are found and a phenomenological theoretical model of the MBE growth process is proposed.

Introduction

The search of new type of semiconducting devices and materials stimulates considerable interest in the heteroepitaxial structures. Barium fluoride is a material most widely used as isolating substrates for epitaxial growth of IV-VI semiconductor films. So the investigation of the PbTe heteroepitaxy onto BaF_2 is of actual importance both from the fundamental science viewpoint and for the practical purposes.

Experiment

Lead telluride films were evaporated by Molecular Beam Epitaxy (MBE) installation EP-1203 equiped with *in situ* quadrupole mass-spectrometer and Reflection High Energy Electron Diffraction (RHEED) system. Because of congruent evaporation, for PbTe growth a compound effusion cell is used and an additinal Te_2 effusion cell to fine adjust the beam flux composition. Due to this adjustment we can obtain both p- and n-type epitaxial films with *apriori* established carrier concentration.

Freshly cleaved single crystal BaF_2 substrates were anealed under superhigh vacuum (10^{-9} Torr) just before evaporation until clear electron diffraction patterns were appeared. Substrate temperatures maintained during the growth process varied in the range 325°C – 425°C. Usual evaporation rate was about 2Å/s. Film thicknesses

0-7803-3223-7/96/$5,00

were determined by the interference microscope.

According to the RHEED monitoring the growth process shall be divided into 3 stages:

1 two-dimentional epitaxial growth of several pseudomorphic monolayers (heteroepitaxy) comes before:

2 islanding and three-dimentional epitaxy. It is caused by the much higher surface energy of PbTe(111) compared to PbTe(001). After deposition of about 1000 Å thickness layer islands coagulate, surface of the growing film becomes again atomicaly smooth and process returns to:

3 a two-dimentional regime (homoepitaxy on the PbTe buffer layer).

Films grown at the optimal substrate temperature (375°C) follows substrate orientation (111), but other samples tend to turn it to (001). Structure quality and orientation were determined by usual Θ-2Θ and by grazing angle X-ray tests. The former method is the most appropriate for the investigation of thin films [1,2] and allows us to record specular reflexes from a large number of crystal planes at a grazing angle of the X-ray incidence. No adaptation of usual X-ray diffractometer is needed for this method. Position of a specular reflex does not depend on the X-ray wavelength, so one can use nonmonochromatic X-rays without any loss of accuracy. As one can see in fig.1 value of the angle φ between the actual film surface plane (in our experiments (111)) and arbitrary crystal plane giving specular reflex is determined by the formula: $\varphi = (\phi - 2\omega)/2$, where ϕ is angle registrated by diffractometer. Comparing of theoretical and experimental values of φ allows us to make identification of reflecting planes.

In fig.2 grazing angle X-ray diffraction patterns from three samples are shown. For the film evaporated onto the substrate mantained at 375°C positions of all specular reflexes coincide with theoretical ones for the perfect (111) oriented PbTe single crystal. Samples obtained at temperatures higher and lower than optimal one show additional reflexes from (001) domains.

Theoretical model and discussion

As a model we consider here a single monoatomic step on the crystal surface. Attachment of each PbTe molecule results in deviation from step-flow regime. From the self-consistent system of equations for flux and force balances we can make conclusion about mechanism of film growth and estimate the optimal growth temperature. During consideration we account the driving force which arises from the capillary force (due to non-step-flow regime) and force due to interaction of the step boundary with adatoms, which opposes boundary motion. The flux balance incorporates the flux along the boundary (due to non-uniform concentration gradient) and the flux which

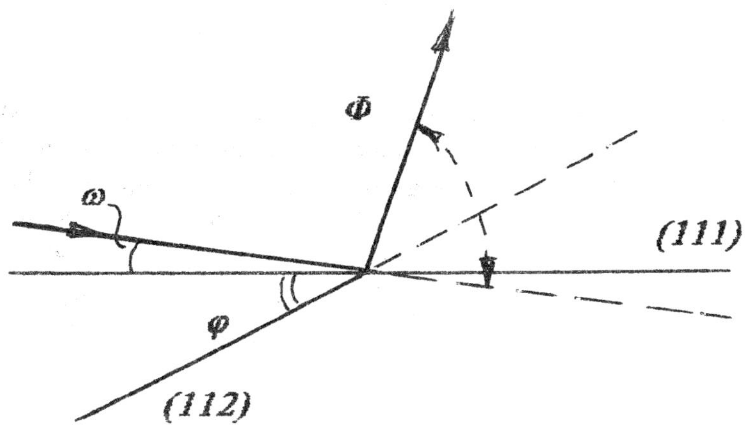
Φ
ω
(111)
φ
(112)

Fig.1

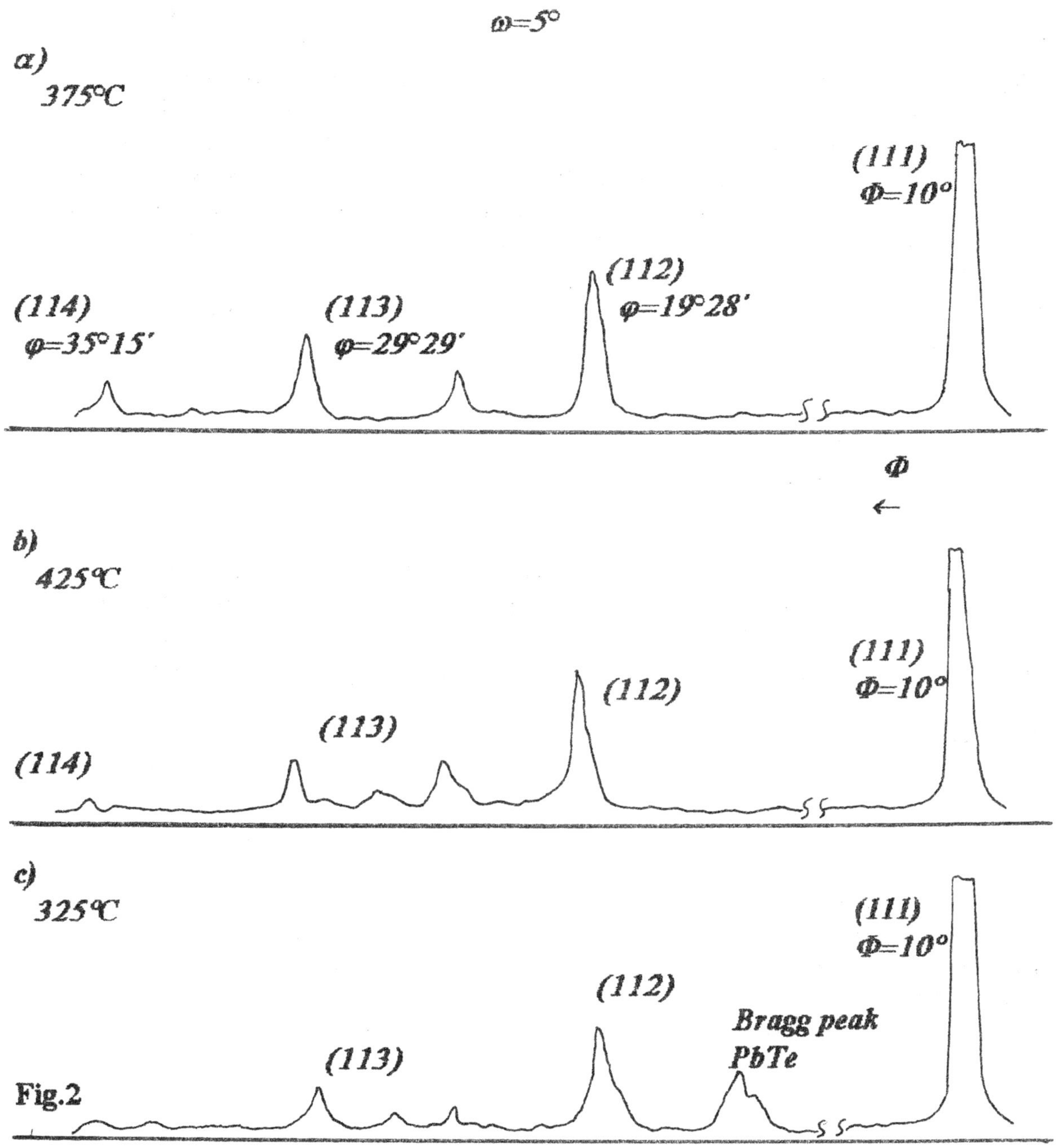
ω=5°
a)
375°C
(111)
Φ=10°
(112)
φ=19°28'
(114)
φ=35°15'
(113)
φ=29°29'
Φ
←
b)
425°C
(111)
Φ=10°
(112)
(113)
(114)
c)
325°C
(111)
Φ=10°
(112)
Bragg peak
PbTe
(113)

Fig.2

tends to accumulate molecules in region containing an above initial concentration C_o of molecules. To give some semiquantative estimations we limit ourselves by steady state case. It is easy to show that steady state amplitude for the step boundary displacement is proportional to $\Delta C_o(D/D')^{1/2}$ where D is the diffusion coefficient along the boundary and D' is some effective value corresponding to the perpendicular diffusion $D' = \mu\gamma/NC_o$, here μ is boundary mobility defined by the molecules supplement to the step, i.e. by boundary kinetics; γ is a surface boundary tension; N is a number of molecules per surface unit.

When D > D' a small change in concentration leads to the significant deviation from step-flow regime and v.v. With account of the different role of linear diffusion (along step boundary) and surface diffusion (at terrace to step edge) we can show that step flow regime occures only in the temperature range T_1 < T < T_2. For T < T_1 the surface diffusion isn't sufficiently fast for the adatoms to find a lattice sites prior to encountering a molecule from the beam. For T > T_2 surface diffusion is too mush faster than the linear one (because of the difference in the activation energies) so that molecule hasn't enough time to incorporate into the nearest kink site before colisions with next income adatoms. For reasonable parameters one can obtain that T_1= 670 K, and T_2=700 K. Despite of the simplified character of the model, calculated results are in good accordance with the experiment data. It should be emphasized that step characteristics are accounted in μ. Consideration of the step configuration in more details (e.g. appearence of additional kinks or set of steps) leads to more complicated temperature dependence.

Conclusions

The epitaxial PbTe films are obtained by MBE method. High crystal quality is confirmed by X-ray analysis. Optimal technological regimes are found. Suggested fenomenological model gives numerical estimations for optimal growth temperatures which are in good agreement with experimental values. The proposed model of the epitaxial evaporation may also be valid for growth description of other congruently evaporated substances.

References

1. Mazur V.A., Jdanova L.I., Keloglu O.Yu. I-st International Conference on Material Science of Chalcogenides and Diamond-Structure Semiconductors, vol.1. Chernovtzy, 1994, p.120.

2. Andronik K.I., Keloglu O.Yu., Mazur V.A., Zasavitski E.A., Ibid. p.66.

ALUMINUM NITRIDE FILMS FOR OPTICAL APPLICATIONS

V.Dumitru[1], Elena Cimpoiasu[2], C.Morosanu[2] , Cristina Nenu[3]
Dana Necsoiu[4]
[1]"Politehnica" University, Dept. of Physics, [2]IFTM,
[3]Zecasin S.A., [4]Institute of Optoelectronics, Bucharest
Romania

Summary:

Hard and adherent AlN layers on glass have been deposited by reactive magnetron sputtering. The deposition rate, the structure and the optical properties have been investigated mainly with respect to possible optical applications.

1. Introduction

Out of the nitride compounds, AlN has a very large optical gap (6.3 eV), a high refractive index in the visible range (>2.1) and a good chemical stability in oxygen. Because of the high interest for this material, we have also studied the possibility of depositing adherent films of AlN on glass with reproducible optical parameters.

2. Experimental details

Films of AlN were prepared in a conventional DC, RF magnetron system using the reactive sputtering in argon and nitrogen. The experimental details have been previously reported /1/. We have deposited layers in the thickness range 0.25-2 μm on quartz, glass and silicon substrates.

The structure of the films was investigated by X-ray diffraction. A DRON-2 diffractometer provided with copper target tube was used.

Optical transmission measurements were performed in the wavelength 200-900 nm by means of a Specord M40 spectrophotometer, in order to estimate the film thickness, the refractive index and the extinction coefficient.

3. Results

3.1. The dependence of deposition rate vs. nitrogen concentration

In the first stage we were interested in the influence of nitrogen concentration upon the deposition rate. We have estimated the deposition rate for the RF deposited layers using different nitrogen concentrations (nitrogen/(nitrogen + argon)) up to 50%, at a constant sputtering pressure of

0-7803-3223-7/96/$5,00

4.2-4.4 mTorr, and two total gas flow rates 50 and 60 sccm (fig.1). For the DC sputtered layers, deposited at various powers, we have calculated the ratio deposition rate vs. deposition power in order to study only the effect of nitrogen concentration (fig.2), as the dependence of the deposition rate vs. deposition power is liniar.

One can notice the decreasing of the deposition rate with the increase in the nitrogen concentration due to the forming of an AlN layer on the surface of Al target at high nitrogen concentrations, which is well known for the reactive sputtering.

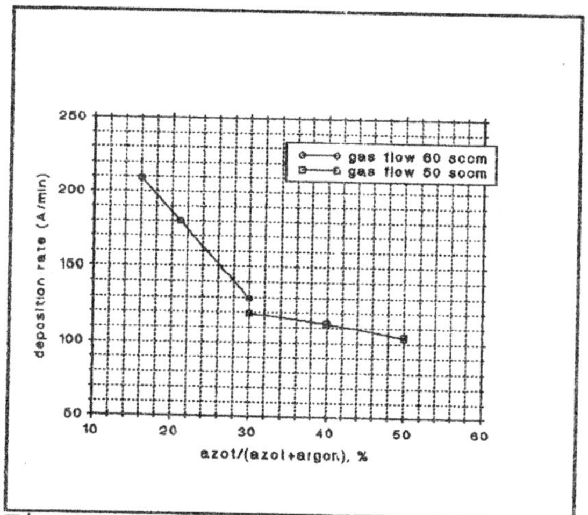

Fig.1 - RF deposition rate vs. nitrogen concentration

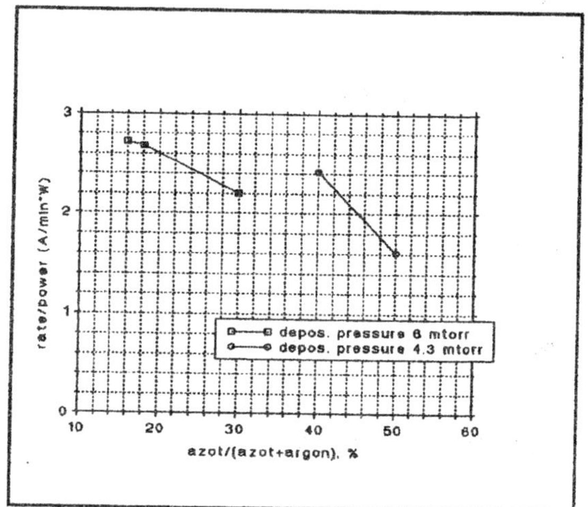

Fig. 2 - DC deposition rate vs. nitrogen concentration

3.2. Structure and optical properties

We have studied the optical properties for AlN layers of 300 nm on quartz substrates, deposited by RF, DC sputtering at 30% nitrogen concentration.

We have first estimated the pre-sputtering time, long enough in order to stabilize the deposition conditions, because in the case of the reactive sputtering the quality of the layers is also influenced by the forming of AlN on the target surface. We have sputtered in the same run three AlN films, each of them, RF deposited for 30 min. Between the second and the third layer, we have measured a difference of optical transmission less than 1%, while between the first and the second it has been of about 2%. That's why a pre-sputtering time more than 45 min was used for all the next depositions in order to obtain reproducible layers.

We have first performed X-ray diffraction measurements on AlN thick layers (1.5 µm) on glass in order to investigate the crystalline structure, necessary for the understanding of the absorption coefficient dependence vs. photon energy.

The RF deposited layers exhibit a hexagonal polycristalline structure with a good orientation of c-axis, normal to the substrate surface (fig.3) while the DC ones, are amorphous. This can be explained by the effect of the positive ion bombardment on the structure of the growing layers, increasing the adatom mobility. We have previously measured that, for our sputtering system, the density of positive ions is about 6-8 times higher in the RF mode respective to the DC one at the same deposition rate /2/.

The optical parameters were obtained from transmission measurements on AlN layers of about 300 nm deposited on quartz, using a computation program based on the procedure of Swanepoel /3/. The refractive index obtained from extremum values of optical transmittance measurements in the low absorption region is extrapolated in the region of high absorption and then the extinction coefficient is obtained (fig.5).

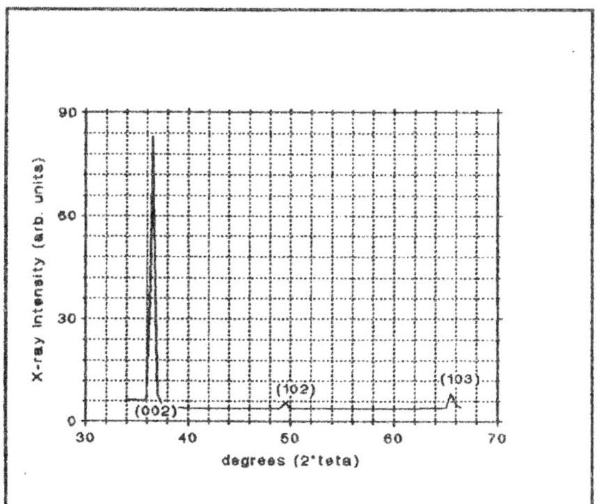

Fig.3 - X-ray diffraction for D20-1 (RF) layer

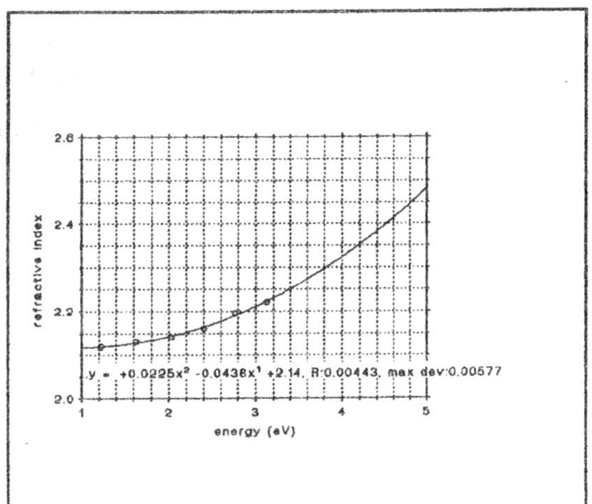

Fig.4 - The refractive index for an RF layer (D13-1)

Using the values of the absorption coefficient, the energy band gap of 4.9 eV for the disordered crystal of the RF polycrystalline layers (fig.4) and the state-to-band absorption coefficient (fig.6) were obtained /4/. For the DC amorphous layers a value of 4.96 eV for the Tauc gap was estimated (fig.7). These values are similar with those reported in literature /4/.

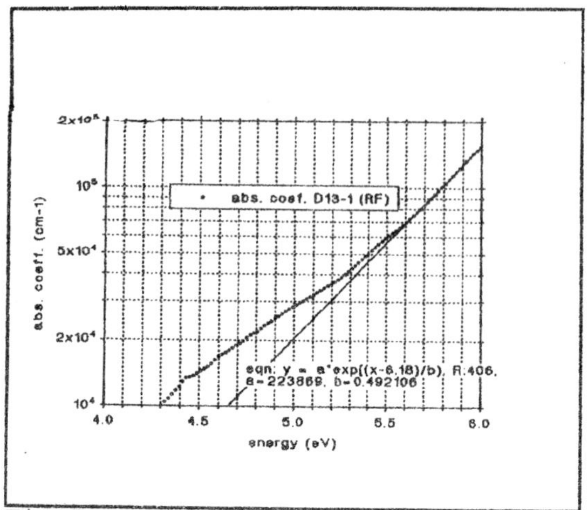

Fig.5 - The absorption coefficient for an RF layer (D13-1)

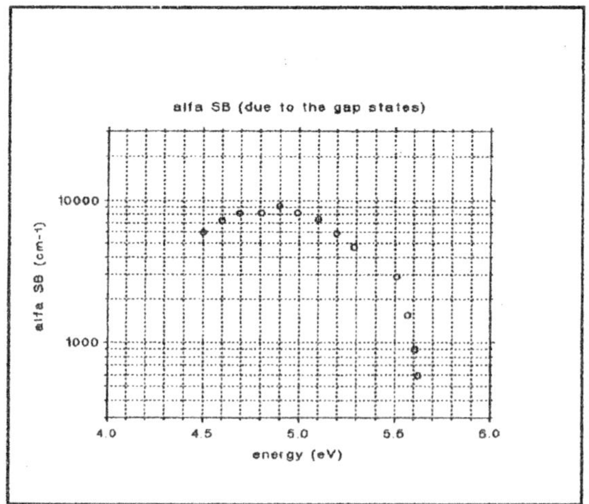

Fig.6-The state-to-band absorption coefficient vs. photon energy for D13-1 layer

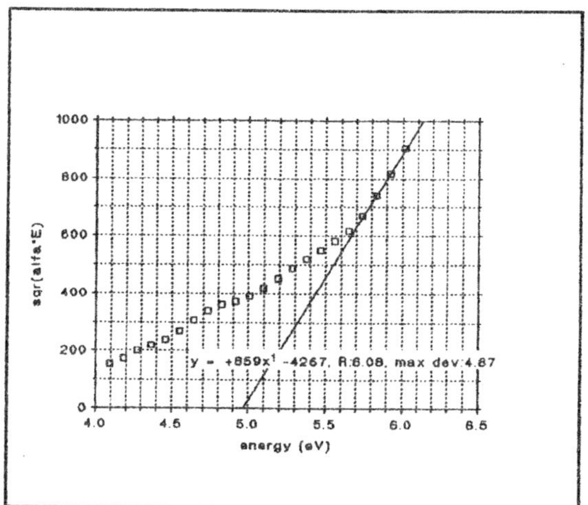

Fig.7 - The absorption coefficient for a DC layer (D15-4)

4. Conclusions

C-axis oriented AlN layers were deposited in a reproducible way by reactive magnetron sputtering. This high refractive index material is interesting for optical applications, particularly when considering the possibility of fabricating various oxinitrides AlO_xN_y for graded index multilayers in the system AlN/Al_2O_3.

References

1. C.Morosanu, T.A.Stoica, T.F.Stoica, D.Necsoiu and M.Popescu - 18th CAS Proceedings, Sinaia, Romania, (1995)
2. C.Morosanu, F.Marica - 3rd National Colloquium of Physics and Technology of Crystalline and Amorphous Materials, Iasi 1988 (p.192)
3. R.Swanepoel - J. Phys. E: Sci. Instrum. 16, 1214(1983)
4. C.R.Aita, C.J.G.Kubiak, and F.Y.H.Shih - J.Appl.Phys.66, 4360 (1989)

TITANIUM STOICHIOMETRY IN ION BEAM INDUCED OXIDATION OF $TiSi_2$.

P. Osiceanu
Institute of Physical Chemistry, Spl. Indep. 202, Bucharest 77208, Romania.

ABSTRACT.

Titanium silicides films grown on Si(100) substrate were investigated by XPS depth profiling after athermal ion beam induced oxidation. The composition and stoichiometry of these films were quantitatively determined as chemical state relative concentrations vs. sputter time.

INTRODUCTION

Refractory metal silicide layers are of great interest for VLSI technology as interconnect material and gate electrods. The chemical stability and reasonably low temperatures of formation and low resistivities are among the most important advantages of these materials. TiO_x films have many interesting (physical, structural, optical) properties/1/, which make them suitable for applications in antireflective coatings /2/ and microelectronics devices /3,4/ due to their high dielectric constant. The athermal oxidation induced by low energy oxygen ion beam is widely applied for sputter depth profiling of thin films, e.g. in AES, XPS and SIMS being generally accepted that this process could lead to near surface compositional changes /5/.

EXPERIMENTAL.

The initial sample was prepared by Ti evaporation on Si substrate and subsequently annealed by means of rapid thermal processing. In this way a $TiSi_2$ polycristalline film was obtained with thickness of about 400nm. The samples were prepared by oxygen ion bombardment at 12 keV and normal incidence and the XPS measurements were performed "ex situ" in a depth profiling experiment. The XPS analyses were done in a UHV apparatus at a pressure ~ $5x10^{-10}$ Torr using a monochromatic AlK_α X - ray source(1486.6 eV and 0.7 eV overall resolution) and facilities for "in depth" ion Ar^+ bombardment.

Ion bombardment of the samples was carried out with Argon ions of current density of 20 mA/cm^2 for different periods up to 10 min.Sputtering conditions were assumed to be comparable to a sputtering rate of 20 A/min. determined by measurement of a standardized layer of SiO_2 / Si. The 1s peak of carbon with a BE of 284.6 eV was used as reference in the as received and briefly sputtered samples in which less than 100A was removed. On further sputtering the N1s was taken as reference assuming no change in its BE value.

0-7803-3223-7/96/$5,00

RESULTS AND DISCUSSION.

The film composition was estimated from the relative area intensities of different high resolution peaks after normalizing with respective relative sensitivity factors after corrections for transmission function and escape depth.The sample $TiSi_2$ after ion beam induced oxidation 12 keV (normal incidence and rastered over 2x2 mm^2) was sputter etched in situ with Ar^+ to remove the oxides for a layer by layer in depth analysis. Spectra of the basic constituents are shown in Fig 1 - 2.

Fig. 1 shows the XPS spectrum for Ti in the oxide state and a small amount of nitride state. We reconstructed the Ti 2p transition from 452 to 468 eV using eight (Gaussian - Lorentzian) components to fit the entire envelope. A Shirley background subtraction was used prior to the curve fit and no restrictions were placed on any of the curve fit parameters.The components required to make the fit to the experimental data are the same in either case selected from the in depth profiling analysis the elemental titanium appearing when the Ar^+ beam proceeds in the depth of the sample. A short comment regarding the presence of a small amount of titanium nitride (we also have recorded the N1s transition) would be desirable in this point.Where from nitrogen in our system with 10^{-10} Torr in the analysis chamber, 99.9999 % Ar purity and after several cycles of purging of the gas handling system? Even if the nitrogen is present in very low concentration in our system (mainly because the purging process was not fully completed) the titanium proves to be very reactive(the sticking coefficient is unity) and traps every atom found in its neighbourhood. Therefore much care must be paid by the experimentalist working with titanium based samples in high and even ultra-high vacuum conditions.Fig.2 shows the O1s transition which can be deconvoluted in two components: one of them exhibits the bonding with titanium having its binding energy at 530.2 eV , while the other being attributed to the bonding with silicon at 531.3 eV.

Performing the data processing of the XPS spectra of Ti 2p, Si 2s, O 1s and Ni 1s levels chosen from different regions of the 3d projection spectra the chemical state relative concentrations for titanium and silicon are obtained.Fig. 3 exibits the chemical state relative concentration vs. sputter time showing titanium in its full oxidized state, titanium suboxides, elemental titanium and a small amount of titanium nitride which is accumulated in the "bulk" of the film. Fig. 4 shows the dependence of the same parameter with the sampling depth but for silicon.

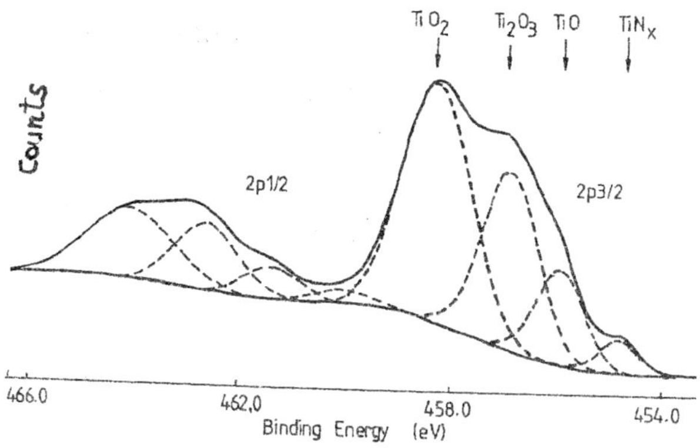

Fig.1 The Ti 2p photoelectron spectrum after 52 sputtering cycles.

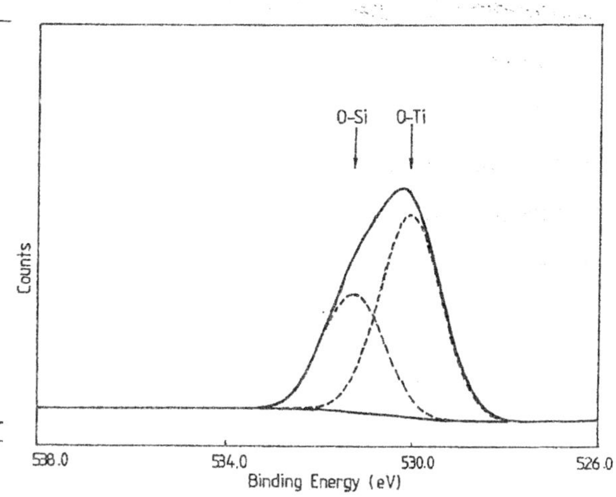

Fig.2 The O1s photoelectron spectrum after 52 sputtering cycles.

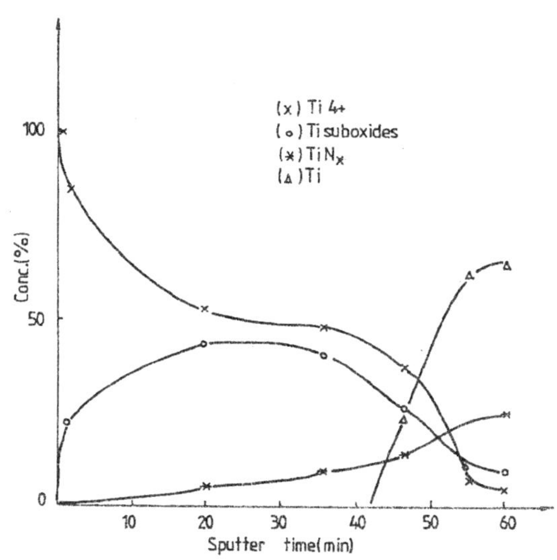

Fig.3 Chemical state relative conc. vs. sputter time for Ti

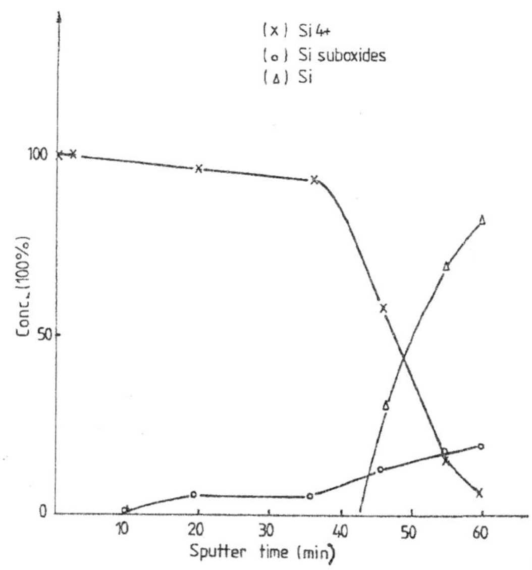

Fig. 4 Chemical state relative conc. vs. sputter time for Si

In the near surface region titanium full oxidized (4+ state) is the dominant feature but proceeding in the bulk of the $TiSi_2$ oxidized film the contribution of the suboxides increases (>40%) over a large range in the depth profile of the sample. In the interface region, which is rather sharp, all the oxidation states are present togheter with elemental titanium and about 8% titanium nitride. The oxidation process of the silicon exibits a different behaviour. The full oxidized silicon (4+ state) can be noticed over a large range of the sampling depth while the contribution of the suboxides in maintained to a low level (~5%) up to the interface region which is rather sharp in agreement with the titanium features. In view of the nitrogen behaviour the existence of an oxynitride of the form TiO_xN_y cannot be ruled out, however.

CONCLUSIONS

1. Chemical state relative conc. of titanium and silicon in ion beam induced oxidation of titanium silicide were studied in a depth profiling XPS experiment

2. While the silicon is found to be full oxidized over a large range in the sampling depth, the titanium oxidation states are mixed, the suboxides being imbeded in the "bulk" of the film.

3. Some trace of titanium nitride is detected revealing the very high reactivity of titanium in titanium silicide even in its oxidized form.

References.

1.M.Gartner, C.Parlog, P.Osiceanu Thin Solid Films 234(1993)561

2. Pettit B., Brinker C.J., Ashley C.J., Solar Cells 15(1985)267

3. Papini M.J., Appl. Phys. 70(1991)777

4. Vopotilov K.A., Orlova E.V. and Petrovsky V.I. Thin Solid Films 207(1992)

5. B.Brijs, W.De Coster, H.Bender, W.Storm, P.Osiceanu, W.Vandervorst Nucl Instr. Meth. in Phys. Res. 99(1995)614

POROUS SILICON PROPERTIES INVESTIGATED BY IR AND UV-VIS SPECTROMETRY

G.Craciun ,M. Bercu[1], Laurentia Marica, Cristiana Bercu[2], A.Dafinei[1], C.Flueraru, V.V.Grecu[1]

Institute of Microtechnology, PO Box 38-160, Bucharest, 72225, [1])University of Bucharest, Faculty of Physics, PO Box MB-11, Bucharest-Magurele, [2])Institute of Chemistry Research, Splaiul Independentei 202, Bucharest, 77208

SUMMARY

IR and UV-VIS spectrometry were used to study the correlation between the anodization conditions and the structure of the porous silicon (PS). The specific surface area increase with the electric charge involved in the etching process. At the same time the total amount of SiH bonds in the PS layer depends on the value of the effective surface. The total electric charge passed through the sample has been related with integral IR absorbances of the Si-H and Si-H_2 bands. An approximately linear function was obtained, especially for SiH. The reactivity of the PS layer related with the anodization conditions have been studied using the behaviour of the Si-O-Si bands after the oxidation in air. The UV-VIS reflection spectrometry shows an interference patterns obtained in the case of a shallow porosity.

1. INTRODUCTION

The high quantum efficiency of visible photoluminescence of the electrochemical etched porous silicon (PS) /1/ depends on the surface morphology and on the chemical structure especially related with Si-O, Si-OH and Si-H_x bond. The PS layers are obtained by anodic electrochemical etching of crystalline silicon in hydrofluoric acid (HF) based solutions and consist in an array of pores that penetrate perpendicularly into silicon substrate. The specific surface aria of PS is very large up to $800m^2/cm^2$ /2/ and surface is covered with silicon hydride as well as silicon oxides and hydroxides /3/. The structural stability of the SiH_x hydrides covering the internal surface is very low especially in an oxidised atmosphere. The bond strength of SiH_x is about 3eV only. The total amount of Si-O and Si-OH bonds increases if the PS is exposed to air even at room temperature. The physical characteristics of PS layer as pore size, specific surface area or morphology of the porous surface are determined by the anodization conditions such as current density, total electric charge passed through the sample, acid concentration, dopant type and dopant concentration in Si. The obtained geometrical structure of the porous surface will determine the reactivity of PS. On the other hand, the corresponding specific surface is related with the amount of Si-H_x bonds of the fresh PS layers. The goal of this study is to relate IR and UV-VIS spectrometric data to condition of the etching process. The amount of SiH_x and SiO bonds in as anodised PS and the total number of Si-O bonds after a long exposure in air have been determined versus the total electrical charge of the anodization procedure.

2. EXPERIMENTAL DETAILS

The porous silicon (PS) samples were obtained on different silicon substrates. All substrates were Czochralski grown, p-type, (100) orientation. The electrolyte was a mixture of hydrofluoric acid HF (49w%) and absolute ethanol in 1:3 volume ratios. The anodization parameters (current intensity and time) are presented in table 1. The porous layers were processed by anodization in a frontal electrochemical cell that was described elsewhere /4/. In the case of 71-74 samples, before anodization on the backside of substrate was realised an ohmic contact by boron implantation ($E=40keV$, $\Phi=2x10^{15}cm^{-1}$) followed by an aluminium deposition. in order to decrease the resistivity of the substrate. For 11-13 samples, before anodization, on these was deposited a boron layer followed by standard diffusion process which results in a p^+ layer of about 6μm thickness and resistivity about 1.2 mΩcm.

The infrared (IR) absorption spectra were recorded in air at room temperature by means of an infrared double-beam spectrometer (PYE UNICAM SP 1100), immediately after PS formation and after exposure in ambient air for 6 days. The UV-VIS reflection spectra have been performed at room temperature (at 22°30' angle of incidence) using a SPECORD M40 double-beam spectrophotometer.

0-7803-3223-7/96/$5,00

Table 1. Anodization parameters for PS processing.

Sample	Current Intensity (mA)	Time (min)	Resistivity Ωcm
73	5		
71	15	10	16-24
74	30		
72	60		
P1	5	10	12-18
P4		5	
13		10	
12	5	15	$1.2x10^{-3}$
11		20	

3. EXPERIMENTAL RESULTS AND DISCUSSIONS

In figure 1 we show the IR spectra of 71-74 PS samples and reference sample (substrate) immediately after anodization. There are three broad absorption bands: 600-750 cm^{-1}, 850-1000 cm^{-1} and 1100 cm^{-1}. In spectral range from 610 cm^{-1} to 675 cm^{-1} we identified the following peaks: 610 cm^{-1} related to substitutional carbon /5, 6/ which is also present in reference IR spectrum; 620-650 cm^{-1} peak related to Si-Si stretching mode /9, 10/, 621 cm^{-1} Si-H_2 deformation mode, and 624 cm^{-1} Si-H bending in Si_3-Si-H /8/; and the peak located at about 675 cm^{-1} related to different Si-H vibration modes. The absorption band located at about 910 cm^{-1} is mainly due to SiH_2 scissors and wagging vibration modes /3,7-10/. The absorption peak from 1100 cm^{-1} is due to a bulk interstitial O-Si-O stretching mode. Fig.1 shows that the increasing in anodization current intensity leads to the increase of the spectral lines intensities, and also to a broadening of FWHM (full width at half-maximum) in spectral range from 610 cm^{-1} to 675 cm^{-1}, and from 850 cm^{-1} to 1000 cm^{-1}. As these bands are due to absorption of different Si-H_x (x=1,2) vibration modes, it means that with increasing of current intensity increases concentration of H contained in PS surface. An explanation is that the PS specific surface increases with increasing of current intensity.

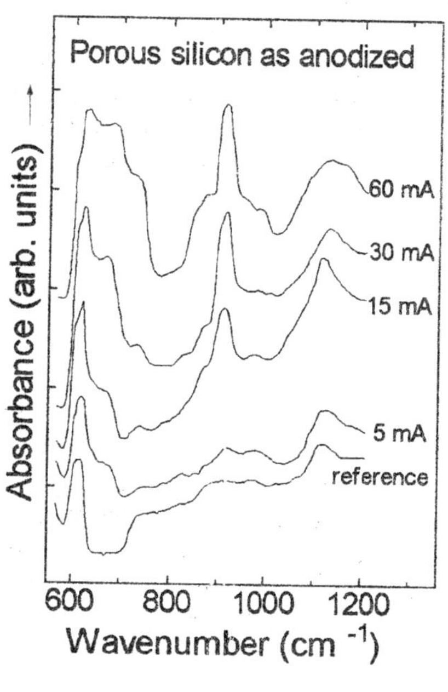

Fig. 1 The IR spectra of 71-74 PS samples and reference sample (substrate) immediately after anodization.

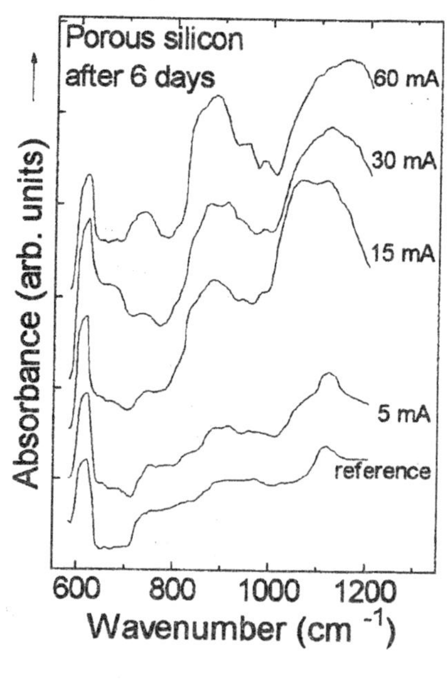

Fig. 2 The IR spectra of 71-74 PS samples and reference sample (substrate) after exposure to ambient air for 6 days.

The IR spectra of PS samples after exposure to ambient air for six days are presented in figure 2. The direct results of the air storage are the replacement of Si-H bonds on the PS surface with the Si-O. This fact can be seen from the decreasing intensity of H-related absorbance peaks (610-675 cm^{-1}), and from the growing intensity and FWHM of O-related absorbance bands. This oxidation effect, which is well known in literature, is due to the large specific surface and, as a consequence, to a high reactivity of the PS layers. Figure 3 shows the integrated absorption of SiH_2 and SiH vibrations versus electric charge of the anodization process. The integrated absorption of SiH bonds (675 cm^{-1}) have been extracted by a deconvolution procedure from the wide band in the range of 600-750 cm^{-1}. The charge dependence of the Si-H absorption is linear. It is interesting to note that the same linear dependence between the SiH stretch mode (2100 cm^{-1}) absorption and electric current density has been found by L.M. Peter /11/. On the other hand, we have obtained the same slope for different resistivities. The charge dependence of the SiH_2 band intensity (908 cm^{-1}, scissors mode) increases at a lower rate as SiH. The behaviour of the oxygen content is also given in figure 4. The number of SiO vibrators (1100 cm^{-1}) increases by exposing to air as a result of the atmospheric pressure oxidation at room temperature. The reactivity of PS depends on the surface morphology. Figure 4 shows that in the case of samples passed by a total charge of about 10 C in the anodization process the surface reactivity reaches its maximum (the connection lines are only a guide for eye).

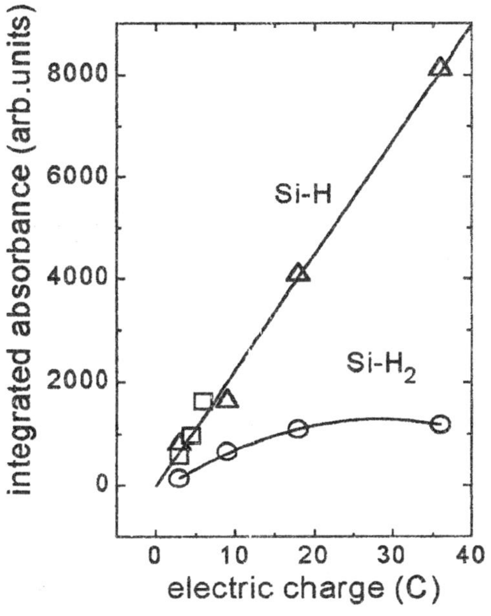

Fig. 3 The integrated absorption of SiH_2 (O) and SiH (□, Δ) vibrations versus electric charge for 71-74 PS samples (□, O) and 11-13 PS samples (Δ).

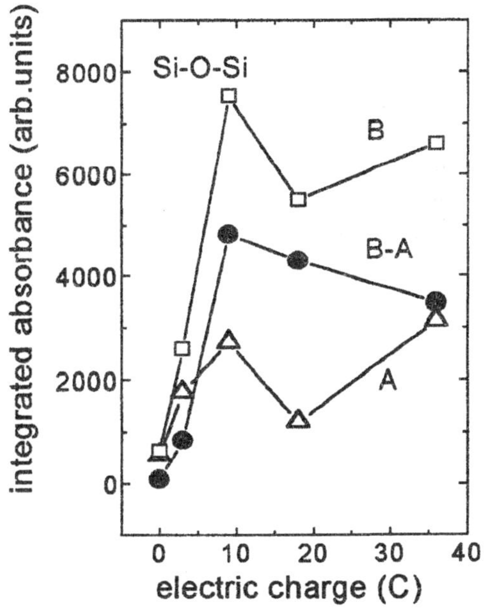

Fig. 4 The integrated absorption of SiO vibrations of 71-74 as anodised PS samples, after 6 days and the behaviour of their difference.

The oxidation starts just after a short exposing in air requested by the manipulation of as-anodized sample (curve A). The most reactive PS layer belongs to the same sample as in the case of oxidised ones. It has been suggested that Si-H and $Si\text{-}H_2$ are important in photoluminescence process of porous silicon surface /10/. The IR spectra of figure 1 and figure 2 show a markedly reduction of Si-H and $Si\text{-}H_2$ bonds from the investigated sample, as a result of the exposing in air for six days.

A preliminary test of the porosity have been done using UV-VIS spectrometry. In fig. 5 the reflection spectra of the three samples correspond at different amount of electric charge used for the porosification of the silicon surface. The reflection intensity at an angle of 22.5 degree decreases for the wave lengths lower than the average dimension of the holes etched on the surface. The diffused light intensity data on the PS layer contains the signature of its morphology. The dimension and geometry of the columns and holes etched on Si surface can be related with light scattering data in the range of 200-1100 nm.

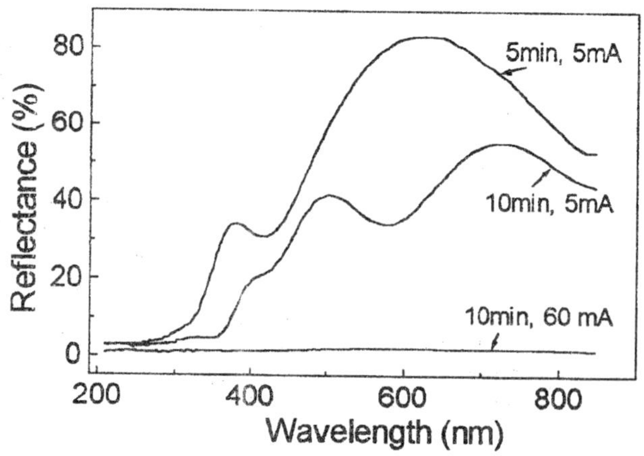

Fig. 5 The reflection spectra of the three PS samples correspond at different amount of electric charges used for the porosification of the silicon surface.

4. CONCLUSIONS

The contribution presents characterisation data on PS samples etched in an electrolyte based on HF for different conditions. We have found a linear dependence between the number of Si-H bonds and the electric charge passed through the sample during anodization process. The integrated absorption signal (675 cm^{-1}) has been calculated by a deconvolution procedure of the IR wide band in the range 600-750 cm^{-1}. The behaviour of SiO band (1100 cm^{-1}) of PS stored in air oxidised samples have been used to determine the dependence between the reactivity of the surface and the anodization conditions. The morphology of the etched Si surface have been related to the anodization process by using a reflectance UV-VIS technique. The investigation of the PS samples exposed in the air shows as other contributions that the Si-H and Si-H_2 bonds are removed by oxygen. The Si-H line (675 cm^{-1}) has been totally removed after the oxidation in air for six days at room temperature.

Acknowledgements

A part of this work has been supported by the Ministry of Research and Technology under contract 121B1/1996

REFERENCES

/1/ L.T. Canham Appl. Phys. Lett. 57, 1046 (1990).
/2/ V.Lehman,U.Gösele Appl.Phys.Lett. 58, 856 (1991).
/3/ R. Anderson, R. Muller and C. Tobias, J. Electrochem. Soc. 140, 1393 (1993).
/4/ G. Craciun, A. Dafinei, C. Vrânceanu, E. Vasile, and C. Flueraru, Int. Semic. Conf., 18th edition (1995), 331, Sinaia, Romania.
/5/ P.Stallhoffer and D. Huber, Solid State Technol. 26, 233 (1983)
/6/ ASTM Data Ser. F 123-81 (1981)
/7/ S.Gardelis and B. Hamilton, J. Appl. Phys. 76 (9), 5327 (1994)
/8/ A.Borghesi, A. Sassella, B. Pivac, and L. Pavesi, Solid State Commun. 87 (1) (1993)
/9/ S.Shih, K.H. Jung, and D.L. Kwong, Appl. Phys. Lett. 62 (16), 1904 (1993)
/10/ L. Lavine, S. Sawan, Y. Shieh and A. Belleza, Appl. Phys. Lett. 62, 1099 (1993).
/11/ L.M. Peter, Phys. Rev. Lett. 62, 308 (1989).

IMPROVED X-RAY SILICON DETECTORS FOR MEDICAL APPLICATIONS

Eugenia Halmagean, Elisabeth Tsoi [*], Konstantinos Misiakos [*],
Madalina Ohanisian [**], Adrian Veron [**], Valerica Cimpoca [***], Doina Lazarovici [***]

Research Institute for Electronic Componenets, [**]Baneasa S.A.
Str. Erou Iancu Nicolae 32B, 72996 Bucuresti, Romania

[*]Institute of Microelectronics, National Centre for Scientific Research "Demokritos",
153 10 Ag. Paraskevi Attikis, POB 60228, Greece

[***]Institute of Atomic Physics,
76900 Bucuresti-Magurele, Romania

ABSTRACT

The paper presents some results concerning the technology and behaviour of X-ray silicon detectors intended for controlling low radiation levels in interventional cardiology. An improved planar process was used for obtaining large area guard-ring test structures. The electrical characteristics, their modification after exposure to radiation and the results of spectroscopic X-ray measurements are discussed.

1. INTRODUCTION

For several years, modern semiconductor detectors have been the primary choice for the measurement of nuclear radiation in various scientific fields /1/. Nowadays the recently developed high resolution silicon detectors find their way in medical applications /2/. As a consequence many efforts have been devoted to the development of high sensitivity radiation hardened X-ray detectors for the energy range 10-100 KeV /3/. The paper presents an improved technology for obtaining low noise silicon detectors. The test structures were electrically characterised before and after exposure to specific X-ray doses between 10^3-10^8 rads. Some detectors were also used for the measurement of the ^{241}Am spectrum in the energy range 10-60 Kev. The results indicated that the proposed process enables the development of reliable silicon detectors to be used in controlling the low radiation levels encountered in medical applications.

2. EXPERIMENTAL

The test structures were fabricated on n-type detector grade silicon wafers produced by Topsil and having a resistivity of 4500 ohmcm. The devices consisted of square geometry pn

0-7803-3223-7/96/$5,00 © 1996 IEEE

geometry pn diodes with diffused guard rings equipped with separate contacts. Two different active areas were analysed: a small diode D1 (3x3 mm^2) and a large diode D2 (5x5 mm^2). The 3 inch diameter wafers were first oxidized at 1050°C in dry oxygen with 2-3% TCE in order to obtain a high quality initial oxide on the silicon surface. The oxidation step was followed by an in situ high temperature N_2 annealing to reduce the interface charges. After the first photoengraving step boron was implanted simultaneously in the active area and in the guard ring. The implantation energy was 40 keV and the dose 1E14 cm^{-2}. A second phosphorous implantation was performed on the back side of the silicon wafers with an energy of 80 keV and a dose of 1E16 cm^{-2}. The implanted impurities were first activated at 900°C in N_2 and subsequently driven in at 1000°C. The diffusion phase comprised two steps: the first one in dry oxygen with 2-3% TCE and the second one in pure N_2. The resulting boron profile in the active area of the detectors as determined by the spreading resistance method is indicated in figure 1. A 5000 A layer of SiO_2 was next deposited on the wafers in an LPCVD hot wall reactor using TEOS as the source. The final thermal treatment was performed at 850°C in N_2, followed by a very slow cooling to 600°C in order to getter the unwanted impurities from the silicon active zones. The contact windows were opened in the central diode area and in the guard ring by standard photolithography. An aluminium layer 10000 A thick was deposited on the wafer surface by thermal evaporation. The reverse electrical characteristics of the detector test structures were measured with a Keithley 616 ammeter and the typical ones are presented in figure 2. The C-V characteristics were recorded at 1 Mhz on a PARC 410 C-V Plotter system and are presented in figure 3. After electrical characterization selected detector test structures were exposed to X-ray radiation with an energy of about 30 keV. The reverse currents of the devices were measured after specific radiation doses in the range 10^3-10^8 rads in order to evaluate the degradation of their electrical characteristics. Some large area detectors were also used for spectroscopic X-ray measurements on an ORTEC set-up described in figure 4. The spectrum of ^{241}Am was recorded at room temperature in the energy range 10-60 keV.

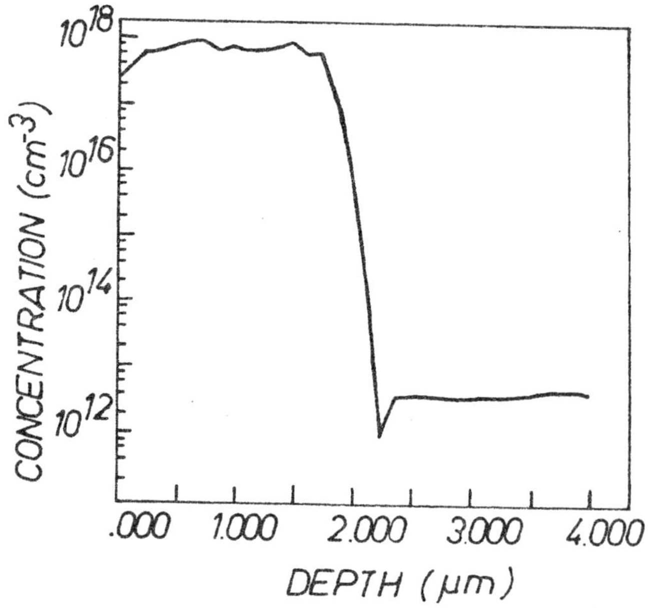

Figure 1. Implanted boron spreading resistance profile.

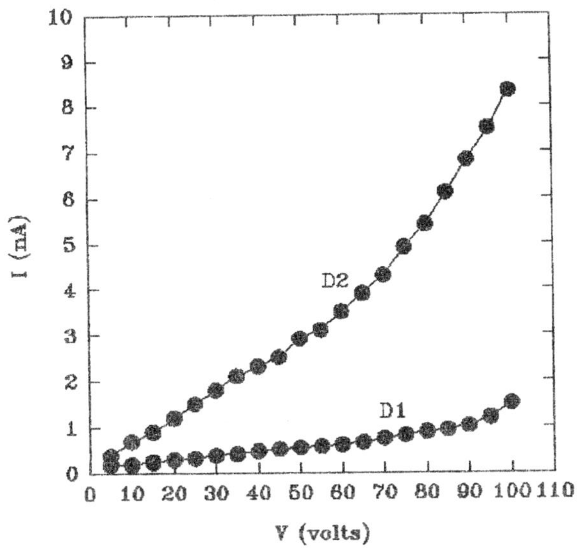

Figure 2. I_R-V_R characteristics.

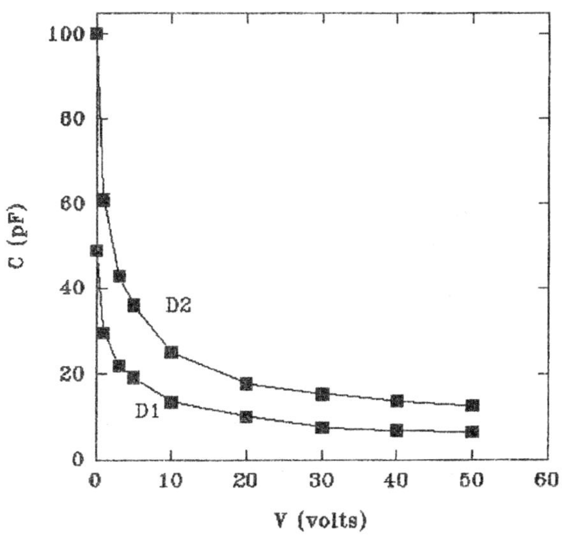

Figure 3. C-V_R characteristics.

3. RESULTS AND DISCUSSION

The measured electrical characteristics of the test structures demonstrated that the proposed processing scheme could provide very small reverse leakage currents on large area devices and is suitable for producing sensitive X-ray detectors. The stability of the reverse leakage currents after detector exposure to different radiation doses is indicated in figure 5. One could observe that after approximately 10^7 rads the leakage current of the tested devices begin to gradually increase. The main reason for this degradation is considered to be the interface states and oxide traps generation due to the ionizing radiation /4/. This dose is high enough to enable the use of the detectors for controlling low radiation levels in medical applications/5/. The recorded spectrum of ^{241}Am is depicted in figure 6. The reverse voltage of 40 V was selected in order to drastically reduce the leakage current of the diode and to correspondingly improve the signal to noise ratio of the detector. The test structure succeeded to distinguish 5 peaks with a resolution of about 3.135 keV.

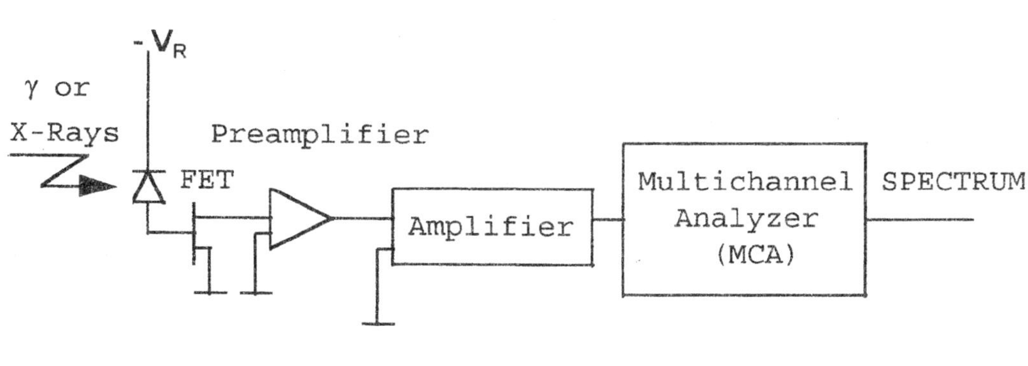

Figure 4. ORTEC spectroscopic set-up for X-ray measurements.

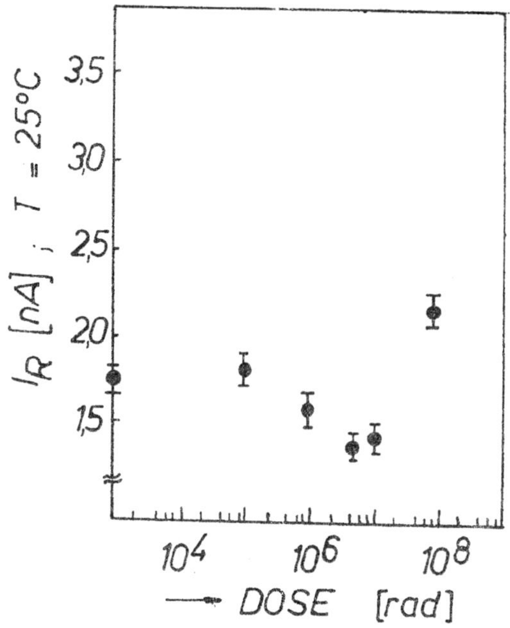

Figure 5. The reverse current versus X-ray doses.

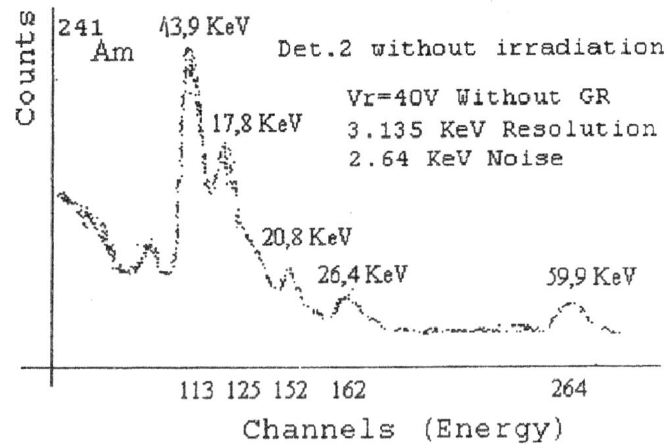

Figure 6. The measured ^{241}Am spectrum.

4. CONCLUSIONS

The proposed processing sequence enabled the development of low noise X-ray silicon detectors intended to be used for controlling low radiation doses specific to medical applications. The critical steps of the planar process are the thermal oxidation of the silicon surface, the annealing of the implanted impurities and the gettering of contaminants. All these factors contribute to the magnitude of the reverse leakage current and consequently affects the stability of the detector after exposure to radiation and the sensitivity of its response. The fabricated test structures proved to be stable in X-ray radiation environments in the dose range 10^3-10^7 rads. The spectroscopic measurements demonstrated the usability of these detectors for analysing the X-ray spectrum of specific sources and for detecting low radiation levels. It is believed that the detector performances could be significantly improved by a more sophisticated device design, including the integration of the preamplifier stage on the same chip.

REFERENCES

/1/. H. Struder, H. Soltau, "High Resolution Silicon Detectors for Photons and Particles", Radiation Protection Dosimetry, Vol. 61, No.1-3, pp.39-46, 1995.

/2/. A. Thompson, "Detectors for Angiography" in Handbook for Synchrotron Radiation, Plenum Press, N.Y. , pp.583, 1994.

/3/. D. Lazarovici, et.al, "X-Ray Dosimetry with Low Capacitance Silicon Detectors", Proc. SPIE, Vol. 2009 - X-Ray Detector Physics and Applications II, pp.56, 1993.

/4/. E. Beuville, et.al., "Measurements of Degradation of Silicon Detectors and Electronics in Various Radiation Environments", Nuclear Instrum. and Methods in Physics Research Vol. A288, pp. 68-75, 1990.

/5/. H.-J. Selbach, "Requirements for Dosimeters Used in Diagnostic Radiology", in Radiation Protection Dosimetry, Vol. 43, No. 1/4, pp. 25, 1992.

LATE NEWS

The Effect of Extrinsic Capacitances on the Microwave Performance of $Ga_{0.51}In_{0.49}P$/GaAs MISFET's (0 nm ≤ t ≤ 10 nm) Grown by GSMBE

Yo-Sheng Lin and Shey-Shi Lu

Room 451, Department of Electrical Engineering
National Taiwan University, Taipei, Taiwan, Republic of China
Fax : 886-2-3638247, E-mail : D82010@cctwin.ee.ntu.edu.twn

ABSTRACT

The effect of extrinsic capacitances on the microwave performance of $Ga_{0.51}In_{0.49}P$/GaAs MISFET's was first studied experimentally and theoretically by varying the thickness of the $Ga_{0.51}In_{0.49}P$ insulating layer. MISFET's with airbridge gate structure showed higher f_t's and f_{max}'s than those of MISFET's with traditional gate structure due to the lower extrinsic capacitances. Moreover, the maximum values of f_t's and f_{max}'s for a 1 μm gate length device all happen when t is between 50 nm and 100 nm. These results demonstrate that $Ga_{0.51}In_{0.49}P$/GaAs airbridge gate MISFET's with insulator thickness between 50 nm and 100 nm were very suitable for microwave high power device applications.

I. INTRODUCTION

An electronic device with high-linearity, high-power and high-speed performance is very important for microwave power applications. A device with good linear characteristics can reduce the intermodulation of high frequency signals, and therefore the distortion of signals at high power level operation is suppressed. It has been demonstrated that if a low-doped or an undoped layer is inserted between the gate metal and the active channel of a MESFET structure, the linearity, current drivability and breakdown voltage characteristics are greatly improved [1-2]. Even though the transconductance is lower than that of the conventional MESFET's, yet the reduced gate terminal capacitance C_g (i.e., gate-source capacitance C_{gs} plus gate-drain capacitance C_{gd}) makes the ratio, g_m/C_g , higher [3-4], and hence the high frequency performance is also improved. Therefore, we would like to study the performance of the metal-undoped $Ga_{0.51}In_{0.49}P$-n GaAs traditional and airbridge gate MISFET's with different insulator thickness t and compare them with similar MESFET's (t = 0 nm).

II. CRYSTAL GROWTH AND DEVICE TECHNOLOGY

The heterostructures were grown by gas source molecular beam epitaxy (GSMBE) on semi-insulating (100) GaAs substrates. They consisted of a 1000 nm undoped GaAs buffer, followed by a 100 nm GaAs n channel layer (6×10^{17} cm^{-3} , Si doped). Then an undoped $Ga_{0.51}In_{0.49}P$ insulator layer was grown on top of the active channel. Finally, a 40 nm n^+ (2×10^{18} cm^{-3} , Si doped) GaAs cap layer was grown to facilitate the formation of good ohmic contacts. The $Ga_{0.51}In_{0.49}P$ insulator thickness t varies from t = 0, 50 - 100 nm, to examine how the insulator thickness affects the the

0-7803-3223-7/96/$5,00

devices microwave characteristics. Schematic diagram of the cross section of MISFET's (t = 50 nm-100nm) and MESFET's (t = 0 nm) is shown in Fig. 1.

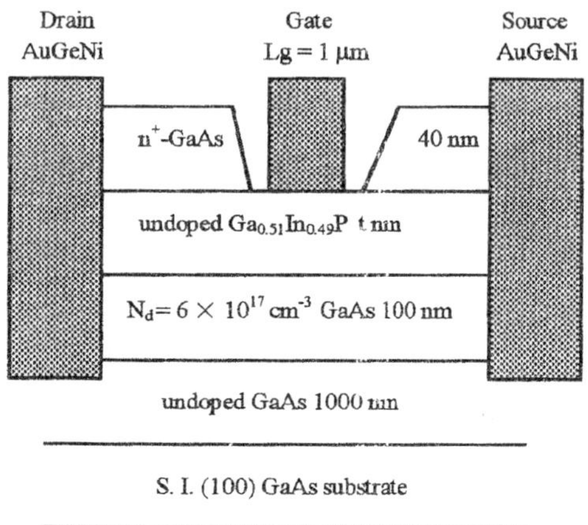

Fig. 1 Device cross section of $Ga_{0.51}In_{0.49}P$/GaAs MISFET's (t = 50nm-100 nm) and MESFET's (t = 0 nm).

III. EXPERIMENTAL RESULTS AND DISCUSSIONS

Microwave on-wafer S-parameters for $Ga_{0.51}In_{0.49}P$/GaAs MISFET's (t = 50 nm and 100 nm) and MESFET's (t = 0 nm) with 1.0 μm-long gate were measured in common-source configuration from 45 MHz to 40 GHz with an HP8510C network analyzer in conjunction with Cascade probes. Fig. 2 shows the f_{max}'s and f_t's as a function of I_{ds} for MISFET's with t = 100 nm and MESFET's biased at V_{ds} = 4V. As can be seen clearly, f_{max}'s and f_t's of MISFET's are much higher than that of the MESFET's. The flat region of f_{max}'s and f_t's of MISFET's are also much wider than that of the MESFET's. Table I is a summary of the results of f_t's, f_{max}'s and extracted C_g (i.e., $C_{gs}+C_{gd}$) from measured S parameters for all devices

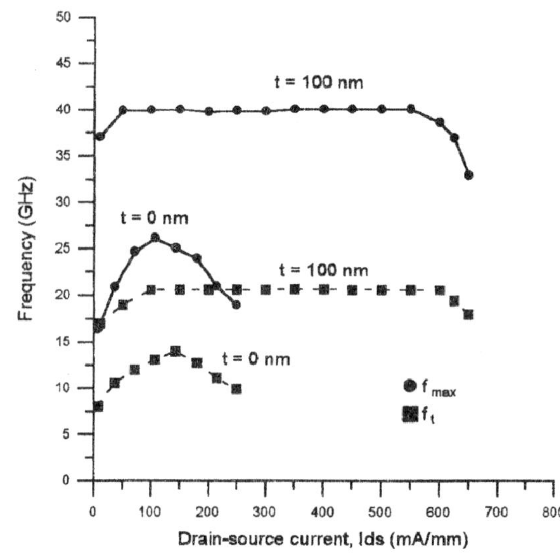

Fig. 2 Microwave characteristics of an airbridge gate MISFET's with insulator thickness t = 100 nm at V_{ds} = 4 V and V_{gs} = 0 V.

biased at V_{ds} =4V, with V_{gs} tuned for a maximum S_{21}. From this table one can find that f_t's and f_{max}'s are greatly improved for all MISFET's and MESFET's with airbridge gate structure compared with that of traditional gate structure. The airbridge gate MISFET's with t = 50 nm exhibited $f_t = 20.5 \pm 0.3$ GHz and $f_{max} = 39 \pm 0.5$ GHz. These valus were roughly 45 % and 49 % higher than those ($f_t = 14.1 \pm 0.3$ GHz and $f_{max} = 26.1 \pm 0.3$ GHz) of the airbridge gate MESFET's (t = 0 nm) but slightly lower than those ($f_t = 20.7 \pm 0.4$ GHz, $f_{max} = 40.1 \pm 0.5$ GHz) of the airbridge gate MISFET's with t = 100 nm. However, the f_t's and f_{max}'s of traditional MISFET's with t = 50 nm were larger than that of traditional MISFET's with t = 100 nm. This can be explained by Fig. 3 and Fig. 4 in which the MISFET model of Ref. [3] was extended to include parasitic gate terminal capacitances C_g^p (i.e., interelectrode capacitance C_g^e plus geometric pad capacitance C_g^{pad}) in the totalcapacitance C_g. That is $C_g = C_g^i + C_g^e + C_g^{pad}$

	t = 0 nm		t = 50 nm		t = 100 nm	
	traditional	airbridged	traditional	airbridged	traditional	airbridged
f_t(GHz)	12.9 ± 0.2	14.1 ± 0.3	17.6 ± 0.2	20.5 ± 0.3	16.2 ± 0.2	20.7 ± 0.4
f_{max}(GHz)	24.6 ± 0.3	26.1 ± 0.3	34.9 ± 0.4	39 ± 0.5	32.8 ± 0.3	40.1 ± 0.5
C_g (fF)	244.1 ± 2	223.9 ± 1.7	139.5 ± 1.4	119.4 ± 1.3	94.1 ± 1	73.9 ± 0.8

Table I Summary of RF characteristics of $Ga_{0.51}In_{0.49}P/GaAs$ airbridge gate and traditional MISFET′s (t = 50 nm and 100 nm) and MESFET′s (t = 0 nm).

. The parameters used are as follows : surface state density N_{SS} = 6 × 10^{11} $cm^{-2}eV^{-1}$, electron mobility : 2000 cm^2/V-sec, saturation velocity 1.2 × 10^7 cm/sec, parasitic resistances R_S = 4.1 Ω, R_d = 4.05 Ω , interelectrode capacitance C_g^e = 11.25 fF, geometric pad capacitance C_g^{pad} = 19 fF, V_{gs} = 0 V and V_{ds} = 4 V. As can be seen clearly from Fig. 3, at small t value (t < 50 nm), even though the transconduntance g_m decreases monotonic with increasing insulator thickness, yet the reduced gate terminal capacitance C_g makes the ratio, g_m/C_g, higher with increasing insulator thickness. This means f_t′s (~ $\frac{g_m}{2\pi C_g}$) will be improved with increasing thickness thickness for small t, as shown in Fig. 4. In theory of Ref. [3-4] only C_g^i is considered, and hence predicted that MISFET′s will have slightly higher or nearly the same values of f_t′s and f_{max}′s with increasing t even if t is up to 500 nm. But, in real case, C_g^p is a not omittable constant and must be considered in the calculation. The C_g^i decreases with increasing t and hence f_t′s and f_{max}′s begin to decay when C_g^p is comparable to C_g^i. The simulated peak of f_t′s of traditional structure appears earlier (50 nm) than airbridge gate structure (75 nm) due to the larger C_g^p. For airbridge gate MISFET′s, the parasitic geometric pad capacitance C_g^{pad} is negligible and hence its total gate capacitance C_g is approximately equal to instead of $C_g^i + C_g^e$. This value of course is lower than that ($C_g = C_g^i + C_g^e + C_g^{pad}$) of traditional MISFET′s as shown in Fig. 3. Therefore, airbridge gate MISFET′s exhibit higher f_t′s than traditional MISFET′s as shown in Fig. 4. In addition, the channel resistance R_i reduce systematically with increasing insulator thickness. Output conductance G_d for MISFET′s fabricated in our laboratory is nearly a constant and about 0.062 mS/mm. This implies the f_{max}′s (~ $\frac{f_t}{2\sqrt{G_d R_i}}$) will be improved more than f_t′s with increasing insulator thickness at small t value, as shown in Fig. 4. The measured f_t′s and f_{max}′s for airbridge gate and traditional MISFET′s were also shown in Fig. 4 for

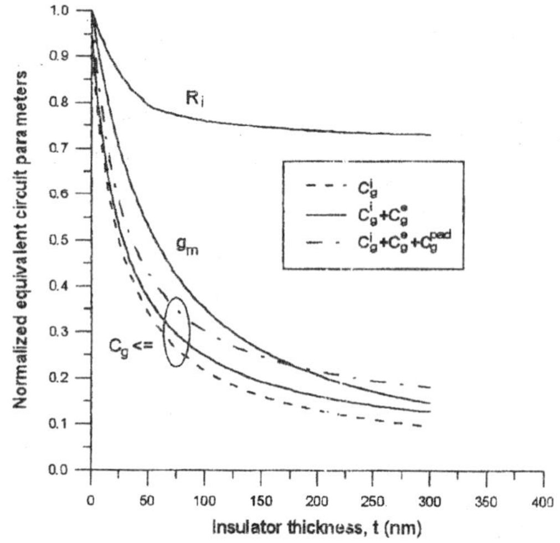

Fig.3 The simulated normalized depedance of the transconductance g_m, gate terminal capacitance C_g and channel resistance R_i versus insulator thickness t.

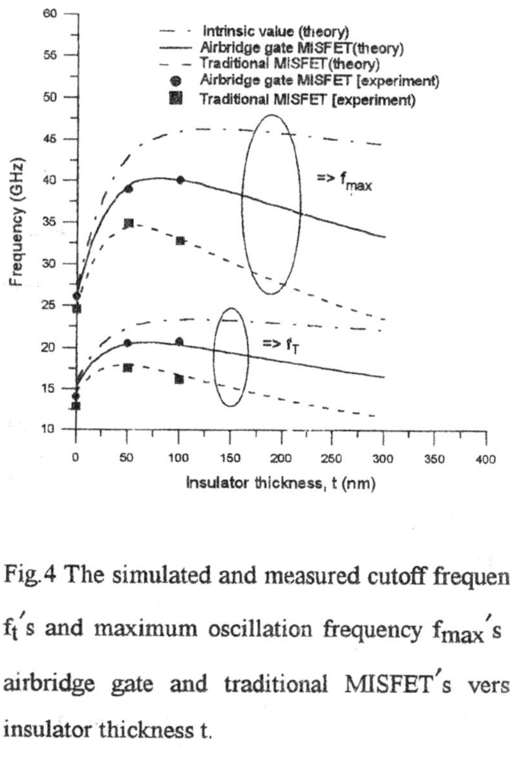

Fig.4 The simulated and measured cutoff frequency f_t's and maximum oscillation frequency f_{max}'s of airbridge gate and traditional MISFET′s versus insulator thickness t.

comparison. For proper gate insulator, e.g. , 50 nm, $Ga_{0.51}In_{0.49}P$/GaAs airbridge gate and traditional MISFET′s clearly show a performance advantage over a similar MESFET′s (t = 0 nm). However, the microwave response (f_t's and f_{max}'s) of airbridge gate and traditional MISFET′s with t = 100 nm are comparable to that of devices with t = 50 nm. This is consistent with the simulated data and demonstrate the validity of the simulation. So, we conclude that the state-of-the-art values of $Ga_{0.51}In_{0.49}P$ insulator thickness should be around 50 nm to 100 nm for 100 nm GaAs active layer.

IV. CONCLUSION

The effect of extrinsic capacitances on the microwave performance of $Ga_{0.51}In_{0.49}P$/GaAs traditional and airbridge gate MISFET′s were evaluated by varying the thickness of the insulating layer. The f_{max}'s and f_t's of MISFET′s using airbridge gate structure were higher than those of MISFET′s using traditional gate structure due to the lower parasitic capacitances. The experimental results suggest that the optimized insulator thickness for 100 nm GaAs active layer is around 50 nm to 100 nm. Due to the $Ga_{0.51}In_{0.49}P$ airbridge gate structure, a high current drivability, high breakdown, high linearity, high speed, and low leakage gate current $Ga_{0.51}In_{0.49}P$/GaAs MISFET was achieved. These results showed that the $Ga_{0.51}In_{0.49}P$/GaAs airbridge gate MISFET′s were very suitable for microwave high power device applications.

REFERENCES

[1] C. L. Chen, W. Smith, B. J. Clifton, L. J. Mahoney, M. J. Manfra, and A. R. Kalawa, *IEEE ED,* vol. 12, pp. 306-308,1991.

[2] Yi-Jen Chan and Ming-Ta Yang, *IEEE EDL.*, vol. 16, pp.33-35,1995.

[3] P. M. Hill, *Tran. ED*, vol. 32, pp. 2249-2256, 1985.

[4] A. Nagashima, S. Umebachi and G. Kano, *IEEE Tran. ED*, vol. 25, pp. 537-539, 1978.